W9-BCK-922

THE CHEMICAL ELEMENTS IN NATURE

The Chemical Elements in Nature

F. H. DAY

M.Sc. Ph.D. F.R.I.C. F.C.S.

Head of Department of Science and Vice-Principal
Carlisle Technical College

REINHOLD PUBLISHING CORPORATION

New York

First published in the United States of America 1964

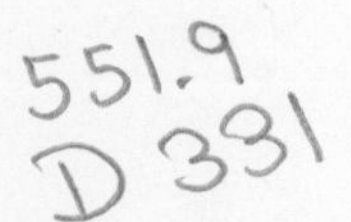

Printed in Great Britain

Preface

Information about the occurrence of what may be called the raw materials of chemistry is obtainable from a number of different sources. The ordinary chemical textbook usually provides notes, incidental to the formal study of an element, and often so brief as to be of almost negligible value. Much of this information is frequently at least obsolescent and sometimes inaccurate. Works on mineral resources contain much valuable matter but not all of it is presented in a manner suitable for non-geological readers. Finally, several admirable texts on geochemistry are now available. However, the discussion in these works is in general related especially to topics like the occurrence of the various elements in conditions not all of which necessarily have any practical bearing.

This work is presented as an attempt to provide information about the chemical elements of a kind and in a form which should be more readily accessible than elsewhere. In particular it deals with the practical sources of chemical materials and the various kinds of transformations they undergo under natural conditions. No attempt is made to include the vast and complex chemistry of living organisms, which has long constituted the study of biochemistry, but some allusion to biological reactions is made where appropriate and desirable.

The behaviour of chemical substances in the laboratory and in industrial operations is very adequately dealt with in the orthodox textbooks of inorganic chemistry, and the ready availability of so many pure chemicals today may well cause the ordinary student to become increasingly the 'slave of the bottle' as far as his materials are concerned. To some extent this book is an attempt to break down this tendency by drawing attention to the themes of 'Where does it come from?' and 'Where does it go to?'.

Many of these topics should be of interest to others besides chemists and although some chemical knowledge is needed for a full comprehension of the text, it contains much that can be referred to without any appreciable chemical background. It is felt that facts and comments about output, consumption, etc., are often of considerable interest and an effort has been made to supply some such matter which may in part be of rather transient importance. Every effort has been made to render quantities strictly comparable by quoting in tons and grammes, and this has required conversion in some cases from more familiar and orthodox units like troy ounces and carats.

British localities have been referred to because of their interest to British residents, even if they are not now of much practical importance. They make no claim to be complete.

No apology is made for the discussion of some relatively unusual sources because they may become of great practical importance as a result of future technological developments.

The attempt to indicate the nature of the end-products of chemical substances after they pass from the sphere of human operations is probably both incomplete and in some cases of doubtful accuracy, but it was included because it was felt that attention was thereby focused on interesting topics which appear to have received little attention in the past.

The first five chapters deal mainly with the principles involved in a study of what may be called ' chemical natural history '. These may be given a general reading while the remainder of the book is referred to independently as required.

While it is hoped that the selected references at the end of the book will be found of some assistance to the reader, they make no pretension to be a complete bibliography of the subject. For comprehensive references to the very numerous papers the reader should consult more specialized works covering the many aspects of the rather diverse topics which are alluded to in this modest work.

The author is indebted to various friends and colleagues and in particular to Mr. N. Thomson for help and suggestions and to Mr. J. K. Hodgson for the execution of the drawings.

F. H. D.

Blackwell Lodge
Carlisle

Contents

CHAPTER 1

The Elements and the Cosmos

For the purposes of this work we can regard the elements as nuclear species characterized by the number of protons they contain. This proton number or atomic number identifies a particular element. Varying numbers of neutrons, which are associated with the nucleus in addition to the protons, lead to the possibility of several nuclides or isotopes for any given element.

The extra-nuclear particles, the electrons, are important for the chemical characteristics of the element and when their number is slightly more or less than the nuclear charge the atom becomes an ion and carries an appropriate net positive or negative charge. Ionic particles are of considerable significance in the structure and interactions of the mineral bodies of which the crust of the Earth is largely composed.

NATURAL UNSTABLE NUCLIDES

Name	*Mass no.*	*Half-life*	*Decay process*
Neutron	1	12·8 min	$\beta \longrightarrow {}^{1}H$
Tritium	3	12·41 y	$\beta \longrightarrow {}^{3}He$
Carbon	14	5568 ± 30 y	$\beta \longrightarrow {}^{14}N$
Potassium	40	$1{\cdot}27 \times 10^{9}$ y	$\beta \longrightarrow {}^{40}Ca$; K-capture $\longrightarrow {}^{40}Ar$
Rubidium	87	$6{\cdot}15 \times 10^{10}$ y	$\beta \longrightarrow {}^{87}Sr$
Lanthanum	138	7×10^{10} y; $1{\cdot}2 \times 10^{12}$ y	K-capture $\longrightarrow {}^{138}Ba$; $\beta \longrightarrow {}^{138}Ce$
Samarium	147	$6{\cdot}7 \times 10^{11}$ y	$\alpha \longrightarrow {}^{143}Nd$
Lutecium	176	$2{\cdot}4 \times 10^{10}$ y	K-capture $\longrightarrow {}^{176}Yb$; $\beta \longrightarrow {}^{176}Hf$
Rhenium	187	4×10^{12} y	$\beta \longrightarrow {}^{187}Os$
Bismuth	209	$2{\cdot}7 \times 10^{17}$ y (?)	$\alpha \longrightarrow {}^{205}Tl$
Radium	226	1622 y	$\alpha \longrightarrow {}^{206}Pb$
Thorium	232	$1{\cdot}389 \times 10^{10}$ y	$\alpha \longrightarrow {}^{208}Pb$
Protactinium	231	34 300 y	$\beta \longrightarrow {}^{231}Th$
Uranium	234	$2{\cdot}69 \times 10^{5}$ y	$\alpha \longrightarrow {}^{206}Pb$
	235	$7{\cdot}07 \times 10^{8}$ y	$\alpha \longrightarrow {}^{207}Pb$
	238	$4{\cdot}5 \times 10^{9}$ y	$\alpha \longrightarrow {}^{206}Pb$

Not all nuclides are stable and the spontaneous disintegration of the unstable ones gives rise to the well-known phenomenon of radioactivity. When the rate of disintegration is relatively low, then radioactive nuclides may occur as natural substances.

Ninety different elements have been found in the Earth and one more, number 43, technetium, is known to exist in certain stars. Number 61, promethium, does not exist in Nature. There are 272 stable and 55 unstable nuclides which are known to exist naturally, but many more have been made artificially by nuclear experiments which have no apparent natural counterpart.

Unstable nuclides. The decay or disintegration of radioactive nuclides proceeds exponentially and the 'half-life' value, which is the time required for half of the initial mass of a sample to decay, gives a measure of the comparative persistence of a particular species. Several kinds of decay processes take place and are discussed in works on nuclear physics to which reference should be made for details.

The Material Universe

Space appears to be populated by galaxies or stellar systems of average diameter about 100000 light-years (1 light-year = *circa* 10^{13} km = 10^{18} cm.) These galaxies are not less than about 1 million light-years apart, the nearest to our own being the Andromeda nebula. The spectra of these extra-galactic nebulae show a displacement of their lines towards the red by an amount about proportional to their estimated distances. This is recognized as a Doppler effect and the conclusion is that the nebulae are receding from us and from each other. This is the basis of the theory of the expanding Universe and in spite of its implications it appears to be the most satisfactory way at present of explaining the observed facts.

By making certain assumptions the rate of expansion can be used to deduce an age for the Universe which turns out to be of the order of 5×10^9 years. This is reasonably consistent with estimates on other bases but nevertheless there is now evidence which places the age of certain types of stars at as much as 10^{11} years.

A galaxy is a vast system containing stars of various types and also gaseous nebulae. Our own Sun is a star and is attended by a planetary system of which the Earth is a member. The Solar System also contains asteroids, meteorites, and comets.

The most important instrument in the study of astrophysics and what may be termed cosmochemistry is undoubtedly the spectroscope in its various forms. Without the knowledge of and the interpretation of spectra very little could be known of the material nature of extra-terrestrial bodies. It should be noted that owing in many cases to the vast distances involved, the objects and events observed are in fact in past time, the extent of which may in some cases be comparable with the geological eras.

The Universe is not static even to the terrestrial observer. Apart from various real and apparent motions which need not concern us here there is the occasional explosive outburst of a star called a supernova. Supernovae are estimated to occur about once every 300 years per galaxy and are probably of great nuclear-chemical significance. The classical example so often quoted is the Crab nebula which is the remnant of an exploded supernova. This was in fact recorded by the Chinese in A.D. 1054 and as the object is 5000 light-years distant, the event actually took place about 6000 years ago.

Origin of the Elements

In a book of this kind, which is in the main factual, it would appear desirable for the sake of completeness to include an outline of present views as to the origin of the elements. This subject, which in fact forms part of cosmochemistry, has developed from a number of speculative views and is necessarily closely connected with general cosmological theory.

However, the present position is much more satisfactory than it was even a few years ago and the improvement may be ascribed to the following points.

(*a*) A fuller knowledge of the age, composition, and distribution of the main types of stars.
(*b*) Developments in nuclear physics, particularly in regard to nuclear reactions and structure.
(*c*) Fuller elaboration of mechanisms of stellar evolution.

It is possible to say very broadly that there are two conceptions in cosmology—catastrophic and uniformitarian. Both of these views have long had their adherents. The modern form of the first is in brief along the following lines. About a decade ago it was widely thought that in view of the theory of the expanding Universe, the known values of half-lives of long-lived radioactive elements, and the estimated age of the solar system by various methods, that the present relative abundances of the elements represented a ' frozen thermodynamic equilibrium ' following from a catastrophic ' act of creation ' occurring about 10^{10} years ago. In the early view of Lemaitre the ' universal atom ' suffered spontaneous disintegration, and the expansive process still continues in the form of the expanding Universe. Among other things the cosmic radiation (very high-energy particles) was postulated as one of the by-products of the operation so that cosmic rays were to be regarded as one of the ' left-overs ' of creation. Later developments adopt various notions of a pre-stellar state of the Universe involving large bodies of dense nuclear matter at a high temperature. Considerable difficulty arises in showing how such massive bodies could disrupt and cool quickly enough to ' freeze ' the elements in their present relative abundances. On another view, from a cold nuclear (chiefly neutron) fluid, polyneutron masses are cast off. Fission, beta-decay, and neutron evaporation are then postulated as the means by which the various elements, up to the heaviest known, were produced (Mayer and Teller).

Alpher and Hermann have developed a neutron capture hypothesis in conjunction with an exploding cosmological model involving high radiation density. Successive neutron captures interspersed by beta-decays provide means for synthesizing heavy nuclei.

The mechanism of neutron capture has been incorporated into other theories of heavy-element production. The second, the uniformitarian theory, is particularly associated with the name of Hoyle and developed by him and his collaborators.

The fundamental postulate of this theory is that matter (presumably hydrogen) is being continuously created in the Cosmos. If this tenet be accepted then it will be found that a consistent theory of galactic and stellar evolution has been built up.

On this view it would appear that the synthesis of the elements is a ' local ' operation in which the *locus* is a star, and stars in all stages of their evolution are available for our observation. The Universe thus maintains a constant average composition in which continuous creation balances continuous expansion. In such a system there is no progress as a whole—a point made many years ago by the late Dean Inge in a somewhat different context.

From the more detailed nuclear-chemical aspect we should observe that consensus of opinion favours the views that:

(*a*) All the elements have been synthesized from hydrogen.

(*b*) The necessary physical conditions for these syntheses can be found in stellar interiors.

(*c*) There is a decided non-uniformity of chemical composition of the Universe.

(*d*) The supernova explosion is a major factor in the synthesis of the heavy elements.

The various problems have been much clarified by the more detailed knowledge of nuclear reactions now available.

Burbidge, Burbidge, Fowler, and Hoyle (1957) find eight different processes necessary to account for the observed features of the abundance curve:

(1) Hydrogen ' burning ' to helium.
(2) Helium ' burning ' to carbon, oxygen, neon, and perhaps Mg, etc.
(3) Alpha-particle processes.
^{16}O and ^{20}Ne to ^{24}Mg, ^{28}Si, ^{32}S, ^{36}Ar, and ^{40}Ca.
(4) Equilibrium ' *e-process* ' accounting for the iron peak.
(5) The ' *s-process* ': production and capture of neutrons at a relatively slow rate.
(6) The ' *r-process* ': neutron capture on a fast time-scale.
(7) The ' *p-process* ': production of proton-rich isotopes.
(8) The ' *x-process* ': production of Li, Be, and B.

The hypothesis that elements are being continuously formed in stars is supported by an impressive array of evidence. Not all elements are manufactured in all stars nor are the proportions of the elements manufactured constant from one object to the next.

All stars initially convert hydrogen into helium. Some may manufacture heavier elements but only a rather massive star can produce the heaviest elements in the Periodic Table.

Certain heavy atoms appear to be produced only under catastrophic conditions. The manufacture of elements appears to be a many-stage process.

An outline of the mechanisms proposed for nucleosynthesis follows.

Commencing with a local condensation of interstellar gas, contraction occurs progressively with release of gravitational potential energy in the form of heat, which raises the interior temperature of the star. At a temperature of 10^6 °K thermonuclear reactions set in, the net result of which is the conversion of hydrogen into helium. Further gravitational contraction and heating ensue and at 10^8 °K the reaction leading to the formation of $^{12}_{6}C$ from three helium nuclei is postulated. Then may follow reactions producing neutrons and neutron capture on a slow time-scale leading to $^{56}_{26}Fe$. At 10^9 °K heavy-ion and photonuclear reactions cause the loss of some and the formation of other heavy nuclides.

A state of thermonuclear equilibrium may now be attained.

With further supply of energy by gravitational contraction, at 5×10^9 °K iron nuclei may break down into helium nuclei. This leads on to the dramatic implosion of the star interior on a time-scale of seconds with the collapse of the envelope also, and emission of immense amounts of energy and ejection of large masses of material into interstellar space. The star is now a supernova and during the explosive phase a neutron blast with density of the order of 10^{32} n/cm^2 s occurs and this provides the necessary conditions for neutron capture by medium nuclei on a fast time-scale forming nuclides up to a mass number of 287. These very heavy particles undergo beta-decay and fission, and in this way are the various heavy nuclides formed. The residue of a supernova constitutes a ' white dwarf '.

The significance of these and other related theories is that they seek to show how the necessary conditions for the synthesis of heavy elements may develop in a star, and they do not require any overall scheme of spontaneous cosmic creation. The use of the conception of neutron capture on a fast time-scale gets over many difficulties, as it was formerly thought that on a purely thermonuclear basis the necessary conditions for heavy element synthesis could never be attained in the course of stellar evolution. Attention may now be drawn to a number of points of some interest to the chemist:

(*a*) All possible stable elements appear to exist on the Earth.

(*b*) The presence of the element 43, technetium, which does not exist naturally on Earth as it has a rather short life, has been established in certain stars, which strongly suggests that nucleosynthesis is therefore taking place.

(*c*) The list of naturally occurring elements is complete and any ' new ' element will be an artificial one and will be inherently unstable.

(*d*) The elements found on Earth are at least as old as the solar system, with the exception of helium, argon, and other disintegration products of the radioactive elements.

(*e*) A considerable residue containing heavy elements remains locked up in the ' white dwarf ' stars.

Temperature Limits of Chemical Change

Stars are classified in order of diminishing temperature as follows:

O	30000–50000°K	
B	12000–28000°K	
A	7800–10000°K	
F	7700– 6100°K	
G	5000– 6000°K	
K	3900– 5000°K	CN, OH, and CH present.
M	– 3800°K	TiO present.

Thus the first evidence of chemical combination appears at 5000°K and there is some analogy between the radicals found and those which are detected in flames.

It is evident that, chemically speaking, the Cosmos is a vast desert, sparsely sprinkled with oases in the shape of class K and M stars and their attendant solar systems, if any.

In these locations alone are chemical reactions possible, all other material concentrations being at too high a temperature for stable electronic systems to exist—in other words no chemical compounds can be formed.

CHAPTER 2

Geochemical Considerations

Geochemistry is concerned with the abundance, distribution, and transformations of the elements in the Earth. The outline of the subject which follows is to enable the reader to understand the principles which are involved in connexion with the individual elements when these are discussed in the chapters which follow.

Several admirable textbooks on geochemistry are available and should be consulted for a fuller knowledge of the subject.

Main Facts about the Earth

The Earth, which is the only part of the Universe of which we have tangible knowledge,[1] is the third planet outwards from the Sun. The Sun itself is rated as a class K star. The origin of the Earth and the solar system is still a matter of keen speculation. Many modern theories tend to the view that the Earth began as a cold body but its interior subsequently became heated, partly by gravitational contraction and partly by radioactivity. We have direct physical evidence that the Earth has a layered structure. The mass of the Earth is nearly 6×10^{27} g or 60 million geogrammes (1 Gg = 10^{20} g).

With an average density of 5·517 g/cm^3 and a surface density of 2·64, it is clear that the Earth is a non-uniform body. Apart from the gaseous envelope, the atmosphere, and the incomplete liquid layer, the hydrosphere, there is definite evidence of discontinuities at certain depths.

At approximately 5–30 km the elastic properties of the solid rocky crust suddenly change—this is the Mohorovičić discontinuity—and the zone beneath it, which is called the mantle, seems to extend down to 2900 km. Below this is a layer which behaves as a liquid, and there is some evidence that at the centre of the Earth there is a solid core. The average density of the mantle is about 5, and of the zone beneath it about 10.

[1] We have in fact also tangible knowledge of meteorites which come to us from (probably) interplanetary space.

The chemical character of the core and mantle is still a matter of speculation but consensus of opinion favours a silicatic mantle of composition somewhat like that of an ultrabasic igneous rock and a core consisting mainly of iron.

At the present time direct sampling of the mantle has not yet been achieved but it seems likely that this will be done in the fairly near future.

It must be emphasized that the whole of our chemical raw materials are derived from the three upper layers of our planet—the atmosphere, the hydrosphere, and the crust—and it is with these that we shall in this book be mainly concerned.

The dimensions of the Earth also determine another very important character—the escape velocity for particles projected upward against the gravitational field of the planet. This is given by the formula

$$V^2 = \frac{2GM}{R}$$

where G is the universal gravitational constant, M is the mass, and R is the radius.

For the Earth, V works out at approximately 11 km/s. The escape velocity is of significance in relation to the gaseous envelope which surrounds a planet, as it determines what gases can be permanently retained in its atmosphere. Planets as small as Mercury and the Moon are unable to retain a permanent atmosphere, planets as large as Saturn and Jupiter will permanently retain all gases, but the Earth, together with Mars and Venus, occupy an intermediate position. This means that loss of lighter gases takes place and three factors are involved—time, temperature, and density of the gases. Therefore the escape velocity is a very important factor in the geochemical differentiation of a planet.

Temperature

The temperatures met with in and on the Earth are of great chemical and biological significance. On the Earth's surface the subject forms an important aspect of the studies of climate and meteorology. In regard to the interior there is some direct evidence and deductions can be made from geophysical data.

It seems unlikely that at the centre the value of 4000 °C is exceeded. Neither energetically nor chemically is the Earth a closed isolated system. It receives massive quantities of radiation from the Sun and other heavenly bodies whilst at the same time losing heat into space.

Chemically there is a steady loss by escape from the upper atmosphere of hydrogen and helium molecules, and a steady accretion of meteoritic material at a rate of perhaps 1000–10000 tons daily. There is in addition presumably a minute intake of matter in the form of primary cosmic ray particles which may be of inter-stellar or extra-galactic origin and which in fact contribute very high-intensity energy as well as matter.

Earth Models

Models of the Earth's internal structure are based on physical and chemical evidence including the above facts about the Earth, particularly the seismic data on discontinuities, the supposed analogy between the Earth and meteorites, and the known behaviour of metallurgical melts and slags.

There is a consensus of opinion in favour of the view that the core of the Earth is metallic and predominantly of iron. This view goes back to at least 1866 when Daubrée proposed it. The alternative of a non-metallic core has been postulated by Kuhn and Ritmann (1941) who considered that the interior consisted of largely undifferentiated solar material having special physical characters on account of the high pressure.

Ramsey (1948) also considered the core was due to a phase change, due to pressure, of the magnesium–iron silicatic material of which the mantle is generally held to be composed.

Neither of these theories has achieved great popularity and that of Kuhn and Ritmann now seems to be out of favour. It has also been shown by Elsasser that quantum theory calculations cannot satisfactorily account for the known value of the density of the core, merely on the basis of a phase change.

In the various models which have been proposed, the general nature of the crust is not a matter of dispute. The sub-crustal layer called the mantle is, however, variously indicated by different authorities. There is general agreement that immediately beneath the crust there is a sub-silicatic layer of dunite, peridotite, or eclogite.

Goldschmidt postulated a sulphide–oxide layer which was thought to be of quite considerable thickness. This conception is largely derived from the practical knowledge of the behaviour of metallurgical melts in which a metallic layer (regulus) may be overlain by an arsenical layer (speiss), a sulphide layer (matte), and finally by the superincumbent silicate layer (slag).

On this basis one writer has suggested the existence of an arsenide as well as a sulphide layer but this view has not been generally accepted. Whilst the presence of sulphides is accepted by most authorities, the existence of a discrete sulphide layer is open to doubt as it seems to imply a much higher value for the abundance of sulphur than is generally believed to be the case.

As previously indicated, it seems probable that actual penetration of the crust and sampling of the mantle may be achieved in a very few years' time, but of course only the uppermost layers are likely to be reached and the general picture of the Earth's structure must still be largely speculative.

It will be observed that it is in general maintained that the structure of the Earth is a layered one, and this implies a process or processes of differentiation. These will be discussed in a later section of this chapter.

Abundance of the Elements in Nature

The determination of the relative amounts in which the various elements occur in Nature has long been one of the major undertakings of cosmochemistry and geochemistry. From the standpoint of this book it will be well to distinguish between the various tables which have been compiled.

(1) Terrestrial abundances deduced from the results of orthodox chemical analyses in conjunction with geological information. This gives, in fact, the composition of the Earth's crust and is the table of most interest in geochemical matters.
(2) Terrestrial abundances for the Earth as a whole deduced mainly from analytical data from meteorites and consideration of the relative proportions of the different types.
(3) Solar abundances deduced from the quantitative study of spectral lines of the elements in the solar atmosphere with due allowance for peculiar conditions of excitation.
(4) Cosmic abundances derived from a carefully considered combination of (1) to (3) together with much additional matter derived from spectroscopic studies of stars, nebulae, and information about primary cosmic ray particles.

It is now recognized that a decided non-uniformity of composition is characteristic of the Universe and this has been taken account of in the development of modern theories of element synthesis as noted in the previous chapter. It may in fact be said that studies of the cosmic

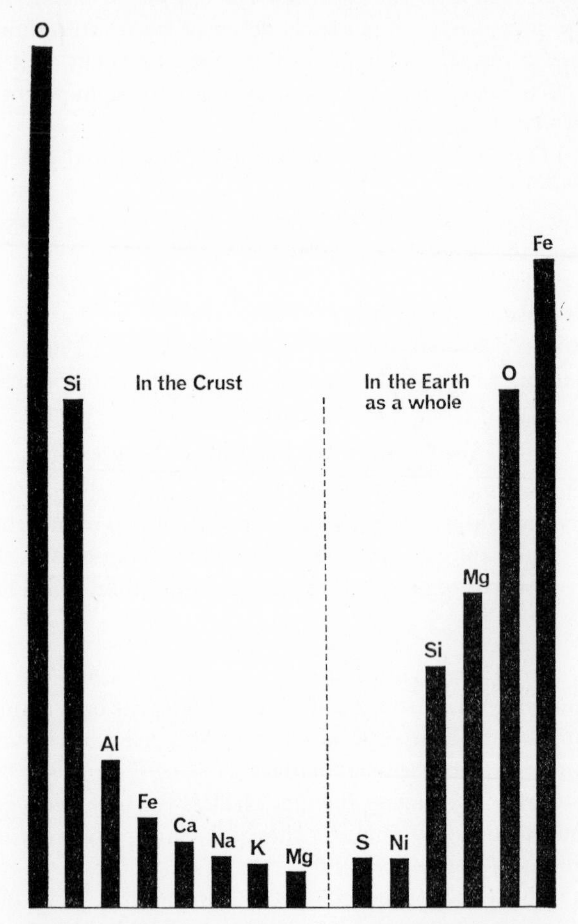

FIG. 2.1. Relative abundance of the commoner elements.

abundance of the elements are mainly of interest in connexion with theories of their origin.

A consideration of the following table of values will show that there are very great differences between solar and cosmic abundances on the one hand, and terrestrial ones on the other.

ABUNDANCE OF THE ELEMENTS

Element	I	II *Cosmic* (log)	III *Earth's Crust* (log)	IV *Solar* (log)	V *Earth's Crust* (g/t)	VI *Meteorites* (log)
Hydrogen	H	10·5		10·5	1400	
Helium	He	9·5		9·5		
Lithium	Li	2·0	2·67	$\bar{1}$·46	30	
Beryllium	Be	1·30	1·35	0·86	2	0·19
Boron	B	1·38	1·45		3	1·58
Carbon	C	7·10	3·43	7·22	320	
Nitrogen	N	6·55	2·03	6·48	46	
Oxygen	O	7·45	6·47	7·46	466000	
Fluorine	F	4·5	3·57		700	2·52
Neon	Ne	7·20	$\bar{4}$·56		<0·001	
Sodium	Na	4·80	5·10	4·80	28300	4·68
Magnesium	Mg	5·90	4·94	5·90	20900	5·96
Aluminium	Al	4·72	5·49	4·7	81300	4·91
Silicon	Si	6·00	6·00	6·00	277200	6·00
Phosphorus	P	3·90	3·59	3·84	1180	3·91
Sulphur	S	5·35	3·21	4·80	520	4·99
Chlorine	Cl	4·75	2·63		200	3·55
Argon	Ar	5·38	$\bar{1}$·00		0·04	
Potassium	K	3·32	4·83	3·20	25900	3·56
Calcium	Ca	4·69	4·65	4·65	36300	4·80
Scandium	Sc	1·35	1·65	1·32	5	1·24
Titanium	Ti	3·39	3·97	3·18	4400	3·36
Vanadium	V	2·32	2·38	2·20	110	2·40
Chromium	Cr	3·88	2·29	3·52	200	3·88
Manganese	Mn	3·62	3·27	3·40	1000	4·76
Iron	Fe	5·07	4·96	5·07	50000	5·86
Cobalt	Co	3·25	1·54	3·14	23	3·28
Nickel	Ni	4·45	1·78	4·65	80	4·58
Copper	Cu	3·00	1·94	3·54	45	2·00
Zinc	Zn	2·72	1·79	2·90	65	1·68
Gallium	Ga	0·95	1·44	0·86	15	1·26
Germanium	Ge	1·70	0·19	1·79	2	1·94
Arsenic	As	0·61	0·43		2	1·17
Selenium	Se	1·73	$\bar{1}$·04		0·09	1·26
Bromine	Br	1·15	0·60		3	1·64
Krypton	Kr	1·71			<0·001	
Rubidium	Rb	0·85	2·14	0·98	120	1·16
Strontium	Sr	1·20	2·72	1·1	450	1·60
Yttrium	Y	0·95	1·51	0·75	40	0·95
Zirconium	Zr	1·00	2·24	0·73	160	2·20

ABUNDANCE OF THE ELEMENTS (*continued*)

Element	I	II *Cosmic* (log)	III *Earth's Crust* (log)	IV *Solar* (log)	V *Earth's Crust* (g/t)	VI *Meteorites* (log)
Niobium	Nb	0·00	1·42	0·45	24	$\bar{1}$·92
Molybdenum	Mo	0·38	0·01	0·40	1	0·90
Technetium	Tc					
Ruthenium	Ru	$\bar{1}$·84	$\bar{3}$·00	$\bar{1}$·93	0·001	0·49
Rhodium	Rh	0·30	$\bar{3}$·00	$\bar{1}$·28	0·001	$\bar{1}$·96
Palladium	Pd	$\bar{1}$·76	$\bar{2}$·56	$\bar{1}$·71	0·01	$\bar{1}$·86
Silver	Ag	$\bar{1}$·36	$\bar{2}$·97	$\bar{2}$·64	0·1	$\bar{1}$·86
Cadmium	Cd	$\bar{1}$·95	$\bar{1}$·13	$\bar{1}$·96	0·2	0·45
Indium	In	$\bar{1}$·25	$\bar{2}$·99	$\bar{1}$·66	0·1	$\bar{1}$·45
Tin	Sn	0·07	0·23	0·04	3	1·42
Antimony	Sb	$\bar{1}$·45	$\bar{1}$·25	0·44	0·2	$\bar{1}$·72
Tellurium	Te	1·55	$\bar{3}$·16		0·002	$\bar{1}$·08
Iodine	I	$\bar{1}$·80	$\bar{1}$·38		0·3	$\bar{1}$·08
Xenon	Xe	0·56			$<$0·001	
Caesium	Cs	1·66	0·58		1·0	$\bar{1}$·98
Barium	Ba	0·58	1·87	0·60	400	0·90
Lanthanum	La	0·60	1·11		18	0·37
Cerium	Ce	0·69	1·59		46	0·37
Praseodymium	Pr	$\bar{1}$·16	0·60		6	$\bar{1}$·95
Neodymium	Nd	1·23			24	$\bar{1}$·53
Promethium	Pm					
Samarium	Sm	$\bar{1}$·39	0·64		7	0·02
Europium	Eu	$\bar{2}$·98	$\bar{1}$·86		1·0	0·49
Gadolinium	Gd	$\bar{1}$·55	0·62		6	0·21
Terbium	Tb	$\bar{2}$·75	$\bar{1}$·76		0·9	$\bar{1}$·71
Dysprosium	Dy	$\bar{1}$·58	0·45		5	0·28
Holmium	Ho	$\bar{2}$·89	$\bar{1}$·87		1	$\bar{1}$·75
Erbium	Er	$\bar{1}$·34	0·18		3	0·21
Thulium	Tm	$\bar{2}$·58	$\bar{1}$·07		0·2	$\bar{1}$·45
Ytterbium	Yb	$\bar{1}$·28	0·20	0·03	3	0·16
Lutecium	Lu	$\bar{2}$·56	1·63		0·8	$\bar{1}$·65
Hafnium	Hf	$\bar{2}$·4	0·23		5	$\bar{1}$·85
Tantalum	Ta	1·25	0·09		2	$\bar{1}$·42
Tungsten	W	$\bar{1}$·10	0·06		1	1·16
Rhenium	Re	$\bar{1}$·40	$\bar{2}$·43		0·001	$\bar{3}$·11
Osmium	Os	$\bar{1}$·90	$\bar{4}$·73		0·001	$\bar{1}$·95
Iridium	Ir	$\bar{1}$·70	$\bar{4}$·73		0·001	$\bar{1}$·69
Platinum	Pt	$\bar{1}$·60	$\bar{3}$·40		0·005	0·38
Gold	Au	$\bar{1}$·16	$\bar{3}$·00		0·005	$\bar{1}$·32
Mercury	Hg	$\bar{1}$·25	$\bar{2}$·60		0·5	$\bar{3}$·85

ABUNDANCE OF THE ELEMENTS (*continued*)

Element	I	II *Cosmic* (log)	III *Earth's Crust* (log)	IV *Solar* (log)	V *Earth's Crust* (g/t)	VI *Meteorites* (log)
Thallium	Tl	$\bar{1}$·05	1·81		1·0	$\bar{1}$·03
Lead	Pb	0·00	0·87	$\bar{1}$·83	15	0·18
Bismuth	Bi	$\bar{1}$·00	$\bar{2}$·99		0·2	$\bar{1}$·08
Polonium	Po					
Astatine	At					
Radon	Rn					
Francium	Fr					
Radium	Ra					
Actinium	Ac					
Thorium	Th	$\bar{2}$·5	0·75		10	
Protactinium	Pa					
Uranium	U	$\bar{2}$·20	0·18		2	

The following observations may be made:

(1) Hydrogen is cosmically much the most abundant of the elements, followed by helium, and the two make up well over 99·8 % of the total number of atoms in the universe.
(2) Cosmically the abundance of the different elements decreases roughly exponentially with atomic number to about 30, after which the abundance is very roughly constant.
(3) Terrestrially the most abundant element is oxygen in the crust, and for the Earth as a whole it is iron which is most abundant.
(4) About 90 % of the Earth is made up of Fe, O, Si, and Mg and about 99·9 % is made up of only fifteen elements.

It is evident that two significant points arise from the above:

(*a*) There is no direct indication that the Earth is derived from solar material—whatever its origin there has clearly been marked differentiation and loss of light elements as compared with the probable constitution of stellar and nebular matter.
(*b*) The relative terrestrial abundances differ very greatly from the abundance of compounds of the elements in common chemical use—*e.g.*, as seen on the shelves of a chemical store. It is clear that practical availability and utility of the elements bears little relation to their natural distribution.

In concluding this section it is to be noted that whilst for the major elements the abundance values are now reasonably well established, the values for the minor ones (and therefore the majority) are subject to almost continuous revision as improved data come to hand. Furthermore it has been shown that the accuracy of the analyses for the less abundant elements in mineral and rock specimens is much lower than has often been supposed. Therefore theories based on abundance values require to be applied with due caution.

Pre-geological Phase of the Earth

The present view is that the Earth has existed as a separate body for about $4 \cdot 5 \times 10^9$ years.

Radioactive dating methods, which have now attained some degree of precision, and which are sufficiently varied to allow some cross checking, enable the age of certain rocks and mineral deposits to be determined.

The oldest formations appear to be of the order of $3 \cdot 3 \times 10^9$ years and it may be assumed that the physical conditions of the Earth's surface layers have not greatly changed during the last 3×10^9 years. In other words during this vast period the Earth has had a crust of silicate rocks, an incomplete watery layer, and an atmosphere not very different chemically from the present one.

In these conditions geological processes of much the same kinds as at the present, have probably been going on—orogeny, intrusion, extrusion, weathering, sedimentation, and metamorphism.

The period of about $1 \cdot 5 \times 10^9$ years before the above is therefore the pre-geological phase and its nature is obviously hypothetical. The older view, in which it was postulated that the Earth condensed from a polytropic gaseous sphere derived from the Sun, is now rivalled by the more recent view that it has been derived from the accretion of a cosmic dust cloud, initially cold. In both theories it is agreed that there has been a high-temperature stage.

However, the relative abundance of the heavy inert gases (which form no compounds) on the Earth is very low compared with the cosmic abundances.

This fact undoubtedly favours the solid accretion and ' cold ' origin theory because a planet formed by hot gas condensation would be likely to have retained elements such as xenon and krypton in something like their solar abundances.

It has been shown that a large dust cloud would contract gravitationally and over a period of the order of 10^9 years would heat up partly from this cause and partly from radioactivity, and attain an internal temperature of upward of 2000 °C leading to fusion and liquid layer differentiation on a physico-chemical basis.

The liquid or partly liquid spheroid would cool on the surface to produce ultimately the proto-crust of the Earth of which no evidence now exists. In addition the embryonic planet would be shrouded in a dense proto-atmosphere formed by release of occluded volatiles from the solid particles. The nature of this atmosphere is speculative but it may well have contained hydrogen, helium, water vapour, nitrogen, ammonia, methane, and carbon dioxide. It is improbable that free oxygen was present.

Chemical Energetics

There is no reason whatever to suppose that the chemical reactions which take place in various natural processes differ in kind from those which are arranged under the artificial conditions of the laboratory. The main ways in which natural reactions seem to be of a different type is that they commonly involve temperatures, pressures, and time intervals which may be greatly beyond those at present attainable experimentally. It is quite justifiable to apply the same laws and principles to the study of geochemical changes as to ' ordinary ' chemical reactions. In particular it is desirable to deal briefly with the topic of energy transformations related to chemical reactions which is usually termed thermodynamics. Thermodynamics pre-supposes no knowledge of the nature of matter, structure of atoms, etc., and the more important aspects can be indicated by means of equations.

The first law, which states the indestructibility of energy, has its counterpart in the law of conservation of matter. (In view of modern knowledge of nuclear reactions the two laws are now in effect combined.)

$$\Delta E = q - w$$

where ΔE is the energy change in a system, q the amount of energy absorbed, and w the mechanical work done. From this can be derived the differential equation

$$\mathrm{d}E = \mathrm{d}q - P\mathrm{d}V$$

where P represents a pressure against which the volume change V occurs.

The second law deals with the somewhat difficult concept of *entropy* which may be described as a measure of the degree of disorder of a system. This law may be stated in many ways, such as

$$dS = d/qT$$

where dS is the change in entropy of a system which is measured by the heat (q) received divided by the absolute temperature. The above applies to a *reversible* process. If the process is *irreversible* then

$$dS > d/qT$$

The internal energy of a system is related to another quantity known as the Gibbs' free energy G such that

$$G = E - TS + PV$$

Where a change takes place at constant temperature and pressure then

$$dG = dE - TdS + PdV$$

so that

$$dG = 0$$

Thus it follows that when a state of equilibrium has been attained, and a close approximation to this is found in many geochemical processes, the free energy of the reactants and that of the products is equal.

When ΔG has a large *negative* value the reaction tends to proceed almost to completion whilst a large *positive* value indicates a tendency to proceed in the opposite direction.

From these conceptions a quantitative statement of Le Chatelier's principle may be derived. This principle, well known in physical chemistry, states that for a system in equilibrium a change in any of the factors which determine the conditions will cause the equilibrium to shift so as to eliminate the effect of the change.

The value of thermodynamics in geochemistry is due in particular to its ability to forecast conditions under which a chemical reaction can or can not take place. It has already proved valuable in indicating the temperature and pressure under which certain minerals can be formed.

There are likely to be greatly increased applications of this kind as thermodynamic data for mineral substances become available to a greater extent.

It should be noted that thermodynamical methods are not concerned with the *rate* or the *mechanism* of a chemical reaction. A further point which will become more apparent later is that a mineral species may have been formed under conditions in the geological past when it was the most stable form then possible. Subsequently, as a result of Earth movements, etc., it may pass into a situation in which it is inherently unstable and therefore is undergoing some process of chemical change —often spoken of as ' alteration ' or ' decomposition '.

For a more adequate treatment of thermodynamics, reference to a standard textbook of physical chemistry is advised.

Geochemical Affinities of the Elements

In the previous section reference was made to the probability of the primary differentiation of the zones of the Earth in the pre-geological stage of its evolution. It is implied that the elements would distribute themselves between a metallic phase, a possible sulphide phase, a silicate melt phase, and a gaseous phase.

To predict the nature of the zones and their process of formation is obviously very difficult in detail because it involves both physical and chemical properties of the elements, and the possible compounds they could form under the particular thermodynamic conditions involved.

In regard to the atmosphere, the substances present are in some cases elements and in other cases compounds.

Some elements like carbon and boron are of low volatility but may readily form compounds which are gaseous like CO_2, CH_4, and BF_3, others like silicon and many of the metals have compounds which are non-volatile or of very low volatility like the elements themselves. A few volatile or gaseous elements readily form compounds of much lower volatility than themselves—*e.g.*, oxygen and the halogens. Hydrogen in the form of water is also more likely to be retained than the free element.

The liquid phases would be essentially the system

Fe–Mg–Si–O–S

Oxygen would of course be greatly in excess of sulphur and in any case the electropositive elements would be in excess, leading to the absence of free silicon, sulphur, and oxygen. Thus the system would in fact be

Metal–sulphide–silicate

Goldschmidt believed that the separation would be analogous to that in certain metallurgical melts. Broadly speaking those elements whose free energy of oxidation exceeds that of iron (59·4 kcal per gramme atom of O) would enter the silicate phase. Elements having a greater energy of formation of sulphides than cobalt (21·1 kcal per gramme atom of S) would enter a sulphide phase.

The relatively small group of elements which lie between these limits are thought to enter the metallic phase. The relationships are indicated in the following table.

Phase	*Constitution*	*Name*	*Type of Element*
Metallic	Metallic bonds	Siderosphere	Siderophile
Sulphide	Mainly covalent	Chalcosphere	Chalcophile Thiophile Sulphophile
Silicate	Ionic	Lithosphere Oxysphere	Lithophile Oxyphile

It can now be seen that in the main the various elements may be expected to distribute themselves in this way and it is important to emphasize that neither relative volatility nor relative density are determining factors. Such could only be the case if no chemical reactions were involved. Confirmation of this is obtained from the case of uranium, which in consequence of its radioactivity can be easily detected and is known to be located in the outer layers of the Earth only. The high density and low volatility of the element would direct it to the core but its electropositive character determines that it is in fact lithophile.

Thus the elements may be classified into groups as given below. Some elements will be noted as appearing in more than one group and this ubiquity is characteristic particularly of certain elements—*e.g.*, iron and carbon.

The table in fact refers to the Earth in its present and not to its

Siderophile	*Thiophile*	*Lithophile*	*Atmophile*	*Biophile*
Fe, Ni, Co, P, As, C, Ru, Rh, Pd, Os, Ir, Pt, Au, Ge, Sn, Mo, W, Nb, Ta, Se, Te, Re	S, Se, Te, Fe, Ni, Co, Cu, Zn, Cd, Pb, Sn, Ge, Mo, As, Ag, Hg, Pd, Sb, Bi, Ag, Ru, Pt, Ga, In, Tl	O, P, H, Si, Ti, Zr, Hf, Th, Sn, F, Cl, Br, I, B, Al, Ga, Sc, Y, La, U, V, Cr	H, O, N, C, Cl, Br, I, He, Ne, A, Kr, Xe	C, H, N, O, S, Cl, I, B, Ca, Mg, K, Na, V, Mn, Fe, Cu, Co, Zn, P, F

primordial state. The main difference is in the atmophile elements and those classed as biophile—present in living organisms.

A somewhat broader and simpler arrangement can be applied to the Periodic Table as shown below.

Lithophile and Atmophile

H																	He
Li	Be	B											C	N	O	F	Ne
Na	Mg	Al											Si	P	S	Cl	A
K	Ca	Sc	Ti	V	Cr	Mn	Fe	Co	Ni	Cu	Zn	Ga	Ge	As	Se	Br	Kr
Rb	Sr	Y	Zr	Nb	Mo		Ru	Rh	Pd	Ag	Cd	In	Sn	Sb	Te	I	Xe
Cs	Ba	La	Hf	Ta	W	Re	Os	Ir	Pt	Au	Hg	Tl	Pb	Bi	Po		
	Ra		Th		U	*Siderophile*				*Thiophile*							

A further table is added which indicates the general forms of chemical constitution of the various terrestrial layers.

Zone	*State of Aggregation*	*Mechanical Condition*	*Chemical Character*
Siderosphere	Liquid	Viscous	Metallic bonds
Chalcosphere	Liquid?	?	Metallic and covalent
Lithosphere	Solid Crystalline	Rigid	Ionic bonds
Hydrosphere	Liquid—solution	Partial convection	Molecules–H–bonds Free ions
Biosphere	All states	Partly free moving	Complex macro-molecules Covalent bonds, etc.
Atmosphere	Gas	Free convection	Simple molecules
Ionosphere	Gas	Static	Gaseous ions

It will be clear from the above that Nature reaches its highest degree of chemical complexity in the biosphere.

Formation of the Crust: Crystallization

Although the primordial crust of the Earth is now no longer to be seen (as far as is known), it may be reasonably assumed that it consisted mainly of silicatic materials, and the following discussion would therefore apply equally to the original crust and the existing assembly of 'igneous' rocks which from geological evidence we believe to have cooled from a rock melt or magma. Such rocks—described more fully in the next chapter—are to a large extent crystalline, formed predominantly of silicates and substantially ionic in chemical structure. A more detailed treatment of silicates will be found in Chapter 14. In this section an outline of the essential principles of crystal formation in its geochemical aspects will be given. Crystals may separate from a melt or from a solution. In geochemistry the distinction between the two is often ill-defined but the controlling factors in each case are not very different.

The structure of crystalline ionic solids is largely determined by geometry and electrical stability. The first requirement, which is governed by the relative sizes of the ions, determines the most effective way in which they can fit into the structure. This will in fact be the arrangement with the lowest potential energy. Geochemically, the radius ratio of the numerous cations in relation to oxygen is the most important factor because, on account of its abundance and large size, oxygen is by far the most important of the anions involved in rock formation. In fact, on a volume basis, 93·77 % of the Earth's crust consists of oxygen —hence the term ' oxysphere ' rather than ' lithosphere '.

The number of oxygen ions which can be packed around a cation gives the co-ordination number of the cation, and this may be as low as 3 and as high as 12. This is deducible by geometry and the predicted numbers agree with those actually observed. It will be noted that the smallest ions have the smallest co-ordination numbers.

A particularly important case is four-fold co-ordination which gives rise to a tetrahedron of oxygen ions of great stability. In Nature it is for example met with as $—SiO_4$, $—PO_4$, $—SO_4$. The size of the ions is related not merely to that of the original atom but also to the valency developed.

A cation has an electron deficit and is naturally smaller than its parent atom. This is not merely from loss of electrons but also because of the contracted electron orbitals due to the relatively increased attraction of the positive nucleus. The same reasons in reverse explain the larger size of the common anions.

The unit employed in these measurements is the ångström ($1\,Å = 10^{-8}$ cm) and the values were originally obtained from X-ray diffraction studies on crystals.

Electrical stability of an ionic structure requires that the sum of the positive and negative charges must balance. Pauling's rule that, ' In a stable structure the total strength of the valency bonds which reach an

RADII OF CATIONS FOR SIX-FOLD CO-ORDINATION

(After Ahrens) (Radii in ångström units)

Ion	*Radius*	*Ion*	*Radius*	*Ion*	*Radius*
N^{5+}	0·13	Co^{2+}	0·72	Cu^{+}	0·96
C^{4+}	0·16	Cu^{2+}	0·72	Bi^{3+}	0·96
B^{3+}	0·23	Ge^{2+}	0·73	Cd^{2+}	0·97
S^{6+}	0·30	Zn^{2+}	0·74	Gd^{3+}	0·97
P^{5+}	0·35	Sb^{3+}	0·76	Na^{+}	0·97
Be^{2+}	0·35	Hf^{4+}	0·78	Eu^{3+}	0·98
Si^{4+}	0·42	Zr^{4+}	0·79	Ca^{2+}	0·99
Se^{6+}	0·42	U^{6+}	0·80	Sm^{3+}	1·00
Al^{3+}	0·51	Mn^{2+}	0·80	Th^{4+}	1·02
Cr^{6+}	0·52	In^{3+}	0·81	Nd^{3+}	1·04
Ge^{4+}	0·53	Sc^{3+}	0·81	Pr^{3+}	1·06
Te^{6+}	0·56	Tb^{4+}	0·81	Ce^{3+}	1·07
I^{5+}	0·62	Pb^{4+}	0·84	Hg^{2+}	1·10
W^{6+}	0·62	Lu^{3+}	0·85	Sr^{2+}	1·12
Ga^{3+}	0·63	Yb^{3+}	0·86	Pb^{2+}	1·20
Co^{3+}	0·63	Tm^{3+}	0·87	Ag^{+}	1·26
Fe^{3+}	0·64	Er^{3+}	0·89	K^{+}	1·33
Mg^{2+}	0·66	Ho^{3+}	0·91	Ba^{2+}	1·34
Mn^{3+}	0·66	Pr^{4+}	0·92	Ra^{2+}	1·43
Ti^{4+}	0·68	Y^{3+}	0·92	Rb^{+}	1·47
Li^{+}	0·68	Tb^{3+}	0·93	Tl^{+}	1·47
Ni^{2+}	0·69	Ce^{4+}	0·94	Cs^{+}	1·67
Sn^{4+}	0·71	Tl^{3+}	0·95		

anion from neighbouring cations equals the charge on the anion ', indicates the tendency for a configuration of minimum potential energy to be assumed, in which the ionic charges are as far as possible neutralized by their immediate neighbours. We thus have in effect a three-fold control on a crystal building process—that it should satisfy geometrical, electrical, and potential energy conditions. It is a consequence of this principle that certain types of compounds do not occur, although valency requirements alone would have suggested their existence.

A classical example is provided by the important rock-forming minerals called feldspars. These are a series of alumino-silicates of mainly sodium, potassium, and calcium. They may, however, contain rubidium and barium, replacing potassium, but lithium cannot replace the other alkali metals nor can magnesium or ferrous iron replace calcium.

The quantitative concept of minimum potential energy is defined by the term ' lattice energy '. This can be calculated for simple binary compounds. It should ultimately be possible to do so also for substances like mineral silicates as adequate and reliable thermodynamic data become available.

It will now be evident that the formation of crystalline minerals is a process in which the size of the ions is all-important. The valencies are much less a determining factor—this being particularly so in the case of the silicates which preponderate in the Earth's crust.

It should be remembered that minerals are in the main ionic assemblies and that the discrete molecules of the organic chemist are not to be met with. The term ' molecule ' is in fact inappropriate. The simple mineral salts of inorganic chemistry come likewise into this category but by long-established custom they are often alluded to as though they were molecular when such in fact is not the case.

When we pass from the ' pure ' materials of the laboratory to the varied assortment of ions with which Nature confronts us it must be appreciated that in the process of crystallization mineral species are formed which seem to ' break the rules ' for distinct chemical substances. It is rather like a military parade where all the soldiers are guardsmen but not necessarily of the same regiment and are assembled without discrimination.

The above remarks translated into more orthodox chemical language imply that the principle of isomorphism is of very great importance in mineral chemistry. We now proceed to discuss some aspects of this topic.

When substances have analogous chemical structures and similar crystal form they are termed isomorphous. The idea is due to Mitscherlich (1819) who employed it to deduce the formula of certain salts, hence the valency of the anion-forming element and so its atomic weight. Isomorphism among minerals has long been studied and used as one of the bases of classification.

With the elucidation of internal crystal structure by X-ray studies the conception has become of increased significance.

Among other things it is now possible to see why analogous compounds of related elements are not always isomorphous because the ionic sizes may constrain them to certain crystal forms. Thus the carbonates of general form MCO_3 are either trigonal or orthorhombic in crystal form. The lower trigonal symmetry applies to barium, lead, and strontium (ionic radii $> 1{\cdot}0$ Å) and the higher orthorhombic symmetry to the carbonates of manganese, ferrous iron, nickel, and magnesium (ionic radii $< 1{\cdot}0$ Å).

Calcium (ionic radius 0·99 Å) is the borderline case and forms two

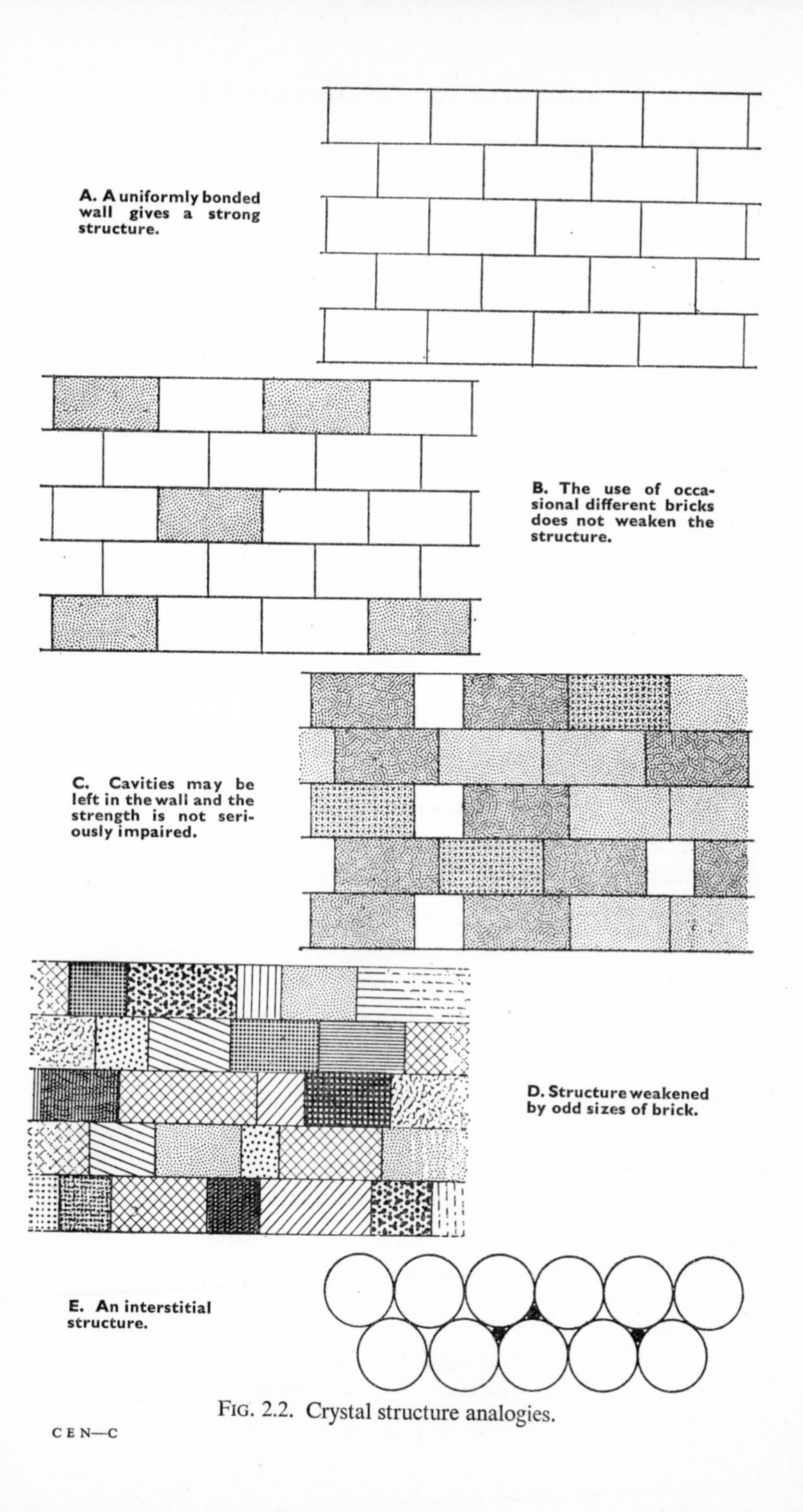

FIG. 2.2. Crystal structure analogies.

carbonate minerals $CaCO_3$. They are calcite (trigonal) and aragonite (orthorhombic). There are numerous other cases of this kind.

The building up of a crystalline mineral closely follows the principles above discussed and it is the rule rather than the exception that there is a variation in the elements whose ions form part of the lattice. When a mineral separates from a magma or solution containing a great assortment of ions then the lattice acts as a kind of sorting mechanism, accepting ions provided that they do not vary greatly in size from the standard. A variation in valency is also possible and can be adjusted in the lattice. Thus we can find many cases of one ion ' deputizing ' for another.

There are certain main types of lattice characteristic of the chief rock-forming minerals and into these lattices numerous minor constituents can enter. There is a tolerance of from 10 % to 15 % for low-temperature minerals to as much as 25 % for those formed at high temperatures.

The following cases should be noted.

(*a*) The ' host ' ion and the ' guest ' ion are of the same valency and of nearly the same size. In this case there is practically equal tendency to enter the lattice.

Examples are Al^{3+} (0·51) and Ga^{3+} (0·62), and Hf^{4+} (0·78) and Zr^{4+} (0·79). These pairs of elements are found closely associated in minerals and are difficult to separate.

(*b*) The ' guest ' ion has the same valency.

(i) Smaller radius in which case it is preferentially accepted.

(ii) Larger radius in which case the bond is weaker with a resultant lowering of melting point. An example is provided by Fe^{2+} (0·74) and Mg^{2+} (0·66) in their orthosilicates.

The magnesium mineral is *forsterite* Mg_2SiO_4 and the iron mineral is *fayalite* Fe_2SiO_4. The second mineral has a melting point 700 °C lower than the first.

(*c*) The ' guest ' ion has about the same size but a different valency. If the valency is higher it is accepted preferentially into the lattice which adjusts itself to maintain electrical neutrality.

Thus calcium replaces sodium in feldspars:

$NaAlSi_3O_8$	$CaAl_2Si_2O_8$
Albite	Anorthite

Ions of lower valency are less readily accepted, thus Li^+ is not normally associated with the other alkali metal ions on account of its smaller size, but it does replace Al^{3+} in such minerals as muscovite.

There are very many cases of isomorphous series in which progressive replacement of one ion by another is encountered.
Examples are:

Forsterite	Olivine	Fayalite
Mg_2SiO_4	$(Mg,Fe)_2SiO_4$	Fe_2SiO_4
Hubnerite	Wolframite	Ferberite
$MnWO_4$	$(Mn,Fe)WO_4$	$FeWO_4$

Crystallization from a Rock Melt

Although the visible and accessible crystalline rocks are to be regarded as localized deposits, however large, they have presumably been formed by processes similar to those involved in the formation of the primary crust of the Earth. The rock melt or *magma* from which they are believed to have separated by cooling is of course far more than a simple silicate melt.

Magmas contain entrapped gases, including steam in large quantity. Silica is the predominant acidic component and the commoner bases include the alkalis, alumina, magnesia, and iron, with of course minor quantities of a host of other lithophile elements.

The crystallization of such a fluid is a progressive differentiation of its components which is governed by physical and chemical properties. The process may be more readily understood by using Niggli's conception of the consolidation of an idealized magma, which represents in all probability a convenient approximation to reality. The melting point–solubility curve for such a system may be shown by means of a diagram such as that in Fig. 2.3 (after Niggli).

With decreasing temperature the system undergoes the following changes:

The presence of even small quantities of component A (water) lowers the melting point of component B and crystallization therefore starts somewhat below the melting point of pure B. As B is removed by crystallization the composition of the liquid phase changes, becoming relatively richer in A, this being initially a slow process. When the flatter part of the curve is reached the rate of crystallization increases and the liquid phase composition becomes very sensitive to temperature changes. At the temperature t_2 the greater part of B has separated and the liquid is now virtually a solution of quite different composition from the original magma. The first stage described above corresponds to the bulk formation of crystalline rocks. The second stage corresponds to the formation of ' pegmatites ', alluded to in Chapter 3. Beyond t_3 the crystallization is slow, and it is in fact the separation of residual components from an aqueous solution. Therefore the whole process is

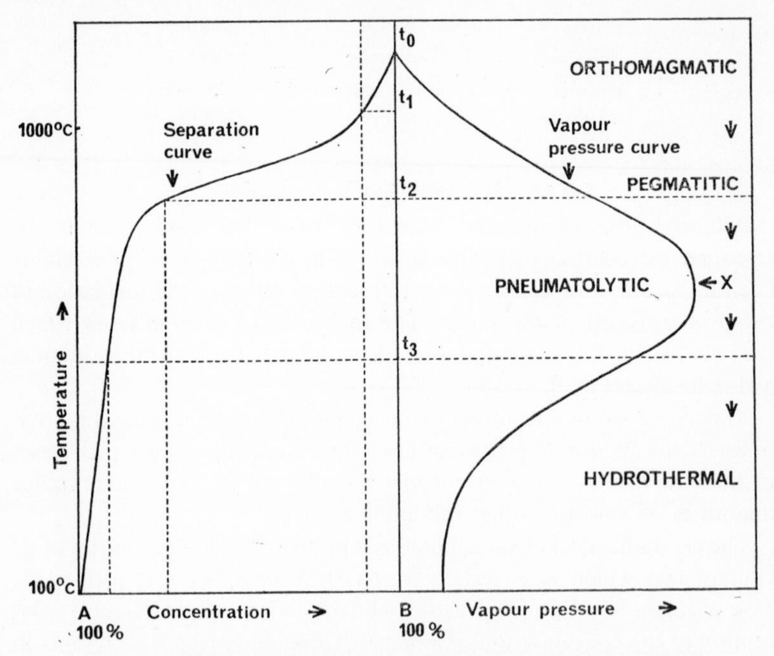

FIG. 2.3. Niggli's diagram for the crystallization of an ideal magma.

envisaged as a gradual change from the solidification of a melt to crystallization of a solution.

It is possible to recognize the following stages in the cooling of actual complex magmas.

(1) Separation of refractory minerals which do not fit into silicate lattices and of the very compact orthosilicate olivine $(Mg,Fe)_2SiO_4$. Some of these by virtue of high density may settle down in the lower levels of the ' magma chamber '.

(2) Main crystallization of the silicates—pyroxenes, amphiboles, feldspars, and residual silica.

(3) Separation of the bulk of the residual silicates and certain other

compounds (rare earths, etc.) under low viscosity conditions with water in its super-critical state.

Following a stage of *increasing* vapour pressure with falling temperature to point X in Fig. 2.3.

(4) Maximum vapour pressure of water and other volatiles. The vapours are mainly acidic and cause dissemination and chemical changes in surrounding rocks.

(5) Below the critical temperature of water crystallization from aqueous solution takes place.

(6) In some cases there is an end-phase of emission of vapours such as steam, CO_2, HCl, etc., often far removed in time and place from the original focus of magmatic activity.

The whole scheme is summarized in the table given below.

CONSOLIDATION OF A MAGMA

Stage	*Products*	*Physical State*
Early magmatic	Orthosilicates, refractory sulphides, refractory oxides	Viscous liquid
Orthomagmatic	Pyroxenes, amphiboles, feldspars, feldspathoids, quartz (silicates)	Viscous liquid
Pegmatitic	Residual silicates, compounds of Li, Be, B, lanthanides, $—PO_4$, Nb, Ta	Low viscosity, rising vapour pressure
High-temperature hydrothermal	Compounds of W, Sn, Mo, Bi, U, etc.	Low viscosity, supercritical
Low-temperature hydrothermal	Compounds of As, Cu, Zn, Pb, Sb, Hg, etc.	Aqueous solution

It should be clearly understood that idealized schemes such as the above are subject to many variations in reality. Furthermore, the separation of various mineral substances also occurs by many other processes than the above. These points will be more apparent in subsequent chapters.

Williams has put forward (1960) a number of tentative conclusions in respect of the origin of sulphide ores.

(1) Metallogenetic provinces are fundamentally due to inhomogeneities in the Earth's mantle and to the ascent of ore fluids along deeply penetrating crustal flaws. Once taken into the crust, particular metals may be involved in numerous cycles of geological processes, and yet remain in the same general geographical region.

(2) The pattern of ore districts may be related to the disposition of crustal lineaments, the ore fluids rising along them from deep-seated magmatic sources. Copper–nickel sulphide ores are genetically connected with certain types of basic igneous rocks.

(3) Many base-metal sulphide deposits are derived from residual hydrothermal solutions emanating from granitic intrusions.

(4) On the whole there is no reason to discard the magmatic-hydrothermal theory of ore genesis.

(5) Granitization may lead to the incorporation of metallic compounds in various silicates or to their segregation as mobile ore fluids.

(6) Metamorphism in some cases may produce ore deposits.

(7) Sulphide ores can be remobilized by hydrothermal solutions and deposited elsewhere as secondary or regenerated ores.

(8) Magmatic water is the predominant solvent for ore-forming solutions.

(9) Sedimentary ores formed by submarine exhalations are probably fairly common.

The Atmosphere

The nature of the original or proto-atmosphere of the Earth is a matter of speculation and ideas are somewhat related to the theory accepted for the origin of our planet. It is reasonable to suppose the atmosphere to have contained light gases and it may well have been devoid of free oxygen. The most abundant gas may at first have been hydrogen but it would soon be largely lost as the r.m.s. velocity of its molecules even at quite low temperatures exceeds the critical value for a gas to be permanently retained in the gravitational field of the Earth. During an initial high-temperature phase water would be entirely present as vapour, but with a cooling of the surface below 374 °C (critical temperature of water) the liquid phase could separate and the evolution of the hydrosphere could commence. The fact that water is still so abundant suggests that in the pre-geological stage the Earth's surface was never hot enough for the velocity of water molecules to exceed the critical value. Jeans calculated that if the average molecular velocity of a gas is a quarter of the ' escape velocity ' then that gas will be completely dissipated into space in 50000 years. If the average velocity is two-ninths of the escape velocity then the gas will be lost entirely in 30 million years and for an average velocity of one-fifth of the escape value a period of 25×10^9 years is required. It is evident that in the latter case the gas is virtually permanently retained.

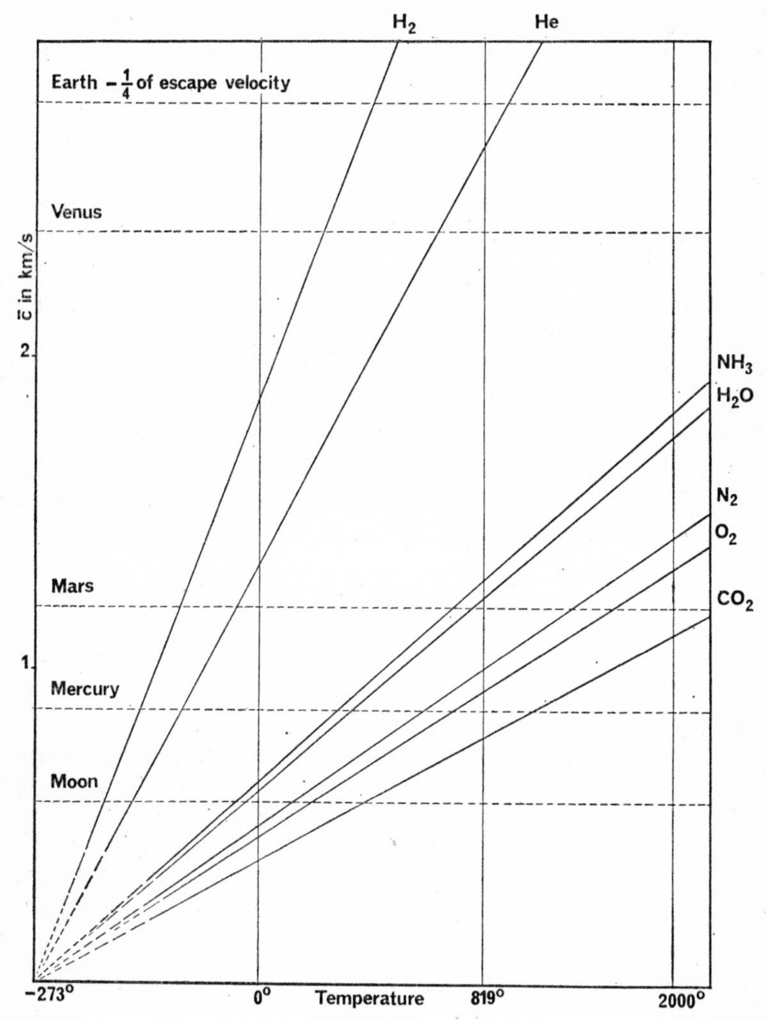

FIG. 2.4. Escape velocity curves for light gases.

A reference to Fig. 2.4 will show that the temperature above which water vapour would not be long retained is of the order of 3000 °C. On the other hand if this is the case there seems no reason why methane and ammonia should not also have been retained. It is therefore clear that other factors are involved. Goldschmidt thought that some of the gases of the primordial atmosphere are now fixed as compounds of chlorine, bromine, iodine, and boron in sea water. It is, however, necessary to consider the possibility of the progressive emission of various gases containing these elements in the form of volcanic emanations during the course of the Earth's history. Poole investigated the nature of the primitive atmosphere from a biogeochemical standpoint. Assuming that the original atmosphere contained no oxygen an attempt was made to calculate its average composition from data supplied from a study of the nature of igneous and sedimentary rocks, and the probable amount of weathering and denudation during geological time. From these considerations it was shown that the total amount of carbon 'fossilized' in the sedimentary rocks could not have been derived from the carbon dioxide present in the primitive atmosphere. The presence of methane in the primitive atmosphere has been suggested by some and discounted by others. The original liberation of oxygen was attributed to the photodissociation of water vapour in the upper atmosphere with escape of the free hydrogen liberated. In this way enough oxygen could have perhaps accumulated to suffice for the needs of plant life.

With the initiation of photosynthesis carbon dioxide would be 'fixed' as organic matter and replaced by an equivalent amount of oxygen. It is important to appreciate that the carbon dioxide biologically equivalent to the amount of carbon now fixed in the form of coal, petroleum, limestone, etc., would represent a partial pressure in the atmosphere far higher than any ordinary form of life could tolerate. It must therefore be assumed that the carbon dioxide has been progressively evolved from the Earth's crust during geological time, perhaps at a rate not greatly different from that maintained at the present time.

It should be observed in this connexion that carbon dioxide is, after steam, the most abundant of volcanic gases. We may now summarize the factors most likely to be responsible for the geochemical evolution of the atmosphere.

(1) The mass of the planet which fixes the escape velocity of gas molecules, with which is also connected the surface temperature. During the cooling phase it may be supposed that the rate of loss of light gases has gradually diminished. A reference to Fig. 2.4 will also show that there is a fairly narrow range of planetary sizes which would favour the retention of the vital gas oxygen and the loss of the lighter gases.

(2) Inorganic chemical reactions leading to the evolution of various

gases from the cooling magma. In many cases these gases would be hydrides of the lighter non-metals such as perhaps H_2O, NH_3, CH_4, B_2H_6, H_2S, and possibly also the halogen hydrides. In many cases these gases would undergo secondary reactions and would ultimately be fixed as a variety of solid compounds in the lithosphere or as soluble ions in the hydrosphere. The methane and ammonia might, however, be the ultimate source of the present atmospheric carbon dioxide and nitrogen. The fact that NH_3 and CH_4 occur in the atmospheres of the giant planets lends some support to this view.

(3) Inorganic reactions leading to the weathering of rocks. These will be more fully considered later. In the meantime it is important to note the significance of the ferrous–ferric oxidation reaction as a probable major factor in the loss of oxygen from the atmosphere.

(4) Biological reactions and in particular the photosynthetic process and the reverse reaction of respiration.

These are further discussed in Chapter 4.

The Hydrosphere

We may assume that this term is literally correct and that we have under consideration a layer of water which covers the greater part (about two-thirds of the surface) of our planet. The hydrosphere must have existed as long as the Earth has been cool enough to allow the condensation of liquid water, but whether it has increased in volume or remained constant is less certain. The present position is that the ocean is a saline solution carrying a large variety of ions and may be considered almost entirely from this point of view. The simplest theory, that the ocean commenced as fresh water and became progressively more saline by extractive processes involved in the weathering of rocks, is not now seriously maintained. It is interesting to note, however, that Joly's calculation of the age of the Earth was based on it. The present salinity of the ocean—and we need not consider purely local variations in enclosed seas—must be the result of a number of factors which are enumerated below.

(1) Solution of the more soluble atmospheric gases when the atmosphere had cooled far enough below the critical temperature of water to allow the main formation of the hydrosphere to take place.

(2) The solution from the solid crust of various ions due to the highly energetic processes which must have taken place during later stages of the pre-geological phase. It is to be noted that the original ocean

may well have been acid in reaction due perhaps mainly to hydrochloric acid. Thus the weathering processes which are now mainly due to the very weak carbonic acid may have initially been extremely rapid. It seems likely that the pH of the ocean has been tending to rise throughout geological time. It is now slightly alkaline and the rate of change during the last 2×10^9 years has probably been very small.

(3) Removal of certain ions by purely inorganic processes. Perhaps the most interesting of these is the virtual elimination of poisonous metals like arsenic and selenium by precipitation on colloidal ferric hydroxide and in other ways. In this sense it may be said that the ocean was being prepared for the advent of life.

(4) Removal of certain ions by direct or indirect biological processes. Most striking here are the massive precipitation of calcium carbonate to form marine skeletal material and its ultimate consolidation as limestone and possibly the precipitation of potassium as the mineral glauconite.

(5) The involvement of some of the ions in inorganic or biological cycles whereby they may temporarily return to the lithosphere or atmosphere.

(6) The progressive addition of some ions by continuous volcanic emanations. This may well be an important factor in the enrichment of the ocean in chloride and carbonate ions and even in water itself, for steam is always a copious volcanic gas. In the carefully reasoned discussion of this subject by Conway and others the general conclusion is that whilst the above factors all operate yet the rate of increase of salinity since the advent of life has been but slight. The following table gives the average composition of the ocean water today and Fig. 2.5 indicates some of the more important processes in its geochemistry.

MORE ABUNDANT IONS IN SEA WATER

Ion	*Percentage*	*Ion*	*Percentage*
Cl	1·8980	Mg	0·1272
Br	0·0065	Ca	0·0400
SO_4	0·2649	Sr	0·0008
HCO_3	0·0140	K	0·0380
F	0·0001	Na	1·0556
H_3BO_3	0·0026		

More than fifty elements have been detected and estimated in sea water in minor quantities down to as low as Au = 0·000008 g/ton.

From its nature and comparative accessibility the geochemistry of the ocean has been the subject of much investigation but there is much which remains to be elucidated. Theories of its evolution are closely related to those dealing with continents and ocean basins. These theories are dependent on geological evidence and as will be seen in the next chapter

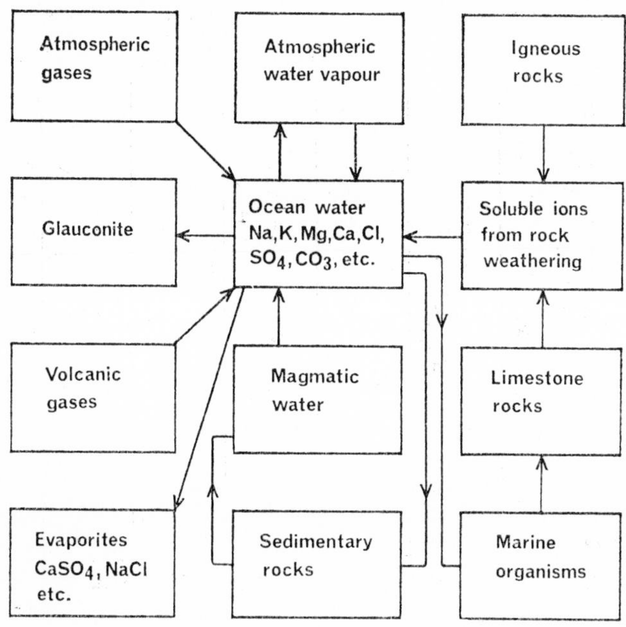

FIG. 2.5. Secondary processes.

the total thickness of sedimentary rocks as a chronological succession represents only about one-sixth of geological time. In consequence the nature, volume, and distribution of the ocean in early pre-Cambrian times is at present a matter largely of speculation. The processes by which the various ions pass into sea water form the subject of the next section.

Weathering Processes

We mean by these the chemical actions between aqueous media and rock materials, and particularly the silicates of the igneous rocks and their related products. As previously noted these silicates were formed at temperatures and pressures in general much higher than those at the

Earth's surface, furthermore the conditions were in general reducing and liquid water was absent. It is not therefore surprising that minerals which are of great stability deep in the crust of the Earth may be quite readily decomposed or ' weathered ' by actions in the presence of water and atmospheric agencies, which take place at or near the surface. The factors which largely determine weathering processes are as follows.

(1) The peculiarities of water as a solvent, particularly its highly polar character which causes it to be attracted to both positive and negative

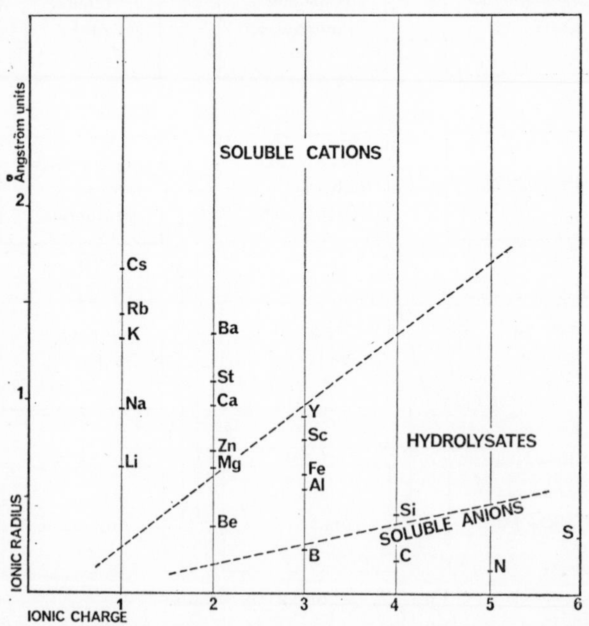

FIG. 2.6. Relation between ionic charge and radius (after Goldschmidt).

ions, its tendency to form complex molecules which give it large latent heats and a long range of liquidity, the ease with which it acts as an ionizing solvent, and its ability under suitable conditions to yield both hydroxonium and hydroxide ions.

(2) Goldschmidt drew attention to the importance of ionic charge in relation to radius in determining the behaviour of ions in aqueous media. The concept of ionic potential thus used is, however, probably better replaced by that of *ionization potential* with which it is closely related. Ionization potentials can now be determined with accuracy by spectroscopic methods and may in fact be used to deduce ionic

radii in certain cases. The table below indicates the way in which ions fall into three groups geochemically. In general those with low ionization potentials tend to form stable cations in solution, those with large values tend to form stable complex anions such as carbonate and sulphate. The ions of intermediate ionization potential tend to form hydrolysates—*i.e.*, to precipitate as hydroxides or hydrated oxides.

IONIZATION POTENTIALS

(After Ahrens)

Ion	*Potential*	*Ion*	*Potential*
Cs^+	3·89	Sc^{3+}	24·8
Rb^+	4·18	Al^{3+}	28·4
K^+	4·34	Zr^{4+}	34
Na^+	5·12	B^{3+}*	37·9
Li^+	5·39	Ti^{4+}	43·24
Ba^{2+}	10	Si^{4+}	45·1
Sr^{2+}	11·03	Ge^{4+}	45·7
Ca^{2+}	11·87	As^{5+}	62·6
Mg^{2+}	15·03	C^{4+}	64·5
Fe^{2+}	16·24	P^{5+}	65
Be^{2+}	18·21	S^{6+}	87·5
La^{3+}	19·2	N^{5+}	97
Y^{3+}	20·5		

* Boron appears to occupy an anomalous position in the series. See Fig. 2.6.

(3) Hydrogen ion concentration which in aqueous media determines the acidity or alkalinity and which is measured by means of pH values. These are on a logarithmic scale and natural media range over at least 11 units of pH. The behaviour of certain ions is much dependent on the pH of aqueous solutions.

(4) Many of the elements exhibit more than one oxidation state in Nature. As a change in oxidation state involves adding or subtracting electrons from the atom certain energy changes are involved and the tendency is measured by the 'redox potential'. In some cases this potential is also controlled by hydrogen ion concentration. A further reference to this subject occurs in Chapter 8.

(5) Colloidal properties which are exhibited by particles in a size range of about 10^{-4} to 10^{-7} cm. In Nature the dispersion medium may be air in the cases of fogs and smokes, but it is more usually water and in special cases liquids like petroleum. The chief significance of the

colloidal state geochemically is that, due to the phenomenon of adsorption, various ions may be attracted to and precipated with certain colloidal materials, particularly ferric hydroxide, hydrated manganese dioxide, and some of the clay minerals.

It will be apparent from the above discussion that the factors which determine the geochemistry of the elements under aqueous conditions are markedly different from those involved in the case of magmatic crystallization. In particular the mobility of aqueous solutions and suspensions can lead to a much greater separation and segregation of the individual elements from each other. Thus it is that economically valuable mineral deposits are often of sedimentary origin whilst concentrations of purely igneous origin are rarer.

In considering the nature of the changes involved in the weathering of igneous rocks it will be understood that basically we are dealing with an aggregate of silicates with or without free silica and in which the ions of the transitional elements tend to be those of the lower valency states —a feature especially characteristic of deep-seated rocks. Thus basalts, gabbros, and olivine rocks are greenish-black in colour in many cases and this is largely due to ferrous compounds. Granites on the other hand are very often pink from exsolution of ferric ions as the oxide Fe_2O_3. In general the silicates which are most stable in deep magmatic conditions are the most susceptible to weathering. Sedimentary rocks containing unchanged feldspars are by no means rare but sands, etc., containing minerals like olivine are exceptions. The weathering process is somewhat analogous to the laboratory treatment of a mixture for analysis by means of acid in order to effect solution. In Nature the acid mainly involved is the very weak carbonic but this is perhaps more than compensated for by the long periods of time which may be involved.

Certain substances are of course completely resistant and remain as an insoluble residue which may frequently undergo physical transport and segregation by means of water. The most important ' resistates ' are quartz, zircon, titania, monazite, magnetite, cassiterite, and the precious metals. As will be seen in Chapter 5 quite a number of elements are extracted from residual deposits of this sort.

Elements whose ions are of intermediate potential can pass into solution initially but they tend to undergo hydrolysis readily. In the case of silica and alumina it commonly happens that the hydrolysates combine together more or less *in situ* to produce the various hydrous aluminium silicate lattices which constitute the clay minerals. The removal of silica—which may occur within a rather narrow pH range—may give residual hydrous aluminium oxides in the form of bauxite.

In the case of iron and manganese, solution as the divalent ions probably takes place first. These ions are fairly stable at moderately low pH values but when they reach the sea the rise in pH into the alkaline region causes oxidation and precipitation as either hydrated Fe_2O_3 or hydrated MnO_2. Important ore deposits of these metals are probably formed in this way in many cases. But the colloidal hydrolysates also tend to adsorb and co-precipitate certain other ions. Thus the positive colloidal $Fe(OH)_3$ adsorbs the anion PO_4^{3-} and the negative colloidal $MnO_2.2H_2O$ adsorbs the cations K^+ and Ba^{2+}. On the other hand the smaller and more readily hydrated cations of Na, Mg, and Ca are not retained in this way.

The ions which reach the ocean in solution are still susceptible of change. Calcium and strontium enter largely into a major biological cycle—they pass into the structure of marine organisms and may ultimately be consolidated as limestones.

The ions still remaining in solution are chiefly those of Na, Mg, and K. Under special localized conditions in inland seas they may separate by a process of fractional evaporation in the general sequence dolomite—gypsum—anhydrite—halite—sylvite—carnallite. These are the evaporites or saline deposits which are of great technical importance and whose origin has been very fully discussed.

The result of the chemical processes of weathering is in most cases the replacement of silicate ions by their equivalent in carbonate ions as far as the more positive metals are concerned. In the case of the metals of intermediate ionic potential like aluminium the hydroxide of the metal would be formed. Owing to subsequent changes the ultimate products as found in sea water may in fact not be carbonates but a variety of other salts, particularly halides. The scheme below represents the reactions which may occur in an idealized form.

$$x(\mathrm{K,Na,Ca,Mg,Fe,Mn})\mathrm{AlSiO_3} + y\mathrm{HCO_3} \longrightarrow$$

$$\begin{array}{lclcl}
\mathrm{K_2CO_3} & & \mathrm{KCl} & & \\
\mathrm{Na_2CO_3} & \longrightarrow & \mathrm{NaCl} & & \\
\mathrm{Mg(HCO_3)_2} & & \mathrm{MgCl_2} & & \\
 & & & \mathrm{(Ca,Mg)CO_3} & \\
\mathrm{Ca(HCO_3)_2} & & \mathrm{CaCO_3} & & \\
\mathrm{Fe(HCO_3)_2} & & \mathrm{Fe(OH)_3} & & \mathrm{Fe_2O_3.}n\mathrm{H_2O} \\
\mathrm{Mn(HCO_3)_2} & \longrightarrow & \mathrm{Mn(OH)_2} & \longrightarrow & \mathrm{MnO_2.}n\mathrm{H_2O} \\
\mathrm{Al(OH)_3} & & \mathrm{Al_2O_3.}n\mathrm{H_2O} & & \\
 & & \text{Clay mineral} & &
\end{array}$$

Fig. 2.7 represents the general scheme of weathering processes which lead to the formation of various kinds of sedimentary rocks, etc.

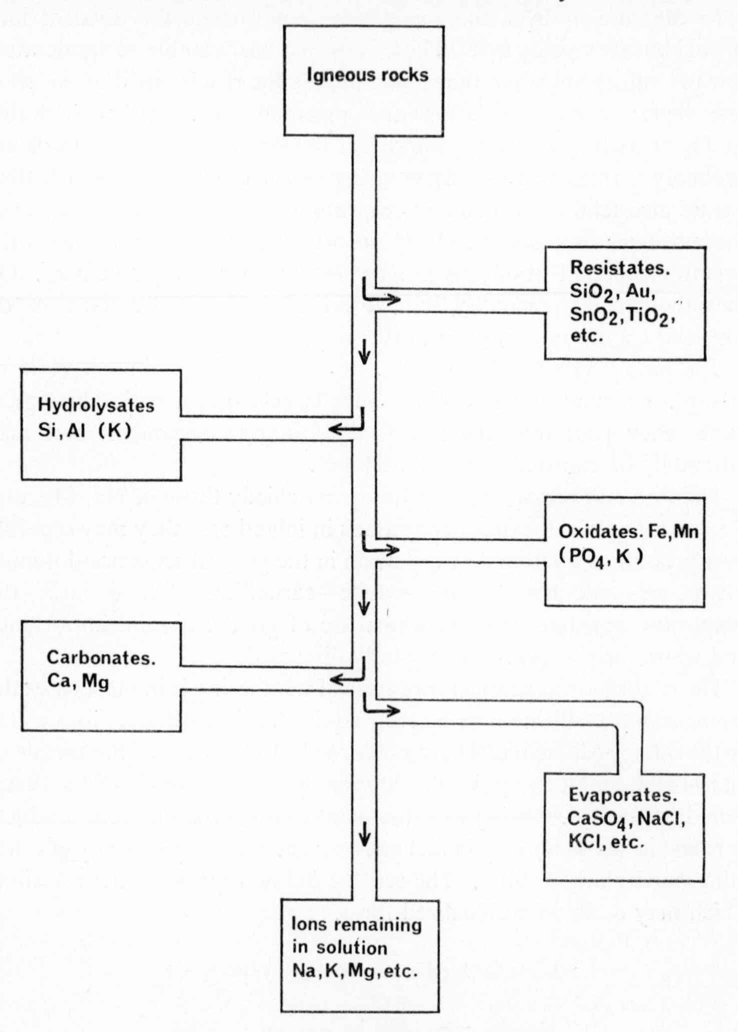

FIG. 2.7. Weathering of igneous rocks.

Biogeochemistry

Under this heading we consider processes in which living organisms contribute to changes in the outer layers of the earth. Some geochemists have postulated a zone of the Earth called the ' biosphere ' but it is in no sense a continuous layer like the others. Organisms inhabit but do not fill the space they occupy. They may be said to live permanently in the ocean and on the solid crust but only temporarily in the atmosphere. We may observe here that limitations of a possible biosphere are related to the very narrow physical and chemical limits within which life is possible.

The most striking aspects of biogeochemistry concern the element carbon but other elements are involved in a less massive but still quite significant manner.

It will now be apparent that the two trends of geochemical processes are either concentration or dispersal and both of these take place as a result of the activities of living organisms. The main effects only will be summarized here. Further consideration is given to the matter under the heading of the individual elements in subsequent chapters.

Carbon. The important photosynthesis process fixes carbon dioxide from the atmosphere or from the ocean in the form of complex organic molecules. Respiration and decay return some of this as carbon dioxide (or methane). There is also a large amount of non-cyclical carbon fixation as solid (or liquid) deposits. These are termed 'bioliths'. Coal, lignite, peat, etc., are called 'caustiobioliths'. Organic limestones are 'acaustobioliths'. Even in spite of combustion of fuels, a large part of this carbon is virtually permanently fixed. However, on a long-term basis, taking into account major orogenic movements, etc., it may be that there is an ultimate release of carbon dioxide to the atmosphere again.

Oxygen. Photosynthesis and respiration involve this element and it appears probable that the processes are complementary and cyclical to a very large extent.

Other elements. Calcium and to a lesser extent strontium are massively involved in carbonate formation.

Potassium is of great significance in connexion with both plant life and with marine deposits of the mineral glauconite which may be indirectly concerned with organic residues.

Phosphorus is involved in some very interesting cyclical processes in some of which there is a high degree of biological concentration.

Other elements which in special cases undergo remarkable biological concentration from high dilution, usually in ocean water, include iodine, bromine, and vanadium. There is also a form of concentration of some of the rarer metals by plants whereby the humus derived from their decayed leaves is enriched in some of the metals which are then retained as insoluble tannin complexes. If the humus ultimately becomes coal, then on combustion an ash may be left in which enrichment is more than a thousand-fold. The classic example is provided by germanium which is now profitably recovered from coal ashes and flue dusts. The sedimentary copper ores, *kupferschiefer*, of Germany are also cited by Goldschmidt as an example of biological enrichment, but more recent views suggest other possible modes of origin. Certain terrestrial plants also adventitiously absorb poisonous elements like arsenic and selenium, and they have been used as a basis of biogeochemical prospecting.

In conclusion the element nitrogen should also be mentioned. Nitrogen is almost entirely involved in cyclical processes between the atmosphere and complex organic molecules, there being only very exceptional examples of solid nitrogen compounds in the Earth's crust. The involvement of the elements in various biological processes is very widespread and it is in fact difficult to say that any one of them is not relevant in some way to the chemistry of living things. For this reason it will be found that the topic is more appropriately dealt with in Chapter 4.

CHAPTER 3

Geological Considerations

Introductory

The three accessible layers of our planet have already been alluded to—the atmosphere, the hydrosphere, and the solid crust. It is the last of these with which we are now concerned. The crust is composed of ordinary rocks which form a layer of somewhat variable thickness and of considerable complexity. Nevertheless the crust consists in fact to a large extent of metallic silicates.

It is the special province of geology to study the solid crust of the Earth. Therefore to understand something of the sources and origins of chemical raw materials it is clearly necessary to have some acquaintance with the subject. The purely mechanical and structural aspects will not be dealt with here and reference may be made, if required, to the many adequate textbooks available. An endeavour will be made to deal with such of the essential facts as are needed for some understanding of geochemical matters.

It is significant that, owing to a sequence of major earth movements throughout geological time, there is now exposed at the surface in the small area of the British Isles an extraordinarily complete range of geological formations and rock types. Because of this the British have largely pioneered the subject of structural geology and much of its terminology is of British origin. On the other hand a tendency to focus attention on physical and stratigraphical phenomena has probably caused some lack of interest in geochemical aspects. Thus geochemistry is not yet studied as extensively in Great Britain as it now is in Scandinavia, Russia, and America.

Kinds of Geological Formations

Geology does not lend itself to the rigorous definitions of chemistry and there is no principle which can be followed like descent in biology. Thus any system of classification of rocks into groups is likely to have anomalies. For ordinary working purposes, however, the difficulties are not necessarily great.

It should be noted that the term ' rock ' is taken to cover all types of solid materials making up the crust and thus embraces the hardest igneous rocks and the softest sands and clays.

It will be convenient to adopt a primary division of the formations into those which are of such a nature as to be called orthodox rocks, and a second group comprising localized and specialized deposits of very variable origin which are of great importance in many cases as sources of chemical raw materials.

Igneous Rocks

The usual description of igneous rocks is that they have been formed from a state of fusion. The rock melt is known (as stated in Chapter 2) as a magma and there is reason to think that initially there were only one or two types of magma from the point of view of chemical composition. It has already been noted that as the cooling of the melt proceeds crystallization sets in, and this is a selective process. A further point is that gravitational settling of early-formed crystals may take place. The magma will now have a different composition from the original and its physical properties may also change progressively. In this way it is possible for a wide variety of silicate complexes to separate from a primary magma. There will also be variations due to rate of cooling, etc., so that igneous rock types are very numerous indeed. However, surveying the Earth as a whole, as far as we are able to get access to the solid crust, it is significant that there are in fact only a very small number of dominant types of igneous rock. The numerous minor types probably represent the result of special local conditions of differentiation and consolidation. As will appear later some so-called igneous rocks may have been formed by a different process and it may in these cases be necessary to abandon the conception of a perfect melt which has completely crystallized. The list below gives the more important rock-forming minerals. The second part of the list includes minerals which are called accessory.

A certain amount of information about the nature of a rock can be gained by macroscopic observations but the bulk of petrological work depends on the examination of a thin section under the microscope aided by polarized light and various optical devices. As the optical properties of rock-forming minerals have been very fully worked out they can be reliably identified and their relationships in the rock determined. Another useful technique is the microscopical study of a polished surface of the rock by reflected light. By these methods it is

LIST OF COMMONER ROCK-FORMING MINERALS

Olivine	$(Mg,Fe)_2SiO_4$
Enstatite	$MgSiO_3$
Hypersthene	$(Mg,Fe)SiO_3$
Augite	$(Ca,Mg,Fe^{\cdot\cdot},Fe^{\cdot\cdot\cdot},Al)(Si,Al)O_3$
Diopside	$MgCaSi_2O_6$
Hornblende	$(Ca,Na,K)_{2-3}(Mg,Fe^{\cdot\cdot},Fe^{\cdot\cdot\cdot},Al)_5(Si,Al)_8O_{22}(OH)_2$
Muscovite	$KAl_3Si_3O_{10}(OH)_2$
Biotite	$K_2(Mg,Fe^{\cdot\cdot}Al,Fe^{\cdot\cdot\cdot})_{4-6}(Si,Al)_8O_{20}(OH)_4$
Orthoclase	$KAlSi_3O_8$
Albite	$NaAlSi_3O_8$
Anorthite	$CaAl_2Si_2O_8$
Plagioclase	Isomorphous series of albite-anorthite
Leucite	$KAlSi_2O_6$
Nepheline	$NaAlSiO_4$
Quartz	SiO_2
Tourmaline	$(Na,Ca)(Li,Mg,Fe^{\cdot\cdot}Al)_3(Al,Fe^{\cdot\cdot\cdot})_6B_3Si_8O_{27}(O,OH,F)_4$
Garnets	*e.g.*, $Fe_3Al_2(SiO_4)_3$
Sphene	$CaTiSiO_5$
Zircon	$ZrSiO_4$
Andalusite	$Al_2Si_2O_5$
Magnetite	Fe_3O_4
Ilmenite	$FeTiO_3$
Apatite	$Ca_5(PO_4)_3F$
Pyrite	FeS_2
Pyrrhotite	$Fe_{0\cdot 8}S$
Fluorite	CaF_2

also quite possible to get a tolerably accurate quantitative estimate of the mineralogical composition of a rock. This work can be supplemented by a physical separation of the minerals in a crushed sample of the rock and also by a full or partial chemical analysis. There are unfortunately technical difficulties, which the chemist will appreciate, in the way of getting a finely powdered fully representative sample of a hard igneous rock and it is much more expeditious to study a good series of thin sections which can be prepared fairly rapidly by means of the diamond slitting disk, etc. There is thus a certain tendency to emphasize the structural and mineralogical aspects of the subject, and interest has often been focused on remarkable and rare rock types. With the further use of modern analytical methods more attention will in future probably be paid to the chemical side of petrology. A recent critical survey (Fairbairn, *Geochim. Cosmochim. Acta* **4,** 143) has shown astonishing discrepancies between rock analyses by different workers.

Classification of Igneous Rocks

From the chemical standpoint petrologists recognize a series of rock types depending on silica content. If there is an excess of silica beyond that required to form silicates the rock is called ' acid '. If under-saturated it is described as an intermediate type. It is also necessary to recognize a group of ' ultrabasic rocks ' rich in iron and magnesium, and a group rich in sodium and potassium which are called ' alkaline '.

It is important to note that these terms are used quite differently from their orthodox chemical meaning. The classification is essentially geological and depends on the manner in which the rock occurs in the Earth and is therefore related to its mode of origin.

Igneous rocks which have consolidated from the molten material or lava poured out from volcanic vents are called ' extrusive '. When a rock melt is consolidated at some depth below the surface ' intrusive ' rocks are formed. The minor intrusions are generally in the form of horizontal sheet-like masses called ' sills ', or vertical masses filling fissures in the crust and called ' dykes '. Under these conditions cooling is relatively rapid, crystals are small, and the rock has some of the features of a lava or extrusive. The major intrusions cooled at great depth in the Earth and are fully crystalline and the crystals may be of some size.

A mushroom-shaped mass of intrusive rock is called a ' laccolith ' and may be up to several miles in diameter. The great intrusions called ' batholiths ' which can be hundreds of miles across are exposed in the cores of denuded mountain ranges as a result of Earth movement and erosion.

Examples of all of these forms of igneous masses may be seen in the British Isles. Very extensive extrusive rocks are encountered in the basaltic lava flows of Antrim and the Inner Hebrides. More ancient and therefore less evident examples are also seen in the Midland Valley of Scotland, in the English Lake District, in Snowdonia, and in many other localities. The most accessible sites of present-day vulcanism for the British observer are to be found in the regions of Vesuvius, of Etna, or in Iceland. Sills are numerous in Britain, one of the best-known being the Great Whin Sill of the North of England. Dykes are particularly abundant in Western Scotland where they radiate from the volcanic foci of Skye and Mull and in parts of the Scottish mainland, and elsewhere. Batholiths, which are not of course seen on the grand scale in British localities, are, however, represented for example by the granite masses of the south-west of England and by many of the Scottish granites. A classification of igneous rocks in tabular form is given below.

COMMON IGNEOUS ROCK TYPES

		Intrusive	
	Extrusive–Volcanic	*Minor–Hybyssal*	*Major–Plutonic*
Acid	Rhyolite Obsidian	Felsite Quartz–porphyry	Granite Granodiorite
Intermediate	Andesite	Porphyrite	Diorite
Basic	Basalt	Dolerite	Gabbro
Ultra-basic	—	—	Picrite Dunite Peridotite
Alkaline	Trachyte Phonolite	Felsite-porphyry	Syenite Nepheline-syenite

In spite of the multiplicity of rock types of which the above is only a fraction, it is found that basalt is by far the most abundant volcanic rock, and granite or granodiorite is the predominant plutonic type. The bulk of the minor intrusions is of relatively small importance.

These facts have led to the idea of all igneous rocks being derived from the differentiation of one or possibly two primary magmas.

The position in regard to granite is somewhat peculiar because there are granites which in some cases suggest by their mode of occurrence that they have originated from a crystallized magma but there are others for which an entirely different origin has been postulated. The theory to account for this will be referred to later in this chapter.

Description of Some Igneous Rocks

Granite. This a coarse-grained and completely crystalline rock. The predominant minerals are the feldspars—orthoclase or microcline (potash feldspars), albite (soda feldspar), or plagioclases (isomorphous mixtures of soda-lime feldspars). Feldspars may constitute up to 60 % of the rock. The other major component is quartz which may make up to 40 %. Mica is the common accessory mineral in granites and may include both muscovite and biotite but many others are found, including hornblende, apatite, zircon, sphene, garnet, magnetite, and tourmaline. Some granites carry large ' phenocrysts ' of feldspar which evidently separated out from the magma at an early stage. In ' normal ' granites the accessory minerals crystallized first, as shown by their perfect shape, ' idiomorphic ', and the non-silicate minerals like magnetite, apatite, and zircon probably preceded the micas. ' Acid ' granites will be likely to carry the light-coloured muscovite but many have both species. In other cases the micas are replaced by hornblende and the rock grades into a granodiorite. Most have orthoclase or its dimorph microcline. Soda or soda-lime feldspars are encountered but rather less commonly. The quartz in granites is usually the last mineral to crystallize and is therefore interstitial.

Diorite. This rock is also coarsely crystalline like granite but only negligible amounts of free quartz are present. The feldspars are mainly of the soda-lime series and may have 30–50 % anorthite (lime feldspar). As much as 40 % hornblende may be present. Diorites do not always show a very distinct sequence of crystallization. They are less abundant rocks than granites but the intermediate granodiorite is a somewhat abundant type.

Gabbro. This basic plutonic rock is found in quite considerable masses which may represent the dissected cores of extinct volcanoes that produced basaltic lavas. A large part of the Cuillin Hills of Skye provide a good British example of a gabbro mass. Gabbros are often very coarsely crystalline and the feldspar contains a high proportion of anorthite. Up to 50 % of the rock may consist of ferro-magnesian minerals belonging mainly to the pyroxene group including augite, enstatite, and hypersthene. There are often notable amounts of ilmenite and magnetite in gabbros.

Peridotites. Rocks of this class, which have several alternative names, make little show on the Earth's surface. They are believed to be essentially of deep-seated origin and may perhaps correspond to the material of the mantle.

These rocks are largely free of feldspars and the predominant mineral is olivine. There may be subordinate amounts of pyroxenes and hornblende. Concentrations of metalliferous minerals occur in some peridotites, as for example chromium as chromite or picotite. Olivine, a mineral of great stability deep in the Earth, is extremely susceptible to chemical alteration near the surface. Consequently many peridotitic rocks are largely modified giving such hydrous magnesium silicates as serpentine, or in some cases pure magnesium compounds such as magnesite. A point of particular interest is that diamond occurs *in situ* in certain peridotites. Thus they are of some chemical and economic interest.

Acid extrusive rocks. These are not very abundant compared with other types of volcanic rocks. The lavas from which they have been formed have been very viscous so that they do not tend to spread out in extensive sheets like basalt. In consequence of rapid cooling crystallization has been wholly or partially inhibited and rhyolites and especially obsidian may be largely in the form of natural glass. In the latter case the rock is almost wholly vitreous. However, acid lavas of considerable geological age are usually largely devitrified.

Basalt. This, the most abundant kind of volcanic rock, has cooled from a very fluid lava. Further, the melting point of iron-rich magmas is relatively low so that basaltic lava flows are often of vast extent. Perpendicular jointing developed during cooling gives many basalts a highly characteristic columnar structure. The high content of ferrous compounds often causes rapid weathering which soon turns the surface to a rusty brown colour. Many basalts are almost entirely crystalline though of fine grain, but sometimes residual glass is present. The minerals of basalts are similar to those of gabbros but they are often vesicular due to gas bubbles in the lava, and in the final rock the vesicles may be filled with secondary minerals—zeolites, etc.

Minor intrusive rocks. These are extremely varied in character both structurally and mineralogically and represent in many cases the results of local differentiation of a magma. As might be expected the very fluid ferriferous magmas of high mobility are most readily able to form widespread sills and lengthy dykes. The viscous, acid magmas tend to more compact types of intrusion. Dolerite, which stands midway between gabbro and basalt in its texture, is perhaps the best-known kind of minor intrusive rock.

We now turn to certain localized formations which are related to magmatic activity but which scarcely build up rock masses in the ordinary sense.

Magmatic segregations. In the cooling of a basic or ultrabasic magma—presumably one of deep-seated origin—various high-melting compounds tend to separate at an early stage. These are particularly iron sulphides and such oxides as magnetite, ilmenite, and chromite. They tend to settle down in the magma so that a considerable measure of concentration takes place. If the rock is ultimately exposed at or near the surface as a result of subsequent Earth movements, then the mineral concentration may become a valuable ore deposit. The iron ores of Kiruna in Sweden, and some of the important chromium deposits of the world are found in magmatic segregations. Magmatic segregations are especially favourable to the concentration of siderophile elements.

Pegmatites. These remarkable deposits represent a late stage in the consolidation of a magma, according to the mechanism discussed in Chapter 2. They contain many of the minerals of ordinary plutonic rocks but are excessively coarse, and crystals measured in feet rather than inches are quite common. The special feature of pegmatites is the tendency for them to contain compounds of the rarer elements which do not readily separate in the main crystallization. They will therefore be frequently alluded to in the course of discussing the individual elements. Pegmatites are but poorly developed in the British Isles. Pegmatites are believed to have crystallized slowly and under conditions of extreme fluidity of the magma and practically represent the end of deposition of silicates in quantity. The later phases of magmatic activity are essentially those of a solution rather than a melt.

Hydrothermal veins. Whilst pegmatites often take a vein- or dyke-like form they do not penetrate far beyond the main intrusion but they are sometimes found to merge into well-defined veins. These in general occupy fissures in the Earth's crust and represent an in-filling by crystallization from aqueous solution of a great variety of minerals.

Whilst vein-type deposits are very common, hydrothermal activity is also found in many varied structural forms; some of these will be referred to in Chapter 5.

Hydrothermal deposits are one of the most important sources of non-ferrous metals and the technique of mining them is very ancient. Characteristically they contain metallic sulphides but many other kinds of compounds are met with. Silica as quartz is very commonly present as an abundant 'gangue' mineral. In the high-temperature and probably super-critical stage there are minerals which may have been deposited by reactions of vapours, particularly those involving compounds of fluorine, chlorine, and boron, and considerable dissemination of minerals occurs into the surrounding 'country rock' as in the tin veins of Cornwall.

The low-temperature stage of hydrothermal activity may be at a great distance from the igneous focus and in fact many mineral veins cannot be definitely related to any specific igneous mass. Low-temperature veins are usually bounded sharply by their walls and carry the minerals which are deposited from the most mobile of the ions in the original solution. There is a fairly uniform sequence of deposition of vein minerals which is in general consistent with the idea of crystallization from solution. It is, however, proper to say that theories of mineralization have also been developed which largely or wholly reject the idea of deposition from a simple aqueous solution and postulate the action of gases or vapours.

Figure 3.1 represents in a very simplified form the possible relationships between the various forms of magmatic activity which have been outlined above.

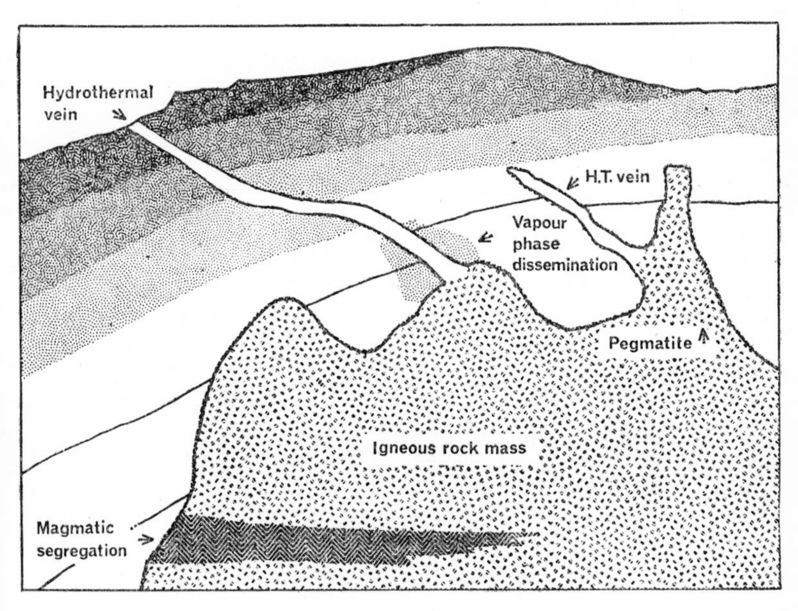

FIG. 3.1. Magmatism and mineralization.

Sedimentary Rocks

These rocks are of secondary origin. They are derived from the products of chemical and mechanical disintegration of the igneous rocks or of other sedimentary rocks. Sedimentary rocks are often very much in evidence at the Earth's surface. A large part of the British Isles for example is covered by them and they have received an amount of attention rather out of proportion to their bulk, which is estimated at about 5% of the 'ten-mile crust'. Sedimentary rocks may naturally contain the minerals of igneous rocks in a fragmentary form but in general the primary silicates are in little evidence on account of their low resistance to chemical weathering; they are largely replaced by compounds of the following types:

COMMONER MINERALS OF SEDIMENTARY ROCKS

Mineral	Formula
Kaolinite, Dickite	$Al_2Si_2O_5(OH)_4$
Montmorillonite	$(Al,Mg)_2Si_4O_{10}(OH)_2.nH_2O$
Halloysite	$Al_2Si_2O_5(OH)_2.2H_2O$
Illite	$K_{2-3}Al_{11}Si_{12}O_{35-36}(OH)_{12-13}$
Glauconite	$K_1\ (Fe^{3+},Mg,Al,Fe^{2+})_{4-6}(Si,Al)_8O_{20}(OH)_4$
Allophane	Amorphous silica-alumina gel
Chlorite (group name)	*e.g.*, $(Mg,Fe^{3+},Fe^{2+},Al)_6(Si,Al)_4O_{10}(O,OH)_8$
Flint, chert, chalcedony	SiO_2 gel or cryptocrystalline
Diaspore	$AlO.OH$
Boehmite, Gibbsite	$Al(OH)_3$
Calcite	$CaCO_3$
Dolomite	$MgCa(CO_3)_2$
Siderite	$FeCO_3$
Hematite	Fe_2O_3
Goethite, Lepidocrocite	$FeO.OH$
Anhydrite	$CaSO_4$
Gypsum	$CaSO_4.2H_2O$
Halite	$NaCl$

The following table of sedimentary rocks will serve for reference, any rigorous system of classification being difficult on account of the very heterogeneous materials and varied modes of formation of these rocks.

SEDIMENTARY ROCKS

Mechanical origin	Scree, gravel, sand, conglomerate, breccia, sandstone, grit arkose, mud, silt, clay, shale, fireclay
Mainly organic origin	Many limestones, siliceous oozes, coprolite, peat, lignite, coal, bog iron ore
Mainly inorganic chemical origin	Some limestones, dolomitic limestones, gypsum, salt beds, flint, some iron ores

The very coarse mechanically formed rocks contain rock fragments rather than individual mineral grains but in the finer grades, commonly called sands, the particles may be mechanically graded and there may be a good deal of sorting of the various minerals due to differences of density and resistance to abrasion, etc.

The sandy or arenaceous rocks are predominantly siliceous due to quartz grains and they also contain the more resistant minerals of igneous rocks. Locally there may be considerable concentration of heavy and often rare minerals in sands. The following list includes some of the more important of these.

Rutile	TiO_2	Zircon	$ZrSiO_4$
Ilmenite	$FeTiO_3$	Thorite	$ThSiO_4$
Magnetite	Fe_3O_4	Olivine	$(Mg,Fe)_2SiO_4$
Cassiterite	SnO_2	Monazite	R.E. phosphates
Garnet	*e.g.*, $Fe_3Al_2Si_3O_{12}$	Gold	Au

Sandstones are compacted sands of various types with various binding materials between the grains—*e.g.*, SiO_2, Fe_2O_3, and $CaCO_3$. Hence they vary much in colour and durability.

The muddy or argillaceous rocks often contain very fine undifferentiated material, ' rock flour ', but they are generally characterized by the presence of one or more of the clay minerals (hydrous aluminium silicates like kaolinite, etc.). These rocks often tend to develop a fissile structure when compacted and are in point of bulk the most important of the sediments, making up about 80 % of the whole thickness. It is estimated that the average thickness of all the sedimentary strata is about 2500 ft although in restricted areas, formerly occupied by geosynclines or troughs of deposition, the thickness may be more than ten times this value.

There is less chemical differentiation in shales than in other sediments so that the average chemical composition of shales in respect of the abundant elements does not differ very much from the average for igneous rocks. Many shales contain notable amounts of carbonaceous matter and this may include liquid hydrocarbons.

In calcareous rocks and the group of chemical evaporites there are frequent examples of enormous masses of material of a high degree of chemical purity and this is because they owe their origin to processes of fractional crystallization or to biological differentiation. The sedimentary deposits forming in the deep oceans are of great interest. It is far from certain at present what proportion of the known sedimentary rocks was formed under pelagic conditions. The estimates vary from as low as 13 % to as high as 50 %. The three main types of deposit are red clay, blue mud, and globigerina ooze, and a recent estimate is that these are in the ratio 8 : 2 : 1 (Kuenen, 1960). The average composition of deep-sea sediments in respect of the common elements is near to that of the other types of sediment but there are many remarkable variations and the subject is at present under active study by many workers.

In this brief survey of sedimentary rocks many special types have not been mentioned. Where they are of some significance in connexion with a particular element they will be found under that heading.

Metamorphic Rocks

This third group of rocks comprises those, originally sedimentary or, less frequently, igneous, which have undergone such profound changes by mechanical, thermal, and chemical processes as to be virtually new types.

In this case also we recognize very many kinds of metamorphic rocks but it will suffice to notice only a few of the more abundant types.

There are minerals characteristic of metamorphic rocks not generally encountered in 'fresh' examples of the others. The following list includes some of the more important.

Mineral	Formula
Andalusite, Sillimanite, Kyanite	Al_2SiO_5
Cordierite	$(Mg,Fe)_2Al_4Si_5O_{18}$
Garnet (*e.g.*)	$Fe_3Al_2Si_3O_{12}$
Talc	$Mg_3Si_4O_{10}(OH)_2$
Serpentine	$Mg_3SiO_4O_{10}(OH)_2$
Brucite	$Mg(OH)_2$
Periclase	MgO
Forsterite	Mg_2SiO_4
Wollastonite	$CaSiO_3$

There are three different types of metamorphism:

(*a*) *Mechanical*, which gives mainly structural changes in the rock.

(*b*) *Thermal*, which is equivalent to a 'baking' process and is somewhat analogous to the operations of the ceramics industry.

(*c*) *Chemical*, in which the original rock is presumably permeated by vapours or solutions. Profound chemical alteration involving complete replacement of one mineral by another is called metasomatism. Actions brought about by the actions of vapours are called pneumatolysis.

Few metamorphic rocks have in fact been formed by the operation of one of these processes singly and the thermal effects in particular often proceed alongside dynamic ones. It may be supposed that masses

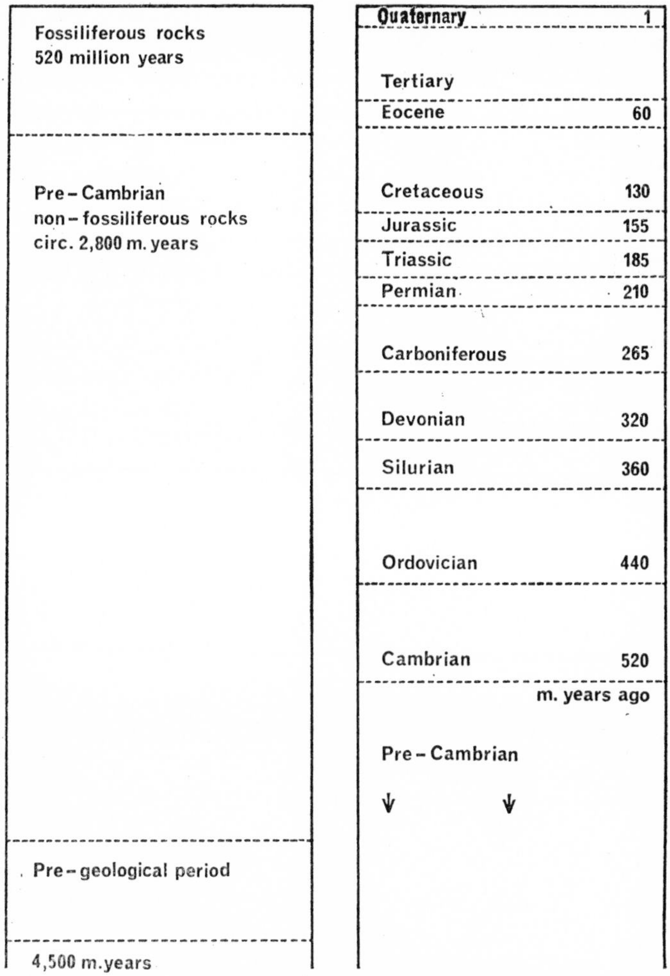

Fossiliferous rocks
520 million years

Pre-Cambrian
non-fossiliferous rocks
circ. 2,800 m. years

Pre-geological period

4,500 m.years

Quaternary	1
Tertiary	
Eocene	60
Cretaceous	130
Jurassic	155
Triassic	185
Permian	210
Carboniferous	265
Devonian	320
Silurian	360
Ordovician	440
Cambrian	520
	m. years ago
Pre-Cambrian	

FIG. 3.2. The geological systems.

of superficial strata can be folded in by major orogenic movements and taken down to great depths where it is probable that all three processes act together. Many interesting chemical changes occur in metamorphism. Dynamic metamorphism is in general of a regional character and associated with major earth movements. Very many of the most ancient rocks of the Earth's crust have suffered dynamic metamorphism on a widespread scale.

Thermal metamorphism is often found in the vicinity of masses of intrusive igneous rocks. The extent to which the surrounding ' country rock ' is heated by an intrusion varies a great deal. Small sills and dykes produce effects extending perhaps only a few inches from the contact.

In some cases a mild degree of thermal metamorphism may extend for several miles outwards from a relatively large plutonic mass. Extreme effects occur when fragments of the roof-rock above a magma chamber become engulfed in it, producing ' xenoliths '. In such cases aluminous minerals may give rise to crystalline corundum.

CHAPTER 4

Biological Considerations

From the geochemical point of view chemical changes related to biological processes are of considerable importance. Living matter is extensively distributed in the Earth, but of course in the outer layers only. The total mass of living substance is doubtless large though very small relative to the other geochemical shells of the planet.

On the other hand, whilst the number of mineral substances known to occur in the accessible parts of the crust is about 2000, the number of plant and animal species exceeds one million. Further, whereas the minerals are substantially ionic in character the greater part of biological material consists of covalent molecules of great complexity. It is evident that the old contrast between ' chalk and cheese ' contains more than a germ of truth.

It will be convenient in this chapter to outline such aspects of the physiology of living organisms as are necessary for an understanding of the part they play in the wider processes of natural chemistry.

The detailed chemistry of living material is not within the scope of this book.

Characteristics of Living Matter

The following are considered to represent criteria for the evidence of life, and all organisms seem to possess some or all of them.

(1) *Assimilation.* The ability to absorb external materials for the purpose of building up the substance of the organism.
(2) *Respiration.* The ability to transform energy by the use of chemical reactions.
(3) *Excretion.* The ability to get rid of substances which are waste products in the chemical changes taking place in the organism.
(4) *Locomotion.* The ability of the whole or part of the organism to change its position by expenditure of energy.
(5) *Irritability.* The faculty of being able to respond to external stimuli out of all proportion to the physical magnitude of the stimulus.

(6) *Reproduction.* The ability of an organism to produce one or more separate and distinct individuals of its own kind.

It would seem that these characters are outside the capacity of any system of non-living matter.

The size and complexity of living organisms vary enormously from some of the viruses which are practically within the limits of molecular dimensions to animals and plants weighing hundreds of tons.

Physically, there must be an upper limit to the size of an organism which is ultimately related to the force of gravity and is therefore a function of the mass of the planet on which the organism resides. From this point of view complete immersion in water is an advantage.

The organisms which we tend to regard as the ' highest ' kinds—the vertebrate animals—are characterized by the ability to carry out large and rapid energy transformations. The mechanism largely responsible is muscular tissue, and as the efficacy of this is largely dependent on its cross-sectional area it is evident that there must be some optimum size for mere physical power.

Essential Role of Water

Before dealing in any detail with structure or physiology of organisms we may conveniently observe at this juncture that, apart from other factors, the operation of any active biological system appears to depend on the presence of water.

As will be seen, the very basis of living substance is an aqueous fluid and water must be looked upon as quite essential both for physical and chemical reasons to all living organisms. The very special and indeed unique properties of water indicate that life evidently involves in this direction a marked degree of chemical specialization. It has been shown that the isotopic compound deuterium oxide cannot replace water in biological systems.

The water content of organisms is large—*e.g.*, Man 60 %.

Temperature is also a very important factor in regard to life. The basic living substance of cells, called protoplasm, is an aqueous fluid which in most cases undergoes permanent alteration (and therefore death) at temperatures much above 40 °C. Clearly such a fluid must also cease to function when the freezing point is reached. Certain special organisms, mainly bacteria, are known to be capable of active life down to about −4 °C on the ocean floor but they can in the form of spores resist temperatures as low as that of liquid air. There are some bacteria actually active in hot springs at 89 °C. Bacterial spores may also be able to resist temperatures of 100 °C for long periods. Nevertheless from the standpoint of the vast cosmic scale of temperature (possibly 10^9 degrees) and the narrow chemical range of temperature (approximately 5×10^3 degrees), then the temperature range for what we call life is very small indeed, say, 10^2 degrees.

It is known that ionizing radiations are very inimical to living organisms. The high-energy gamma-rays are the most powerful but the effect is noticeable in the shorter ultra-violet and it remains to be discovered whether any sort of organisms can permanently resist the high radiation in interplanetary space.

The special feature of all but the lowest forms of life is the existence within them of a multiplicity of units called cells which tend to develop particular forms and functions suitable to the requirements of the organism to which they belong.

Finally we observe that living things are susceptible of a classification based essentially on the principle of descent and with a tacit recognition of the conception of evolution.

Organisms are commonly classified as plants and animals and it will be found convenient to deal with typical plants and animals first, and then to consider some of the peculiarities of the lower types of organisms which are scarcely sufficiently specialized to be claimed beyond any possible doubt by any one of the two kingdoms.

Plants

In this section a typical green plant will be considered. Such a plant is an essentially autotrophic organism—*i.e.*, it absorbs nutritive material from simple inorganic sources. It will be found that, very largely, the plant absorbs *ions* and uses them to build up complex molecules. Plants therefore form an important link between the inorganic (and largely ionic) structure of the lithosphere and the organic (and largely molecular) structure of the biosphere.

The typical plant can be considered as consisting of roots and leaves. The food the plant requires consists in the main of carbon, nitrogen, and water, and various mineral substances. All of these nutrients except the carbon are absorbed by means of the roots (or root hairs).

Geochemically the marine flora is at least as important as the terrestrial one and it consists mainly of algae—a simpler type of plant, largely undifferentiated, and absorbing all its nutrient materials from the aqueous solution in which it resides. Microscopic algae will be further alluded to later in this chapter.

Carbon Assimilation

Carbon is taken in as carbon dioxide and it is the special function of the leaves to perform this task. In the case of aquatic plants the carbon is absorbed from solution as the HCO_3^- ion. (This is also strictly true for terrestrial plants in regard to the internal mechanism.)

In the main the assimilation of carbon takes place by the process called photosynthesis. The basic reaction of photosynthesis may be represented by:

$$6CO_2 + 12H_2O = C_6H_{12}O_6 + 6O_2 + 6H_2O \quad (\Delta G = +690000 \text{ cal})$$

The reaction may be regarded as the reduction of carbon dioxide, and water is the hydrogen donor. The high energy requirement of this reaction is derived from the light absorbed by the agency of the group of pigments generically called chlorophyll.

There do not appear to be any important variations in the chemical constitution of the chlorophylls from different plants, even simple unicellular algae.

The product of the reaction is some form of carbohydrate, possibly a hexose, but the matter has been the subject of much research and discussion by biochemists and is not yet completely understood. Works on plant biochemistry should be referred to for more details.

It is of importance to observe that the oxygen evolved in photosynthesis is derived from the water only. This fact has been established by isotopic tracer techniques using ^{18}O for the purpose.

The chemically opposite change to photosynthesis is of course respiration, and for both plants and animals is essentially the same:

$$C_6H_{12}O_6 + 6O_2 = 6CO_2 + 6H_2O \quad (\Delta G = -690000 \text{ cal})$$

However, in plants, whose energy requirements are relatively low, respiration, although essential, is a much less evident phenomenon than in animals.

Much of the assimilated carbon in plants is ultimately elaborated into cellulose which forms the greater part of the rigid structure of the plant and the material of its cell walls—a very important difference from animals.

The carbon dioxide content of the atmosphere surrounding the plant is an important factor and up to a point it appears that an increase favours increased photosynthetic activity, provided there is adequate light intensity. However, biochemists seem to agree that there is an upper limit and ultimately an increase in CO_2 content produces a toxic effect on the protoplasm of the plant cells involved in the photosynthetic process. At the other end of the scale it appears that there is a threshold value for CO_2 (of about 0·009 vol. %) for land plants below which photosynthesis cannot occur. The initial presence of at least a small amount of oxygen is probably also necessary for ordinary types of green plants. The extent of fixation of carbon by plants is further dealt with in connexion with that element.

Nitrogen Assimilation

Nitrogen enters into the constitution of proteins and other complex substances which are essential components of protoplasm, and is in general absorbed by autotrophic plants from simple substances in solution by means of the roots, or in the case of algae by the whole surface of the plant. The usual source of nitrogen is in the form of nitrates but in some cases it appears that ammonium and nitrate ions may also be absorbed.

The nitrogen content of soils is usually less than 1 % and of this only about 1 % is immediately available as inorganic salts. These necessary compounds are continuously produced in the soil by the degradation of complex nitrogenous materials of protein type which ultimately give ammonia. Oxidation of ammonia, first to nitrite and then to nitrate, is brought about by the agency of soil bacteria like *Nitrosomonas* and *Nitrobacter*, the presence of which is therefore essential to a fertile soil. Plants of the Leguminosae family can also obtain their nitrogen in another way. These plants have on their roots nodules, in which live in a symbiotic relationship certain bacteria, for example *Rhizobium*, which are able to fix atmospheric nitrogen, possibly in the form of aspartic acid, which can then be absorbed by the host plant.

The marine algae must obtain their nitrogen from the small content of dissolved material in ocean water. This is less than 1 g/ton and may be as little as about one-thirtieth of this value; only about 5 % of this nitrogen is in the form of nitrates. Whilst the mechanism of absorption and utilization of nitrogen is not appropriate to discussion here, it may be noted that there is evidence that the nitrate must be reduced by way of nitrite to ammonia which can then be combined in the form of amino-acids, for example by reaction with α-ketonic acids derived from a simple carbohydrate.

It thus appears that the process of nitrogen assimilation is roughly the reversal of the process of degradation of protein in the soil and therefore a large part of organic nitrogen cycles in this way (see Fig. 16.1).

The major part of nitrogen assimilation appears to be associated with bacterial activity but we must add that a small amount of nitrate ion is made available by way of the formation of nitrogen oxides by atmospheric electric discharges, etc., and this passes ultimately into solution.

Mineral Constituents of Plants

All the elements referred to below are probably absorbed by the plant in ionic form from solution. This absorption is partly, but not entirely, an osmotic mechanism, by way of the root hairs or other appropriate parts of the plant.

It has been found that a very corrosive environment developed near the root tips causes the ready extraction of mineral matter from soils.

The average land plant contains upwards of 90% of the elements C, H, O, and N based on the dry weight so that the numerous additional elements are present in quite small amounts. There is of course considerable concentration of mineral constituents in the ash of the plant and many analyses of plant ash have been recorded. The results show quite remarkable variation. It may, however, be stated that of the various metals potassium tends to be present in greater amount than the others, although occasionally exceeded by calcium. Among the non-metals sulphur, phosphorus, and silicon can be present in notable quantity.

The following elements may be said to be commonly present in perceptible amounts: potassium, calcium, sodium, magnesium, iron, aluminium, manganese, sulphur, phosphorus, and chlorine. About forty elements have been detected in plants. Some occur rarely; it is by no means certain that they are all functionally necessary. In fact biologists are still by no means certain of the actual function of some of the quite common elements.

There is some evidence that the assimilatory system of a plant is not particularly selective and that a number of ions may be absorbed adventitiously if they happen to be present in the medium in which the plant grows.

On the other hand it has been found that several toxic elements—*e.g.*, lead, vanadium, and uranium, appear to be immobilized by precipitation on the cell walls of the roots and thus do not reach the active centres of growth.

The particular characters of an ion which determine whether or not it will be absorbed are not yet fully understood. For the major elements there is a marked degree of selection as shown by the analysis of the cell sap.

Ion	*Content in* Vallonia[1]	*Sea water*[1]
Cl^-	21·2	19·6
K^+	20·14	0·46
Na^+	2·1	10·9
Ca^{2+}	0·07	0·45
Mg^{2+}		1·31
SO_4^{2-}	0·005	3·33

[1] In parts per thousand.

Vallonia is a marine alga and is therefore fully surrounded by a solution rich in sodium ions but in fact it takes up about ten times the weight of potassium ions.

Forms and Functions of the Elements

Potassium. This ion appears to be present in the plant in a soluble form and may be present in the cell sap in high concentration. There is much conflicting evidence, but it may be that one of the main functions of the element is to further catalytic activity in the formation of starch by the plant. However, it may well have other functions.

Calcium. The functions of this element are by no means clear. In many plants calcium oxalate is found and it has been suggested that the element acts by removing in an insoluble form excess of oxalic acid which would otherwise be toxic to the plant. Some calcium is needed in the protein of the cytoplasm and for calcium pectinate which keeps cells together as tissues. The classes of plants called by ecologists calcicoles and calcifuges are so termed on account of their supposed need for or rejection of a calcareous soil. The differences are probably in the main due to different pH requirements which are controlled by calcium carbonate and also the physical conditions created in such a soil. Calcium ions could of course be quite readily absorbed from even an acid soil and plants of the calcicole type are not necessarily rich in calcium.

Magnesium. The particular function of this element must lie in the formation of chlorophyll, the complex molecule of which contains a magnesium atom. There is also some evidence of the possible use of magnesium in phosphorus metabolism, particularly in connexion with lipoid substances.

Iron. The function of iron is probably in the main catalytic, as is perhaps the case with some of the other metals. Although not a constituent of chlorophyll, iron seems to be a necessary element in its formation.

Other metals. A number of metals which exist in plants in small or trace quantities are now known to be present as atoms in the molecules of certain of the enzymes which are essential to the various chemical operations of plants. This is true of copper, zinc, manganese, and molybdenum. The case of sodium is somewhat peculiar. It is often present in very large quantities in halophyte plants and would appear in some cases to be able partially to replace potassium. It does not appear to be an essential element to most plants. The nature of its function, if any, is obscure.

Phosphorus. Phosphorus is usually present in lower quantity than sulphur in plant material but it performs vital functions. In particular phosphate enters into some of the coenzymes connected with carbohydrate metabolism and other important processes. We note in particular the nucleotides which are compounds derived from phosphoric acid, a sugar, and a pyrimidine or purine. They are in fact phosphoglycosides. Adenylic acid or adenosine monophosphate (AMP) can take on one or two more phospate groups to form adenosine diphosphate (ADP) and adenosine triphosphate (ATP). The pyrophospate bonds which are present in the two latter have the high energy of 12000 cal/mole and play a major part in the various energy transformations involved in metabolic processes. It is remarkable that phosphorus is present in these and all other natural compounds as simple or derived phosphate groups —PO_4. The only exceptions are perhaps the formation of phosphine PH_3 by the action of certain anaerobic bacteria and the occasional presence of metallic phosphides in meteoritic material.

Sulphur. This element is often present in fairly large quantity and it is an essential component of some of the amino-acids of protein. Some families—*e.g.*, Cruciferae and Liliaceae, have species particularly rich in sulphur compounds which are glycosides and contain isothiocyanates (mustard oils). Some marine algae contain carbohydrate sulphates.

Sulphur is absorbed as sulphate ions which may be present in the cell sap but there are examples of the absorption also of elemental sulphur.

Silicon. Some plants, particularly the grasses and horsetails, often contain large amounts of silicon. Its exact function is not certain but it appears that its presence confers increased resistance to parasites on account of the rigidity and toughness of cell walls rich in silica. The element is taken in presumably as silicate ions.

Boron. The importance of boron as an essential trace element has been fully established but its function or functions are as yet somewhat obscure. It has been suggested that it affects the water-absorbing capacity of the protoplasm, that it favours the absorption of cations and retards that of anions, and that it is particularly concerned with the absorption of calcium and again with that of potassium.

Chlorine. This element is generally present in the plant, probably as chloride ion. It may be adventitious only, but there is not yet sufficient evidence at hand to decide.

Concluding Note on Plants

Plants are evidently predominantly accretionary organisms. They absorb large amounts of a variety of materials of simple constitution, including water, and they give out only oxygen and the water of transpiration (and a very small amount of carbon dioxide). Plants have in general no means of excreting any waste products they produce except by such processes as the decay and fall of leaves, etc. It therefore follows that plants are much more important than animals as accumulators of organic material. There are for example no deposits of animal origin which are analogous to coal—a vegetable product.

Animals

We now pass to the consideration of the chemical characters of animals in relation to their external surroundings. The characteristic of animals, as distinct from green plants, is their inability to manufacture their essential complex structures from simple materials. The animal is mainly a protein structure and it must build up its own special proteins by using others for the purpose. The same remark in general applies to most of the other types of organic substance needed. The only simple substances taken in are water, oxygen for respiration, and some of the simple inorganic ions. Animals are thus either directly or indirectly dependent on plants.

The essentials of nutrition of higher animals are well known and may be stated as follows: carbohydrates, fats, proteins, special organic compounds like vitamins, some mineral salts, and water.

The special molecular species of these types can in general be elaborated by the animal but only from relatively complex starting materials.

The characteristic carbohydrate of animals is glycogen which is produced from glucose, and the glucose, if not taken directly, is derived from other carbohydrate foods. It is also possible for fat in some cases to be converted into carbohydrate and the reverse process also takes place. In the case of proteins it is necessary for the food to be such as will provide all the amino-acids which are needed to synthesize all its own proteins and these must include cyclic acids like tryptophane. Not all the vegetable proteins contain the necessary complement of the required amino-acids, but there are some which do so. Because an animal has an enormously greater energy output than a plant, and because there is a continual wastage of its structure which has to be made good, there is a very large intake of food relatively to the weight of the animal. In the case of a large animal like Man the energy requirement is of the order of 3000 kcal per day which is provided by the carbohydrate food mainly. The tissue wastage is made good by consumption of protein and the requirements in this respect vary greatly as between adults and growing immature animals.

Essential Elements in Animals

The major elements are of course the same as in plants but the minor elements are apparently much more numerous.

Hutchinson lists the following as probably functional in Mammalia: H, O, C, N, S, P, K, Na, Cl, Ca, Mg, Fe, Mn, Zn, Cu, Al, Co, I, F, and Br.

The following have been recorded in Mammalia and may be non-functional: Li, Rb, Cs, Sr, Ba, Hg, Cd, B, Ga, Sc, Y, La, Ce, Nd, Gd, Dy, Ti, Ge, Mo, Sn, Pb, As, U, Se, Ni. In the case of some marine invertebrates we must also include vanadium.

There is no doubt that animals possess in a high degree the power to concentrate relatively rare elements. The presence of iodine in the substance thyroxin is a good example as the content of iodine in the ordinary diet of animals is very low indeed. Another case is the production of a dibromindigo—the famous Tyrian purple—by a marine mollusc, which must extract the necessary bromine from sea water.

Further treatment of the various elements in organisms will be found under the appropriate headings in the later chapters of this book.

Excretion in Animals

Another peculiarly animal characteristic is the ability which it possesses to get rid of the waste products of its metabolism. The elimination of carbon dioxide as a general part of the respiratory process takes place in all animals as in plants, but is vastly more important. It constitutes a geochemical factor of some significance. The main excretory process, however, is concerned with the elimination of waste nitrogenous material and also smaller amounts of other elements. It will suffice to take note of the various forms in which these elements are excreted. There is a good deal of variation with different types of animal.

Nitrogenous compounds, the waste material from breakdown of proteins, are eliminated by the higher animals both in solution and in the solid state as faeces. The most important compounds are: ammonium salts, amino-acids, urea, purines, insoluble protein, etc., and creatinine. It is interesting to observe that ammonium salts and amino-acids tend to be excreted by aquatic animals, and urea and purines by terrestrial forms. Insoluble matter in solid faeces includes undigested material and usually considerable masses of bacteria from the gut.

Few animals can digest cellulose and lignin to any extent, so the faeces of herbivores are very bulky. On the other hand carnivores excrete a meagre quantity of solid matter. Chitin which is common in the structure of arthropods is a nitrogenous carbohydrate and it is usually excreted unchanged.

The mechanism for the disposal of waste nitrogenous matter also serves for eliminating most of the other elements. Most of these probably come away in ionic form. Thus phosphorus is usually excreted as phosphate and sulphur as sulphate. Calcium as carbonate is got rid of by animals which cast their shells, etc., and this in effect is a form of excretion.

For most animals excretion is essentially a process of dispersion and the material which is of value in plant nutrition can of course be readily absorbed at low concentrations. It is well known that in closed systems like small aquaria it is possible to maintain a balance between plant and animal life which are largely complementary in their requirements. This may also be approximately true for the Earth as a whole, but there are, however, other factors involved.

There are some animals, particularly civilized Man and some sea birds, whose excreta may be considerably localized and concentrated.

Lower Organisms

In spite of their size and impressive appearance it is doubtful if the higher types of plants and animals are as important geochemically as the humbler and often microscopic organisms.

In this section no attempt is made to discriminate between plant and animal and in fact in numerous cases the terms become rather meaningless. It is now a recognized fact that the micro-organisms which are free-floating in the sea and are called plankton are in the aggregate responsible for a much greater amount of carbon fixation than the aggregate of terrestrial plants. This is of course in respect of the phytoplankton. The zooplankton are equally important as a source of food for the higher organisms. In the following remarks no rigid order is used, but some groups are referred to which make important contributions to the chemical cycles in nature.

Protozoa

Some of these have mainly animal and some mainly plant characteristics. They are exceedingly abundant in the soil, in the water, and in muds at the bottom of lakes and the sea. One example will suffice, the foraminifera. These are generally of small size and mainly live in the bottom mud of the sea where they no doubt employ organic residues as a means of nutrition. Their special character is that they form quite complicated calcareous shells and they are therefore of some importance as accumulators of calcium carbonate. The remains of foraminiferal shells constitute globigerina ooze which is one of the typical bottom deposits of the ocean. Foraminifera have been abundant in past geological time as shown by the fossil remains in many limestones.

Algae

We now refer more particularly to the microscopic forms which constitute phytoplankton to a large extent. Green algae are of course autotrophic and they are of vast importance in the economy of Nature as providing a primary food source for higher organisms. In spite of their relative morphological simplicity they are able to elaborate all the characteristic organic molecules of larger plants. Many of them produce oils which are stored as globules in their cells. Marine microalgae are recognized as probably the major instrument for the fixation of carbon in a form which ultimately becomes liquid petroleum. One of these groups of algae possesses the ability to absorb silica from water to produce complex skeletal structures. These are the diatoms. Diatoms are very numerous in lakes as well as the ocean and they thrive at relatively low temperatures and so are abundant in sub-arctic regions. The remains of diatoms form an important siliceous deposit—diatomaceous earth—which under different names has a great range of uses. There are also calcareous algae which can build up structures simulating coral. These have also contributed with their remains to the formation of limestones. Their seasonal absorption of calcium carbonate in the case of small lakes can have the effect of causing pH variations in the water. In a similar manner there are seasonal variations in the silicate content of lake waters due to the activities of diatoms.

Aquatic algae of some species such as *Spirogyra* have the capacity to remove surprisingly large quantities of elements from the water in which they grow. It has been found that this alga, growing in a mine water containing a total of 16 p.p.m. heavy metals, contained 2900 p.p.m. zinc, 6600 p.p.m. lead, and 920 p.p.m. copper (on dry weight). It is also stated that a large proportion of the copper content of certain lake waters is contained in the body structures of the plankton population.

Fungi

These plants are not of course necessarily minute but in their nutrition they are quite different from orthodox green plants as they do not manufacture any carbohydrate by photosynthesis. Fungi are holozoic or animal-like in their absorption of food. Geochemically perhaps the most interesting are the types which absorb organic matter from solution and in general are parasitic or saprophytic on animals and on plants. The group known as moulds live on organic residua and are undoubtedly one of the major factors in its decay. In many cases the food of fungi, such as proteins, fats, and cellulose, is insoluble. The fungus secretes enzymes into its surroundings which break down the food material, render it soluble, and so enable it to pass through the fungal cell walls.

The enzymes seem to be secreted in larger quantities than are actually needed and must contribute in large measure to the decay of animal and vegetable residues in the soil and elsewhere. In this sense fungi perform a useful function. Fungi include yeasts which are of interest on account of their capacity to build up protein when fed with carbohydrate and suitable mineral salts. They are also of great interest in connexion with fermentation. Fungi do not require light and some will grow without a regular supply of oxygen.

It is desirable to allude also to the remarkable plants called lichens. These, so common and familiar on rocks and tree trunks, represent a symbiosis or partnership between a fungus and an alga. The alga by photosynthesis provides the nutrition and the fungus is thought to conserve moisture and supply the necessary mineral salts. Lichens produce acids which are able to attack their substratum and this probably contributes to the decay of rocks and the beginning of soil formation. Whilst lichens are extremely resistant to temperature changes and long periods of desiccation yet they seem to be very sensitive to atmospheric pollution, probably by sulphur dioxide. Thus lichens are rarely seen in industrial areas.

Bacteria

Though practically the smallest of living organisms, the bacteria, by reason of their abundance and extraordinary fecundity, undoubtedly constitute an important geochemical factor. Bacteria are of interest on account of their varied and peculiar modes of nutrition and their adaptability to different physical and chemical conditions. The great function of bacteria in Nature is as destroyers, breaking down complex organic material and ultimately returning it in an inorganic form which may again be used by autotrophic plants. We note here that the pathogenic species which medically are all-important are in our case of little or no interest.

Bacteria are unicellular and their cell walls appear not to be of cellulose, as in plants, or of protein, as in animals, but instead of remarkable complexes which combine both carbohydrate and protein structures. These minute organisms are unique in the fact that they can make use of mechanisms for nutrition and energy sources which are quite different from those used by other living things. These will now be discussed, first observing that many species are facultative—*i.e.*, they can adapt their nutritive mechanism to changed conditions.

A few bacteria are capable of photosynthesis but they use a pigment which is not identical with the chlorophyll of higher plants. These species use hydrogen sulphide as the hydrogen donor in the reduction of carbon dioxide to carbohydrate according to the equation

$$12H_2S + 6CO_2 = C_6H_{12}O_6 + 12S + 6H_2O \quad (\Delta G = 104{\cdot}4 \text{ kcal})$$

The energy is obtained from sunlight but no oxygen is given off and none is required by the bacteria.

More remarkable are those which carry out the above synthesis but obtain the energy necessary for the purpose by the reaction

$$2H_2S + O_2 = 2S + 2H_2O \quad (\Delta G = -97{\cdot}4 \text{ kcal})$$

It will be seen that this provides an adequate source of energy, particularly as the hydrogen sulphide is likely to be readily available in the medium.

In some species the sulphur is stored as a reserve material and used when required as follows:

$$2S + 3O_2 + 2H_2O = 2H_2SO_4 \quad (\Delta G = -80{\cdot}2 \text{ kcal})$$

It is then supposed that some of the sulphur can act as a reducing agent for the carbon dioxide thus:

$$6CO_2 + 10H_2O + 4S = C_6H_{12}O_6 + H_2SO_4$$

Such organisms are entirely self-sufficient, requiring neither light nor organic material.

Many species of anaerobic bacteria can get their oxygen from chemical substances and hence they act in effect as reducing agents. Thus nitrate may be reduced to nitrogen or ammonia, sulphates to sulphides, and carbonate to methane.

Some of the ' methane bacteria ' carry out the interesting operation of using carbon dioxide as an oxidizing agent to convert ethyl alcohol to acetic acid:

$$2C_2H_5OH + CO_2 = 2CH_3COOH + CH_4$$

Methane formation is typical of anaerobic decay of carbonaceous matter in swamps, etc., but a similar action goes on in the rumen of herbivorous animals. It is stated that the domestic cow, for example, may produce over 100 litres of methane per day in this manner.

The end products of such anaerobic actions generally tend to be hydrides. Thus nitrogenous matter gives ammonia, sulphur proteins give hydrogen sulphide and mercaptans, and complex phosphorus compounds give phosphine. Hence it follows that anaerobic bacteria tend to produce offensive odours and with the general preponderance of nitrogen, the ammonia formed tends to raise the pH of the medium. This type of bacterial action is independent of light and does not need a high temperature but it is relatively slow. In aerobic actions the organism obtains its energy by oxidation of a suitable substrate by means of atmospheric oxygen (or the gas dissolved in water). In this case carbon dioxide will be the end product from carbon compounds and the mineral elements will generally be fully oxidized ions—*e.g.*, sulphate, etc.

The iron bacteria can oxidize ferrous ions to ferric. The numerous bacterial population of soils is essentially aerobic.

BACTERIA OF CHEMICAL INTEREST

Name	*Habitat*	*Characteristics*
Azotobacter	Soil	Aerobic. Uses N_2 to form protein. Tolerates pH down to 4
Rhizobium	Legume root nodules	As above. Symbiotic in nodules
Thiobacillus thio-oxidans	Soil	Oxidize S as energy source. Tolerates pH down to 1
Thiobacillus denitrificans	Soil	Anaerobic. Oxidize S using NO_3 to give N_2
Hydrogenomonas	Soil	Oxidize H_2 as energy source
Nitrobacter	Soil	
Nitricystis		Oxidize $-NO_2$ to $-NO_3$
Nitrosomonas	Soil	
Nitroscoccus		Oxidize NH_3 to $-NO_2$
Gallionella	Ferruginous	Oxidize Fe^{2+} to Fe^{3+}. Sheath of $Fe(OH)_3$
Leptothrix	waters	
Chlorobacteriaceae	H_2S waters	Reduce CO_2 with H_2S photochemically
Thiorhodaceae	H_2S waters	As above but store S as reserve
Athiorhodaceae	H_2S waters	$H_2S + O$ for energy but reduce CO_2 with H_2. Can use light
Beggiatoa	H_2S waters	Fully chemautotrophic. $H_2S + O$ for energy. Reduce CO_2 to carbohydrate by S
Desulphovibrio	Black muds	Anaerobic. Reduce $-SO_4$ to H_2S
Methanomonas		Oxidize CH_4 to carbohydrate

It will be seen from the above table that bacteria are clearly the most versatile and ubiquitous of living organisms. Few of them require light and it is possible for some species to develop actively under conditions which no other form of life could tolerate—*e.g.*, in the brines of deep oil wells. In respect of different species there is a range of toleration of pH values from 1 to nearly 13, and as previously mentioned they have the widest toleration of temperature of any known living forms. They are, however, sensitive to short-wave radiation and they cannot actively develop in the absence of water.

Chemical and Physical Conditions for Life

The foregoing outline of some of the physiological characters of different kinds of organisms will focus attention on the varying requirements of different forms. It is noteworthy that the living forms which are accepted as the ' highest ' are those which require the most specialized conditions, both physically and chemically.

A high degree of activity requires a high rate of energy transfer and this apparently can be achieved only by an oxidative reaction using carbohydrate and oxygen as in all animals. But the special mechanisms of metabolism in some of the bacteria show clearly that the ordinary ' oxygen–carbon dioxide ' system of life—common alike to animals and higher plants—is not the only possible mechanism. In all cases, however, a high degree of molecular complexity appears to be necessary and it is clear that the framework of living substance is a carbon framework. In view of the unique chemical character of carbon it does not seem possible that any other element could replace it.

It would thus appear that life outside the Earth, whilst it ought not to be ruled out, is not likely to be of a very advanced type unless, as would seem improbable, the physical and chemical conditions approximated very closely to those on our own planet.

Speculations by Haldane, Bernal, and others have led Firsoff to develop an ingenious theory for the existence of a life system in which liquid ammonia replaces water as the basic fluid medium. It is suggested that a whole scheme of carbon compound reactions suited to vital processes could be developed on this basis. Such organisms are envisaged as ' ammonia drinking ' and ' nitrogen breathing ' and might therefore live in the conditions thought to exist on Saturn and Jupiter. It must be observed, however, that liquid ammonia with its low polarity has an analogy with liquid water which is more apparent than real.

Possible Origin of Life

This subject, naturally of a speculative and controversial character, has been of interest to mankind at least since times of classical antiquity. It is unnecessary here to deal with its philosophical implications but development of views in recent years bears some relation to topics dealt with in this book and thus requires consideration.

Those biologists who consider the problem from a non-vitalistic standpoint, tend to subscribe in general to such views as developed by Haldane, Bernal, and particularly Oparin. It is maintained that living organisms have evolved from non-living systems and some notion of this kind may in fact have occurred to Darwin a century ago. It is now possible to go some way towards suggesting tentative mechanisms for the process.

The first postulate is that the development occurred in water. The pre-geological stage of the ocean is conceived as a solution containing a great assortment of ions and some simple molecules.

Owing to the virtual absence of oxygen from the primitive atmosphere it is held that strong ultra-violet radiation reached the Earth's surface and supplied the necessary energy for numerous reactions involving water and the atmospheric gases such as methane, ammonia, carbon monoxide, etc., which are thought to have been present. Interaction between some of these primary products in aqueous solution, perhaps catalysed by a variety of mineral substances, also took place. It is also claimed that a number of hydrocarbons could have been formed by action of water on primary metallic carbides. By such processes simple to moderately complex alcohols, aldehydes, acids, ketones, and even some heterocyclic compounds could have been synthesized.

The sponsors of these views claim that the occurrence of these reactions is now no longer possible because of the changed nature of the atmosphere and the screening off by the oxygen of most of the short-wave radiation. However, electric discharges, possibly more numerous in pre-geological times, have also been postulated as a source of energy. Miller, as a result of experimental work, has suggested reactions by which quite complex organic molecules including amino- and hydroxy-acids, fatty acids, and some polymers could be formed. The next stage is the combination of the appropriate amino-acids, etc., to form peptides and polypeptides. This is thought to have occurred by agency of adsorption on colloidal clay particles.

The problem now becomes increasingly complex and difficult because there enters into it the characteristic molecular dissymmetry of most of these complex substances.

Most biologically important molecules are in effect 'left-handed'. Ordinary chemical syntheses always give mixtures of both right-handed and left-handed forms. It has been suggested that the orientation was initially determined by the surfaces of, for example, laevo-crystals like quartz acting as catalysts or possibly as a result of the partial polarization of the light which provided the energy for the reactions.

Oparin has developed the idea of a ' coacervate ' system as first put forward by Bungenburg de Jong as a result of experimental work. This phenomenon occurs with hydrophilic colloids and is essentially the separation of droplets of colloid-rich sols in an aqueous solution almost free of colloid.

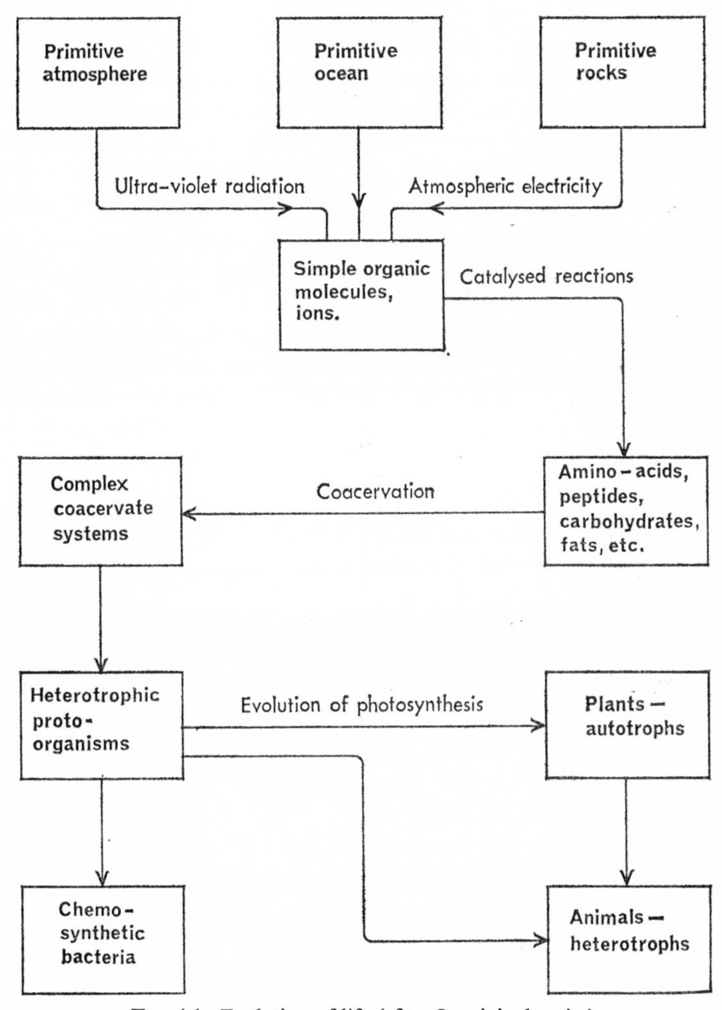

FIG. 4.1. Evolution of life (after Oparin's theories).

Such systems can be established with purely inorganic materials, but they are of special interest with organic colloids including proteins and in which several components are present. It is suggested that such systems are analogous to a rudimentary form of protoplasm. Oparin has elaborated the theory so as to provide a course of evolutionary development of these drops until they have attained an enclosing film and have developed through interchange with the external medium a system of reactions which can cause actual growth of the coacervate. A special point is made that when eventually a reaction co-ordination in time and space has been attained the system, which is an ' open ' one, is so vastly more efficient as a chemical synthesizer that it overwhelms all other competing systems. At some such stage it can be regarded as a proto-organism and ' life ' may be said to have been initiated.

Thus the primitive organisms are envisaged as heterotrophic, deriving their nourishment from the variety of organic molecules then abundant in the water of the ocean.

Only at a later stage did photosynthesis develop and a vast step forward in the nutritional process was achieved. The evolution of more elaborate heterotrophs dependent on the green ' proto-plants ' paved the way to the development of the animal kingdom. The chemosynthetic bacteria may have been off-shoots from the primitive line or possible degenerate forms.

It is evident that whilst the theory is in many ways attractive there are still weak links in the chain of reasoning. Nevertheless it is now sufficiently elaborated to merit the interest of biologists and geochemists.

CHAPTER 5

Occurrence and Extraction

We now proceed to deal with practical sources of chemical raw materials and are therefore concerned only with terrestrial locations for it is manifest that no others are accessible to us. The three possible sources of materials are:

(1) The atmosphere.
(2) The oceans and other waters.
(3) The solid crust of the Earth.

This appears to correspond with the three states of matter but it will in fact be found that whilst the atmosphere can supply only gases, the two other spheres can be sources of gases, liquids, and solids also. Thermodynamically the highest degree of entropy is met with in the atmosphere, somewhat less in the oceans, and least in the Earth's crust. This means that the greatest degree of dispersal is met with in the first two locations and only in the third is there any measure of order or concentration of materials.

Atmosphere as a Chemical Source

The composition of the air is nearly constant in the troposphere due to convection and is as follows (in per cent by weight):

Nitrogen	75·51	Neon	0·00125
Oxygen	23·15	Krypton	0·00029
Argon	1·28	Xenon	0·000036

In the above table we ignore the small variable amounts of carbon dioxide, water vapour, nitrogen compounds, ozone, etc.

It is of great interest to observe that every one of the component gases of air is extracted as a regular industrial undertaking. The processes are essentially physical and depend mainly on fractional liquefaction and distillation. In the case of xenon, the extraction of 0·36 g/ton of air compares favourably with the extraction of perhaps 1 g of gold per ton from very low-grade ores, a process which is barely economic.

There is a further aspect of the use of the atmosphere as a source of chemicals and that is in the use of its oxygen in combustion processes. An estimate of the probable annual world consumption of oxygen in this way can be based on the estimated fuel output which is about $3{\cdot}5 \times 10^9$ tons. The oxygen needed to burn this will therefore be about 9×10^9 tons. Allowing for unaccounted combustion—*e.g.*, forest fires, etc., the total is probably of the order of 10^{10} tons or 10^{16} g. Now the total oxygen content of the atmosphere is estimated at $11{\cdot}841 \times 10^{20}$ g (11·841 Gg). Therefore the oxygen of the air could be expected at the present rate of consumption to last about 11000 years. However, the oxygen output by photosynthesis by green plants (terrestrial and marine) is about 36×10^{16} g so that the effect of human activities is evidently negligible.

Oceans as a Chemical Source

As previously noted (Chapter 2), the water of the oceans is a complex solution containing ions and some undissociated molecules and the major components were tabulated (p. 42). We now re-state the composition for the major ions in grammes per ton of sea water. There are variations in density and concentration in waters from different parts of the ocean but it has been established that the ratio of the different inorganic ions as quoted below is almost invariable.

Ion	*Content* (g/ton)
Cl^-*	18980
Na^+*	10556
SO_4^{2-}	2649
Mg^{2+}*	1272
Ca^{2+}	400
K^+	380
HCO_3^-	140
Br^-*	64·6
H_3BO_3	26
Sr^{2+}	13·13
F^-	1·3

The ions marked with an asterisk are at the present time being commercially extracted on a substantial scale. The methods which are presumably available are:

(*a*) *Evaporation*, used mainly for common salt NaCl and for smaller quantities of certain other chemicals.

(*b*) *Precipitation*, used for magnesium as $Mg(OH)_2$ and possibly also for potassium.
(*c*) *Volatilization*, used for bromine.

The technical requirements in the three cases are somewhat different. Where the product is a relatively low-cost material like common salt then evaporation is possible economically when it can be carried out mainly by solar heat. The alternative of removing most of the water by freezing out is also possible and is practised—*e.g.*, in Russia and Sweden, where the climate is appropriate. ' Solar salt ' production is carried out efficiently on the Pacific coast of the U.S.A. where the conditions are particularly favourable. These include strong lateral currents bringing in fresh ocean water continually, high insolation with warm dry winds and low rainfall during the working season, and proximity to markets. Under these circumstances controlled evaporation is possible enabling the less soluble compounds (mainly calcium sulphate) to crystallize first and the main product (sodium chloride) to separate out in a pure state apart from the mother liquor (' bittern ') from which may be recovered, when economic conditions are favourable, salts like magnesium sulphate and chloride and potassium chloride, and also free bromine.

Production of 1000 tons ' solar salt '
requires approximately 40000 tons sea water,
also yields 28 tons potassium chloride,
27 tons magnesium chloride,
16 tons magnesium sulphate,
2·4 tons bromine.

The second technique is at present in use for the extraction of magnesium. In this case the sea water is treated with either slaked lime or with slaked calcined dolomite whereby the magnesium ion is precipitated as the sparingly soluble hydroxide which can be collected by sedimentation and filtration, and subsequently converted into whatever magnesium compound is required. The precipitant here is a low-cost material.

An alternative technique is possible for potassium and a process is being developed and is believed to have reached a commercial scale in Norway. For this, the precipitant, which is dipicrylamine, is expensive and is recovered by decomposing the potassium complex with nitric acid so that the end product is potassium nitrate.

In view of the many organic complexing agents now available it may well be that further possibilities exist for separating some of the other ions which occur in smaller quantity in the ocean. The problem is to a considerable extent an economic one and it is probable that the highest efficiency will be attained when several salts can be extracted at the same plant, thereby reducing civil engineering, pumping, etc., costs to the minimum.

The third technique, volatilization, is used for bromine. It depends on adjusting the pH of the water to 3·5, treating with chlorine and liberating the free bromine by steaming out. There does not appear to be much likelihood of employing volatilization methods for any other component of sea water except perhaps boric acid.

Other Waters

There are special cases of the use of springs, deep well and mine waters, geysers, lakes, etc., as sources of particular chemicals. These will be referred to where appropriate in connexion with the elements concerned.

Solid Crust as a Chemical Source

The solid crust of the Earth is by far the most important source of chemical raw materials both in quantity and variety of chemical forms and physical states. There are so many different kinds of mineral deposits in the wider sense and so many different mineral substances that a rigid classification presents difficulties. For the purpose of this book the following simplified table, largely based on the classification of Bateman, is presented.

The various kinds of deposit can be roughly classified on a non-technical basis into:

(1) Those which extend more or less uniformly in three dimensions—*e.g.*, bodies of igneous and other rocks.
(2) Those which extend essentially in two dimensions approximately horizontally—*e.g.*, some sedimentary slates, coals, gypsum, and salt beds. Many bedded ore deposits.
(3) Those which extend essentially in two dimensions approximately vertically—*e.g.*, a very large number of mineral veins or lodes, simple or modified.

The methods of extraction are related closely to the type of deposit. The following table provides a summary.

METHODS OF EXTRACTION FROM EARTH'S CRUST

Surface Methods

Pitting and trenching, quarrying, stripping, open cutting, panning, hydraulicking, dredging.

Underground Methods

Vein-mining
Shrinkage stoping, cut-and-fill stoping, square setting, top slicing, rill stoping.

Flat beds and coal seams
Room and pillar, longwall.

Drilling
Gas, oil, sulphur, salt brines.

The various surface methods of extraction are used for solid materials when practicable because of generally lower cost and they are particularly appropriate to low-grade deposits in sedimentary formations of fairly soft rocks when the deposit is at no great depth. With the extensive use of modern excavating machinery it is now possible to work such deposits economically to depths of 100 ft or more and in the case of large open pitlike workings, as for diamonds in South Africa and copper in Nevada, depths of ten times this value are attained.

The exact techniques used necessarily vary with the nature of the deposit.

EXAMPLES OF OPEN OR SURFACE MINING

Material	*Locality*
Coal	England
China clay	Cornwall, England
Pyrite	Rio Tinto, Spain
Asbestos	Quebec, Canada
Iron ores	Lincolnshire, England
	Lake Superior, U.S.A.
Copper ores	Chile, S. America
	Nevada, U.S.A.
Bauxite	British Guiana
Diamonds	Kimberley, S. Africa

The underground mining methods are of course well known and very ancient. The process called ' stoping ' consists in cutting out mechanically or by explosives the vein material from below so that it falls down to the floor of the ' level ' by which it can be removed to the surface.

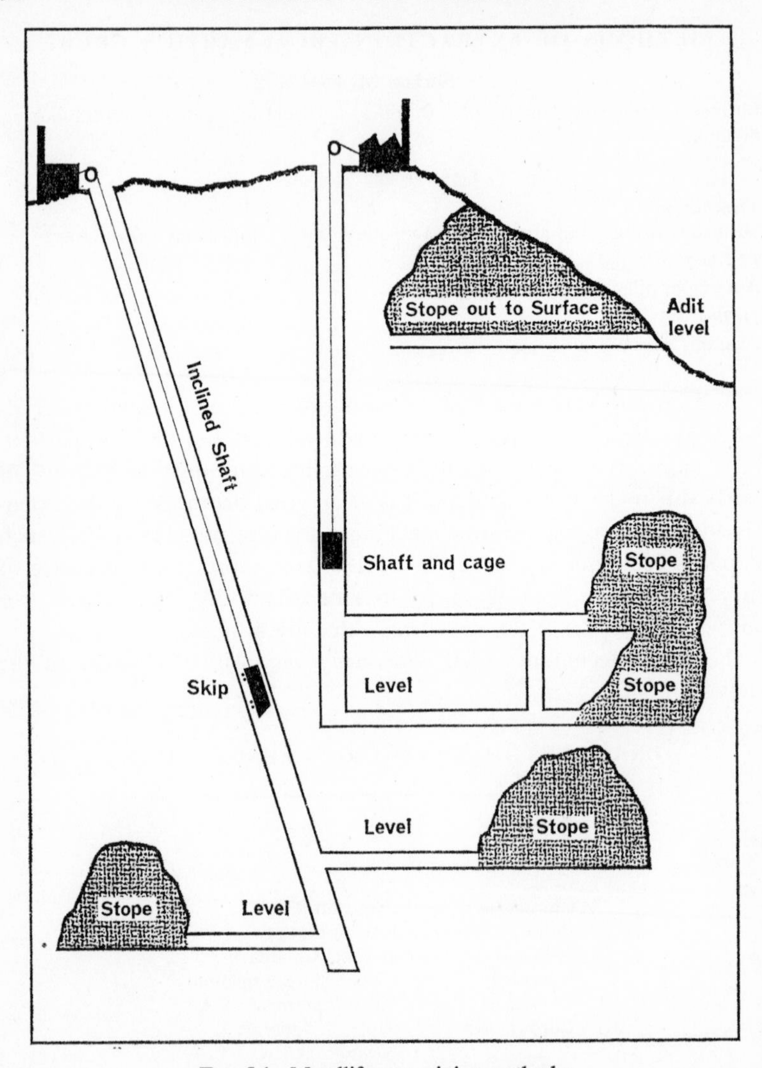

FIG. 5.1. Metalliferous mining methods.

Where the deposit is concentrated and valuable as with some vein minerals, then stoping-out of the veinstone can be done in many cases fairly economically. In this way a great deal of the ancient mining was carried out, sometimes with much profit. Many of the important deposits worked today are of low to medium grade but very extensive and necessitate removal of large amounts of rock and mineralized material.

There is a nice economic balance between the various factors which determine whether or not a deposit can be profitably worked and this may vary from time to time, hence the alternations of activity and stagnation in some kinds of metalliferous mining. There is an increasing tendency, due to modern methods of concentration or ' milling ' of ores, to increase the number and kind of deposits which can be economically worked.

The tenor of an ore is the lowest concentration which can be economically worked. Values of the tenor, which are clearly variable, will be found under the heading of the various elements concerned.

The meaning of the word ' ore ' may perhaps seem somewhat indefinite. It refers traditionally to natural compounds of the common metals which were at least originally reduced by thermal smelting processes. In fact there does not appear to be any real distinction between such a compound as galena or lead sulphide which is the usual ore of lead and, for example, common salt or sodium chloride which is just as certainly the ore of sodium as it is equally the ' ore ' of chlorine. It will be found that in practical mining the term often applies not necessarily to a pure mineral but to the whole mass of material extracted including the relatively valueless ' gangue '. Many of the so-called gangue minerals—*e.g.*, barytes, fluorspar, calcite, may in some circumstances be valuable minerals in their own right. The reader is referred to works on mining for technical details of methods of extraction and concentration.

We now pass to a brief consideration of the methods of procedure for materials in the liquid phase.

The raw materials concerned may be initially liquid and this applies to petroleum and of course to water. The techniques of drilling have now been very highly developed so that wells may be taken down to more than 15000 ft.

In a large number of cases natural gas pressure suffices to raise the fluid to the surface, and otherwise pumping may be used. There are two kinds of raw materials which exist *in situ* as solids—these are common salt and sulphur—and they are capable of being removed in the liquid phase. In England it has long been the practice to extract salt by introducing water to the deposit which is generally of roughly lenticular form and enclosed in impervious rocks. The resultant brine can then be pumped out.

Sulphur in the famous Gulf Coast deposits of the U.S.A. is obtained by the well-known Frasch process. This involves the fusion of the sulphur by injecting superheated water and then blowing the molten material up to the surface by means of compressed air.

An extension of such methods to other mineral substances would no doubt be welcome but unfortunately there are few if any other raw materials which have melting points low enough to be liquefied below ground in this way. It does not seem very likely that any extension of liquid phase mining is probable in the foreseeable future.

The extraction of natural gases, mainly fuel gases, is clearly an analogous operation and it is becoming of increased importance. The techniques are in general analogous to those used in the extraction of petroleum.

One interesting but probably obsolescent application of gaseous phase extraction is the recovery of boric acid from the *suffioni* of Tuscany. Here the volatility of the acid in steam is made use of as a means of separating it but the process seems to be on a diminishing scale.

Location of Deposits

The search for valuable minerals and other important substances of economic interest has been familiar to mining practice for many centuries. The medieval prospector probably employed methods akin to those of the water diviner but from quite early times considerable skill in the observation of metalliferous regions must have been attained. There are few mineralized areas in Europe which have not a long history of mining activity and the multiplicity of workings and trials on the surface in these districts shows the assiduity and degree of success achieved by the ‘ old men ’ by purely empirical methods where experience and an ‘ eye ’ for the country must have been the chief factors.

Modern prospecting methods aim not merely at the location of the deposit but also on an evaluation as far as possible of its extent and quality. Many of the deposits now worked are of large size but of low grade and the economics of mining are of major importance. Much of the primitive mineral prospecting must have been in almost complete ignorance of the geology. It is now usual for geologists and prospectors to work with each other and, in the case of petroleum prospecting in particular, the geology of the region requires to be well known.

Modern prospecting methods are at present mainly geophysical with some growing attention to geochemical techniques. They are summarized below.

(1) Magnetic methods. These involve the study of variations in direction or intensity of the Earth's magnetic field and whilst they may be useful in locating ore bodies of magnetic minerals they are mainly suitable for a limited number of ores of iron, cobalt, and nickel. Magnetic surveys are also useful for structural studies because different rock types show considerable differences in magnetic susceptibility.

(2) Electrical methods. These are chiefly valuable for locating metallic mineral deposits. They mainly depend on differences in conductivity between country rock and mineral deposits of various kinds. The actual measurements made may be of resistance, current, or potential. There are also methods dependent on induced currents which are of superior accuracy but more laborious. In all these methods adequate knowledge of the geology of the area is needed in order to interpret the results of field measurements.

(3) Gravity methods. These are in effect regional surveys of the Earth's gravitational field by one of several methods. Variations in density of rocks cause minute variations in the force of gravity which can be measured with sensitive spring balances called gravimeters. Alternatively the pendulum or the torsion balance may be employed. Plotting of the values on a map of the area may serve to locate an ore body or concealed geological structures.

(4) Seismic methods. These depend on studying the reflection or in some cases the refraction of shock waves from an underground explosion. Hard and soft rocks behave differently as reflectors and refractors of such waves and the technique is widely and successfully used in oil prospecting.

(5) Radioactive methods. Initially these were used for locating radioactive ores—*e.g.*, those of uranium and thorium. They employ a suitable detector of radiation such as a Geiger–Müller or scintillation counter and a high enough sensitivity is now attainable to use the method for differentiating between various kinds of igneous rocks.

(6) Geochemical methods. These have been developed extensively in Russia and they generally depend on the determination of elements in soils, vegetation, ground water, etc.

It has been established that certain elements determined in trace quantities may be indicative of a zone of mineralization—they are called geochemical indicators. The ratio between such pairs as (Ni : Co) and (Ba : Sr) is also found to be related to the nature of adjacent rocks and therefore in some cases of a possible mineralized area. In mapping an area geochemically a large number of determinations of a selected element (usually a metal) is necessary and suitable analytical methods include spectrography and colorimetry. Investigations may be applied to the bedrock, to the top cover of loose material or overburden, to ground waters draining from the area, and to vegetation.

Certain plants are confined to areas where particular metals are present in appreciable quantity in the soil. The following are examples: *Viola calaminaria zinci* (zinc); *Trientalis europea* (tin); *Agrostis alba* (copper).

Some metals cause certain peculiar characters to develop—*e.g.*, blossoms of *California emolcia* are bluish in the presence of copper but lemon yellow in the presence of zinc. Uranium and thorium are said to cause pathologic forms to develop (perhaps mutations caused by gamma-radiation). The plant *Gypsophila patrini* appears to be a remarkable copper indicator.

It would seem that there is a considerable scope in the application of biogeochemical and geobotanical techniques. It is worthy of note that biological evidence for the existence of veins is cited by Agricola as early as the sixteenth century and the Spaniard A. A. Barba described botanical techniques for prospecting for ores in Latin America in the early seventeenth century. Needham has also drawn attention to the use by the Chinese of biogeochemical knowledge and it appears that the presence of certain plants was employed for locating ores as early as the third century A.D.

EXAMPLES OF GEOCHEMICAL PROSPECTING PROJECTS

U.S.S.R.	Fersman	1939	Soil analysis, tin
	Segeev	1941	Soil analysis, tin
	Tkalich	1939	Vegetation
	Maliuga	1947	Vegetation
Cornwall	Palmqvist	1936–	Plant ash analysis for Sn and W
Wales	Brundin	1939	
Finland	Rankama	1940	Plant samples for Ni
Norway	Vogt	1938	Weathering products of sulphides in soils, etc.
Canada	Lundberg	1941	Spectrographic analysis of vegetation
U.S.A.	Geological Survey	1946	Various
Africa	Roberts (Nigeria)	1953	Soil surveys
Japan	Kimura	1950	Soils, waters, rocks

CHAPTER 6

Hydrogen

Abundance

The relative abundance of this element, of about 90% of all matter in the Universe, means that hydrogen atoms are numerically far more important than any of the others. Whilst variations are observed in different stars, only helium is ever met with in major proportions. All the other elements are cosmically of comparative insignificance. Some stars, such as HD 124448, HD 160641, HD 168476, do not seem to contain any hydrogen.

By contrast hydrogen is not quantitatively a major element in the Earth's crust. The abundance, estimated at 1400 g/ton, places it well down in the list—about the same abundance as phosphorus. However, on a relative atomic basis the value is of course appreciably higher.

Isotopes

There are three isotopes of hydrogen, all of which occur naturally, but in very different amounts. Alone of the elements, hydrogen has been dignified by three separate names for its isotopes.

$^{1}_{1}H$, ordinary hydrogen—protium, (H).
$^{2}_{1}H$, deuterium, (D).
$^{3}_{1}H$, tritium, (T).

The normal isotopic composition of terrestrial hydrogen is H = 99·9844% and D = 0·0156%, or a ratio of 6409 : 1.

Evidence available suggests that deuterium is of lower abundance cosmically than terrestrially. The thermonuclear reactions postulated for stellar interiors can involve deuterium, but it tends to be eliminated at further stages, thus

$$^{1}H + {}^{1}H \longrightarrow {}^{2}H + \beta^{+} + \nu$$

then

$$^{2}H + {}^{1}H \longrightarrow {}^{3}He + \gamma$$

At very high temperatures reactions like

$$^{12}C + {}^{16}O \longrightarrow {}^{26}Mg + {}^{2}H$$

might also provide a source of deuterium.

It has been suggested that terrestrial deuterium may form in ocean waters by cosmic ray neutron capture:

$$^{1}H + n \rightarrow {}^{2}H$$

The third isotope of hydrogen, tritium, is very scarce in Nature and it is stated (Faltings and Hartuk) that the total quantity in the entire atmosphere is only about one gramme molecule (6 g).

It appears to be formed by action of high-energy cosmic ray neutrons on nitrogen:

$$^{14}N + n \rightarrow {}^{3}H + {}^{12}C$$

With its brief half-life (12·41 years) it is clear that an equilibrium between formation and disintegration

$$^{3}H \rightarrow {}^{3}He + \beta^{-}$$

will have been established.

Later estimates (Kaufmann, Libby, Buttlar, 1954, 1955) place the figure much higher, at 1·8 kg.

The tritium content of the atmosphere, mainly as T_2O in the stratosphere, is believed to have been increased on a massive scale (*circa* 30 kg) as a result of the tests of thermonuclear devices. This is probably the most significant example to date of a profound disturbance of the amount and balance of an element in Nature as a result of human activity.

Terrestrial Occurrence

(1) Atomic hydrogen. This appears to be produced by the photo-dissociation of water vapour:

$$H_2O \rightarrow H + OH$$

The most effective wavelengths are in the ultra-violet, particularly below 1800 Å, but as molecular oxygen absorbs strongly in the same range, appreciable formation of free H atoms must necessarily be determined by O_2 concentration. The free atoms are believed to be involved in a number of fast reactions:

$$H + O_3 \rightarrow OH + O_2$$
$$H + O_2 + M \rightarrow HO_2 + M$$
$$H + O + M \rightarrow OH + M$$
$$OH + O \rightarrow O_2 + H$$

(M is an inert 'third body'). Hence only at great altitudes, above

60 km, is atomic hydrogen important; it is estimated to be of the order of 10^9–10^{10} atoms per cubic centimetre and at the greatest altitudes it is probably the dominant species in the atmosphere. Recent rocket techniques in connexion with the scattering of the Lyman-α-line at 1216 Å seem to indicate that most of the hydrogen is above 200 km.

The loss of this hydrogen into outer space as a result of kinetic molecular motion must necessarily be balanced by its rate of production. It is therefore evident that the atomic hydrogen content of the atmosphere is determined by physical and chemical equilibria.

(2) Molecular hydrogen H_2. Free hydrogen is met with on the Earth in a variety of conditions.

(*a*) *Volcanic emanations.* Free hydrogen has frequently been reported in volcanic gases. It may well have been formed in such cases by chemical reactions between water and various components of the heated rock material associated with the volcanic regions concerned. Many of the older analyses show significant amounts of free hydrogen in these gases.

Source	*Date*	H_2 (% vol.)	*Analyst*
Reykjalidh	1853	25·14	Bunsen
Monte Pelee	1902	8·12	Moissan
Kilauea	1912	10·2	Day and Shepherd
Santorini		29·43	Fouque

(*b*) *Other natural gases.* Gases associated with petroleum are mainly hydrocarbons but free hydrogen has been occasionally reported in amounts up to 22·5 %.

Very high hydrogen content has been reported in occluded gas from the salt deposits at Stassfurt—93·5 % (Precht). Occluded gases from rocks and meteorites generally appear to contain free hydrogen but it is not clear how far it was originally present or possibly formed in reactions of various rock minerals with water or acids during the process of extracting the occluded gas for analysis.

(*c*) *Biological sources.* A number of species of anaerobic bacteria produce hydrogen gas and particularly those found in the lower intestinal tract and in the faeces of animals. Thus *B. coli* generates a gas containing about 50 % H_2 and 50 % CO_2 whilst *B. aerogenes* gives a gas containing about 33 % H_2 and proportionately more CO_2.

(3) Hydrogen in combination. The forms in which hydrogen occurs in chemical combination in Nature may be summarized as follows:

(*a*) As water, in all three states of aggregation.
(*b*) As water of crystallization in minerals.
(*c*) As metallic hydroxides and basic salts.
(*d*) As hydrocarbons.
(*e*) As complex organic molecules (other than (*d*)).
(*f*) As the solvated hydrogen ion in natural acids.
(*g*) As the hydrides of a number of other elements.

These will now be considered in some greater detail.

(*a*) *Water*. Possibly the most abundant single compound on the Earth is water. The mass of the hydrosphere (usually taken to include the polar ice-caps) is quoted as approximately 14100 Gg or $14{\cdot}1 \times 10^{17}$ tons. The relative amount of ' fresh ' water is perhaps about $\frac{1}{2000}$th of this and the water vapour content of the atmosphere about 0·13 Gg or $1{\cdot}3 \times 10^{13}$ tons. Practically all natural waters may be regarded as solutions of assorted inorganic ions with variable and usually minor amounts of non-ionic matter, colloidal or otherwise, together with dissolved gases, which are usually those of the atmosphere.

' Fresh ' water from lakes and rivers may have salinity as low as 20 p.p.m. and water of enclosed seas like the Dead Sea may contain nearly 260000 p.p.m.

The chemical reactivity, solvent properties, and ionizing power of water are reflected in the large number of ionic and molecular species met with in different kinds of natural waters. The most important molecular species are probably boric acid H_3BO_3 and silica (perhaps as a collosol).

The ionic species reported in waters are typified by the list for sea water which exceeds forty ions determinable by appropriate analytical methods.

Whilst in its major constituents the water of the ocean shows little variation, the minor constituents show very marked variation in concentration both in depth and in geographical location. Thus the ocean regarded as an aqueous solution exhibits a high degree of inhomogeneity.

The variations, both quantitative and qualitative, between the mineral contents of other natural waters are very great indeed. As might be expected this effect is most marked in the case of waters of deep wells and mineral springs.

The isotopic constitution of natural waters is of some significance and has been investigated to a considerable extent.

In respect of the hydrogen isotopes alone, the following molecules may be expected: H_2O, D_2O, T_2O, HDO, HTO, and DTO. However, as there are also three oxygen isotopes in Nature (^{16}O, ^{17}O, and ^{18}O) it follows that eighteen possible isotopic molecules of water can exist.

Isotopic fractionation occurs particularly in evaporation, and deep ocean waters are impoverished in deuterium due to preferential evaporation of protium oxide from the surface layers. The various analyses of water for deuterium content are not at present sufficiently conclusive of any perceptible differences between water of the hydrosphere generally and the so-called ' juvenile ' water postulated by geochemists as coming up in volcanic and other emanations from the deeper layers of the lithosphere. It may be supposed that during geological time there has been a perceptible contribution to the hydrosphere from this source but its magnitude cannot at present be assessed with any accuracy.

It is important to draw attention to the presence of water as an essential part of living organisms and the content may be about 50% as in mammals to almost 99% as in some marine invertebrates.

The presence of water vapour in the atmosphere of Venus has now been established. The presence of a little water vapour in the Martian atmosphere has been inferred but not positively confirmed. The polar caps of Mars are generally thought to be of ice and it has also been claimed that the rings of Saturn have the same composition.

(*b*) *Water of crystallization.* A large number of minerals are formulated as containing water of crystallization which in general is largely lost by heating to temperatures not much above 100 °C. This is quite distinct from the much larger number of species which are referred to in (*c*) as compounds containing OH ions.

The most important and abundant types of hydrous minerals are met with among the following:

(i) A few oxides—*e.g.*, chalcophanite $ZnMn_3O_7.3H_2O$.
(ii) Most of the borates—*e.g.*, borax $Na_2B_4O_7.10H_2O$.
(iii) Some carbonates—*e.g.*, lansfordite $MgCO_3.5H_2O$.
(iv) Silicates, particularly in the zeolite group and some of the clay minerals.
(v) Many of the phosphates and arsenates—*e.g.*, vivianite $Fe_3(PO_4)_2.8H_2O$
(vi) Sulphates—*e.g.*, gypsum $CaSO_4.2H_2O$

It will be seen that these minerals are either those of the upper weathered layer of the Earth's crust or are mainly derived from it. Loosely combined water is largely absent from igneous rocks and metamorphic rocks, and the most important sediments containing it are the clays, shales, and some of the evaporite deposits.

Hydrous vein minerals are essentially those of the upper zones of sub-aerial weathering.

(*c*) *Metallic hydroxides and basic salts.* The OH ion is a unit very frequently encountered in the structure of crystalline minerals. There is a limited number of simple metallic hydroxides and a very large number of species which contain some OH ions in addition to the various common anions such as carbonate, silicate, phosphate, sulphate, etc. On account of close similarity in size the OH ion often replaces F and O ions in minerals.

The hydroxides are for the most part layered-lattice structures. Thus brucite $Mg(OH)_2$ has double (OH) sheets with Mg ions between the sheets. As the (OH) ions in the sheets are held together by weak bonds this type of mineral has a well-developed cleavage.

The common hydroxides are

Brucite	$Mg(OH)_2$	Boehmite	$AlO(OH)$
Pyrochroite	$Mn(OH)_2$	Manganite	$MnO(OH)$
Lepidocrocite	$FeO(OH)$	Gibbsite	$Al(OH)_3$

The second class of hydroxidic minerals are in effect basic salts of various kinds. They are so numerous that only selected examples will be referred to. Excluding the silicates we may note that the following metals are very prone to form basic salts in Nature, *viz.*, Cu, Zn, Al, Bi, Fe, Cr.

On the other hand some anions, particularly borate, nitrate, iodate, and tungstate, are rarely met with as basic salts. The following may be regarded as examples of minerals which contain OH ions—it will be seen that they are typically compounds produced by oxidation under hydrous conditions of ' primary ' minerals of various kinds.

Atacamite	$Cu_2Cl(OH)_3$	Fluoborite	$Mg_3(BO_3)(F,OH)_3$
Laurionite	$Pb(OH)Cl$	Brochantite	$Cu_4(SO_4)(OH)_6$
Hydrozincite	$Zn_5(CO_3)_2(OH)_6$	Alunite	$KAl_3(SO_4)_2(OH)_6$
Malachite	$Cu_2(CO_3)(OH)_2$	Clinoclase	$Cu_3(AsO_4)(OH)_3$
Hydrocerussite	$Pb_3(CO_3)_2(OH)_2$	Hydroxyapatite	$Ca_5(PO_4)_3(OH)$
Gerhardtite	$Cu_2(NO_3)(OH)_3$	Wavellite	$Al_6(PO_4)_4(OH)_6.9H_2O$
Salesite	$Cu(IO_3)(OH)$	Turquoise	$CuAl_6(PO_4)_4(OH)_8.5H_2O$

Silicates with OH ions. Examples of these are best considered under separate headings.

(i) Minerals of igneous rocks:

Tremolite	$Ca_2Mg_5(Si_4O_{11})_2(OH)_2$
Muscovite	$KAl_2(AlSi_3O_{10})(OH)_2$
Biotite	$K(Mg,Fe)_3AlSi_3O_{10})(OH)_2$

(ii) Minerals of sedimentary rocks:

Kaolinite	$Al_4Si_4O_{10}(OH)_8$
Chlorite	$Mg_{10}Al_2(Si_2Al_2)(O_{20})(OH)_{16}$

(iii) Minerals of metamorphic rocks:

Talc	$Mg_3Si_4O_{10}(OH)_2$
Serpentine	$Mg_3Si_2O_5(OH)_4$

(iv) Other silicates:

Chrysotile	$Mg_3Si_2O_5(OH)_4$
Hemimorphite	$Zn_4Si_2O_7(OH)_2.H_2O$

The above examples indicate the very large number of basic salts occurring as minerals but from the point of view of hydrogen content they are not very important. Thus a mineral such as kaolinite contains only about 1·55% of hydrogen. It would seem from the average mineralogical composition of igneous rocks that the hydroxidic and hydrous minerals are not very significant and in the ultrabasic rocks of which the sub-crystal zone is thought to be composed they are practically absent.

(*d*) *Hydrocarbons.* These occur in Nature in great variety and very considerable quantity. As they are chiefly of interest as compounds of carbon they will be dealt with more fully in Chapter 13. In this chapter their general modes of occurrence will be summarized.

Some volcanic gases have been reported as containing small amounts of methane and other simple paraffins but it appears that the vast majority of occurrences of hydrocarbons are directly or indirectly of biogenic origin.

The most important hydrocarbon of direct biological origin is methane CH_4 and it is known to be produced by the action of anaerobic bacteria on carbonaceous matter. Hence it is found in marshes and swamps, is freely formed in the rumen of herbivorous animals, and is the chief constituent of the firedamp of coal mines.

It has now been shown that hydrocarbons comparable to those of petroleum are present in recent marine sediments and it may be supposed that they have originated largely from bacterial decay of the organic matter of the bottom muds known as sapropels. It seems that there is little doubt that the primary stage of petroleum formation occurs in these circumstances.

The special feature of liquid and gaseous hydrocarbons is capacity for migration from their original site of formation, and when they come to be retained by domes of impervious cap-rock the conditions for gas and oil accumulations exist.

On account of great variation in the modes of migration and accumulation it is common to find that sometimes liquid and sometimes gaseous hydrocarbons occur alone, but in the case of very many oil-producing areas they occur together. Important gas-producing areas have been exploited in many parts of the world, examples being in the western parts of North America (California, Alberta), in Northern Italy, and in the Pyrenees foothills in Southern France.

Hydrocarbon content of natural gas consists mainly of methane but paraffins up to hexane may also occur in it.

In the case of the liquid material—*i.e.*, crude petroleum, the composition is extremely variable but it is essentially a mixture of hydrocarbons of various types with smaller amounts of other materials such as sulphur compounds. The hydrogen content is likely to be in excess of 10 % and therefore considerably greater than that of coal (about 6 %).

(*e*) *Organic substances.* The hydrogen content of organic matter is relatively small and on a basis of dry weight may be taken as around 6 %. It may be supposed that such hydrogen—present particularly in carbohydrates, proteins, and fats—has been derived originally from water by means of the basic photosynthetic reaction. In the processes of decay some of this hydrogen finds its way into various hydrides such as ammonia, hydrogen sulphide, and methane.

(*f*) *The solvated hydrogen ion.* H_3O^+ is the basis of acidity, and is encountered in Nature in a wide variety of conditions.

(i) Free acids in volcanic emanations. The most commonly occurring acids are HF, HCl, and H_2SO_4. They may have been formed by the action of water on various substances originally present in the magma, particularly the reactive halides of elements. The common presence of boric acid in volcanic gases suggests the possibility of formation of HF by decomposition of BF_3 by water or steam.

Sulphuric acid could be derived from such sulphur compounds as

H_2S and SO_2 which are commonly present in volcanic gases. An estimate of the output of the two hydracids in certain localities has been made. Thus in the Valley of Ten Thousand Smokes in Alaska it is

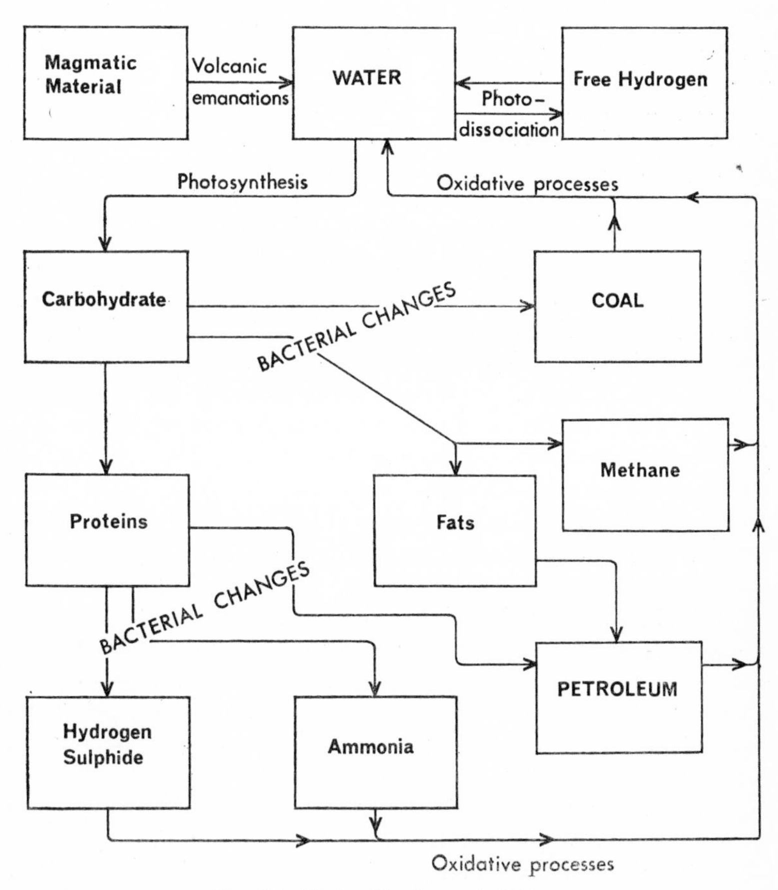

FIG. 6.1. Cycle of hydrogen in Nature.

stated that the annual emission of HF is 200000 tons and of HCl 1300000 tons.

(ii) Solid minerals which are acid salts. Only a very small number of minerals of this type exists in Nature. Apart from the sodium and potassium carbonates, which are of course alkaline and are associated with saline deposits, the acid sulphates are mostly closely connected with fumarolic activity, or oxidative conditions in sulphide ore deposits.

Some of the more important mineral acid salts are listed below.

Nahcolite	$NaHCO_3$	Rhomboclase	$FeH(SO_4)_2.4H_2O$
Kalicinite	$KHCO_3$	Monetite	$CaH(PO)_4$
Mercallite	$KHSO_4$		

Boric acid may also be mentioned here. It occurs as the solid mineral sassolite H_3BO_3.

Hydrogen ion concentration in Nature. It is convenient to adopt the well-known notation of pH values, a logarithmic scale in which the hydrogen ion concentration varies inversely as the pH number. pH 7 exists in perfectly neutral solutions and higher values represent alkaline conditions whilst lower values than 7 represent acid conditions. The highest acidity met with in Nature seems to be pH 1·0 which is due to strong mineral acids (HCl and H_2SO_4) in solution in volcanic waters—*e.g.*, Crater Lake, Java and Rio Vinagre, Mexico—and occasionally in waters draining from mineral deposits where it is due to H_2SO_4 formed by oxidation of pyrite. More localized acid waters with pH falling nearly as low (2·0–1·5) can be due to action of those sulphur bacteria which are able to oxidize sulphur to sulphuric acid. Peat waters may show pH as low as 4·0–4·5 due to various complex organic acids. Rain water due to solution of carbonic acid is normally slightly acid—pH 5·9. Surface waters, rivers, and lakes in regions of non-calcareous rocks approach neutrality with pH 6·0–7·0 but in limestone regions the water becomes alkaline with pH 8·0–8·4.

The reaction of sea water is also slightly alkaline at pH 8·1–8·3 but changes of pH with depth are characteristic. In restricted seas and inlets the action of bacteria in the bottom muds lowers the pH due to formation of H_2S and values of 7·2–7·3 are recorded at 1000 m depth. Active photosynthesis in surface waters has the opposite effect by rapidly removing carbonic acid so that quite strongly alkaline (pH 9·6) waters are met with in small freshwater ponds with much vegetation, and a similar but smaller effect is noticed in shallow pools of sea water.

The highest natural pH of about 10·0 seems to be that found in the peculiar alkaline (Na_2CO_3) soils of certain desert regions—*e.g.*, Sudan and Utah, U.S.A.

(*g*) *Hydrides of elements* (*other than oxygen and carbon*). These constitute a very limited number of compounds, some of which have already been mentioned. They are gaseous but may pass into aqueous solution. The three acid gases HF, HCl, and H_2S are all found in volcanic emanations and may have been formed by the action of water or steam on binary fluorides, chlorides, and sulphides, as stated above.

Hydrogen sulphide is formed freely by the reducing action of anaerobic bacteria particularly in the bottom muds, or sapropels in sea and estuarine waters. Ammonia is formed also by anaerobic decay of nitrogenous material mainly in soils.

Extra-terrestrial Hydrogen

The preponderance of hydrogen in the stars and nebulae has already been noted. We now refer to the solar system. Spectroscopic observations show that the atmospheres of the planets (Mercury and the Moon are devoid of atmospheres) contain a variety of gases. On Venus it appears that a little water vapour exists in a dense CO_2 atmosphere. The polar caps of Mars are generally thought to be of ice so that water is present there also.

In the case of the outer planets it is found that strong absorption bands for CH_4 exist and there is evidence of high-pressure H_2. Ammonia is almost certainly present in varying amounts (possibly solid) in the outer planets. It is of interest that in these large bodies—Jupiter, Saturn, Uranus, and Neptune—the atmospheres are to a large extent of hydrogen compounds or free hydrogen whereas those of Earth and Venus are rich in oxygen or its compounds.

Industrial Sources

There appear to be only two sources of hydrogen in use and these are water and methane.

A high proportion of hydrogen comes from water and the two main processes are electrolysis (directly or as a by-product of other electrolytic processes) and reduction by carbon—*e.g.*, the water gas process. The production of hydrogen from hydrocarbons (mainly methane) depends either on thermal cracking (obsolescent) or more usually on the reaction

$$CH_4 + 2H_2O = CO_2 + 4H_2$$

so that in fact half of the hydrogen comes from the water and half from the methane.

Output

Approximately 66% of hydrogen produced is used in synthetic ammonia production and on this basis the annual output is about 1 800 000 tons.

The actual output of elemental hydrogen as gas is difficult to ascertain but the British production is in excess of 1 million ft^3 per annum and the U.S.A. figure, on basis of June 1961, is about 60000 million ft^3, equivalent to 150000 tons.

CHAPTER 7

The Inert Gases

General

These six elements, one of which, radon, is a short-lived radioactive nuclide, are all alike in their virtually complete lack of chemical properties. Their distribution in Nature is therefore determined entirely by nuclear reactions and physical properties.

Helium

This element is second only to hydrogen in cosmic importance and its abundance relative to hydrogen is as quoted below (after Aller):

Stellar atmospheres, $He : H = 0{\cdot}16$.
Planetary nebulae, $He : H = 0{\cdot}18$.

This is a complete reversal of the position in regard to the Earth, where it ranks among the rarer of the elements. In view of modern theories of stellar evolution it is probable that the helium content of the interiors of some types of stars is high relative to hydrogen.

Terrestrial Abundance

Helium is one of the very few of the elements now on the Earth which did not form part of its primordial composition. It is virtually certain that all terrestrial helium is of radiogenic origin.

The nuclei of helium, known as α-particles, are formed in the decay of several members of the three natural radioactive families and of ^{147}Sm. The isotope ^{3}He which is present to the extent of $1{\cdot}3 \times 10^{-4}\%$ is the stable decay product of tritium ^{3}H so that the origin of all terrestrial helium can be accounted for.

The great part of the helium in the Universe, however, has presumably been formed in the so-called ' hydrogen burning ' stars, including our own Sun, by the following reactions:

(1) $^{1}H + {}^{1}H \rightarrow {}^{2}H + \beta^{+} + \nu$
(2) $^{2}H + {}^{1}H \rightarrow {}^{3}He + \gamma$
(3) $^{3}He + {}^{3}He \rightarrow {}^{4}He + 2{}^{1}H$

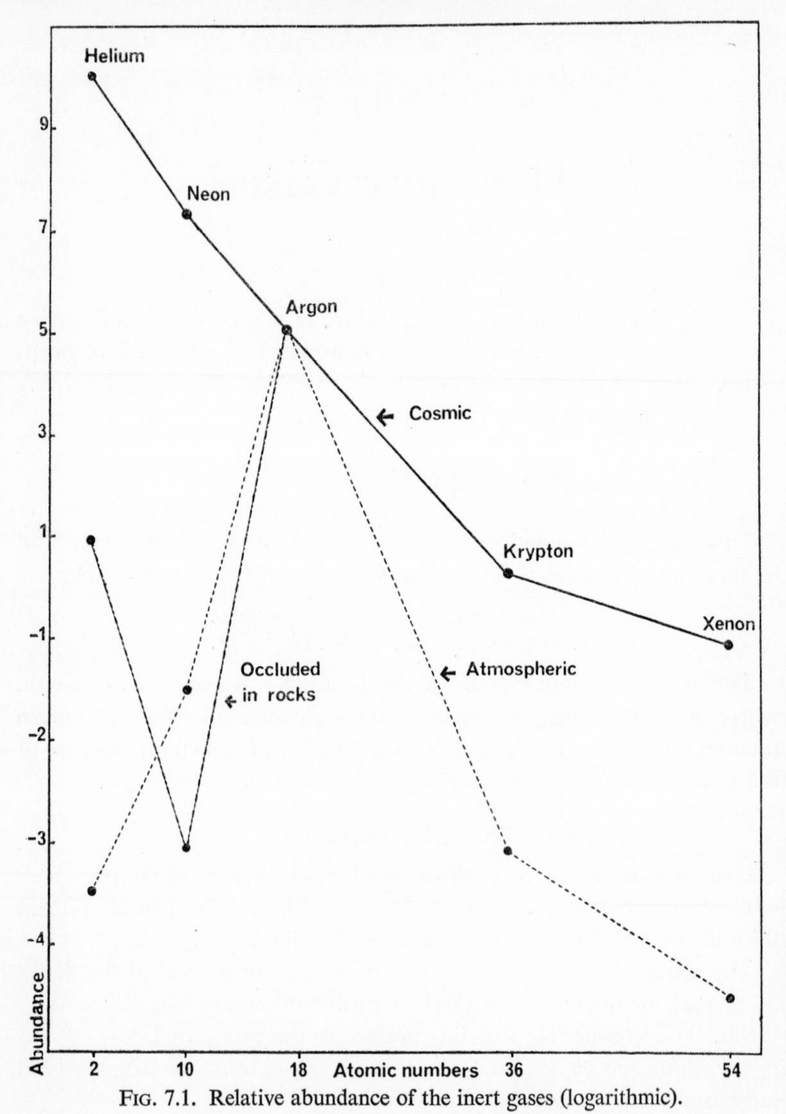

FIG. 7.1. Relative abundance of the inert gases (logarithmic).

These occur in the early stages of hydrogen burning. If the temperature does not exceed 8×10^6 °K the process stops at (2). At about $1{\cdot}3 \times 10^7$ ° K the following set in:

$$(4)\quad {}^3\text{He} + {}^4\text{He} \longrightarrow {}^7\text{Be} + \gamma$$
$$(5)\quad {}^7\text{Be} + \beta^- \longrightarrow {}^7\text{Li}$$
$$(6)\quad {}^7\text{Li} + \text{p} \longrightarrow 2\,{}^4\text{He}$$

It is thought that the reactions in the Sun proceed mainly as above. If some carbon atoms are present then the well-known carbon–nitrogen cycle takes place

$$
\begin{aligned}
&(1)\quad {}^{12}C + {}^{1}H \rightarrow {}^{13}N + \gamma \\
&(2)\quad {}^{13}N \rightarrow {}^{13}C + \beta^{+} + \nu \\
&(3)\quad {}^{13}C + {}^{1}H \rightarrow {}^{14}N + \gamma \\
&(4)\quad {}^{14}N + {}^{1}H \rightarrow {}^{15}O + \gamma \\
&(5)\quad {}^{15}O \rightarrow {}^{15}N + \beta^{+} + \nu \\
&(6)\quad {}^{15}N + {}^{1}H \rightarrow {}^{12}C + {}^{4}He
\end{aligned}
$$

Terrestrial helium. It is estimated that the total radioactive material in the Earth's crust will produce $2 \times 10^{7}\,m^{3}$ of helium per year. It is thus clear that the small actual amount of helium in the atmosphere (0·0005 %) represents an equilibrium between rate of production and rate of loss by escape from the upper atmosphere because the Earth's gravitational field cannot permanently retain a gas as light as helium.

The extraction of helium from the air on a large scale is not very practicable, particularly due to the presence of neon which is difficult to separate from it.

Alone among the inert gases, there are, however, considerable sources of helium in connexion with various natural gases.

The most prolific sources occur where geological conditions favour the accumulation of gases in underlying strata. It would appear that helium occluded in rock material can be leached out by flow of large amounts of various kinds of natural (including petroligenic) gases. Volcanic gases and gases from thermal springs may have perceptible helium. The following table includes some of the most interesting sources.

HELIUM SOURCES

Country	*Source*	*He* (%)	*Yield* (m^3/year)
(1) England	King's Well, Bath	0·15	
(2) France	Santenay, Carnot, Côte-d'Or	9·97	17·845
(3) France	Bourbon-Laney	1·84	10·074
(4) France	Neris, Allier	0·97	33·99
(5) France	Lesquin, Lille	0·924	1359
(6) France	Vaux-en-Bugey	0·095	20000
(7) Germany	Nuengamme, Hamburg	0·025	25000
(8) U.S.A.	Petrolia, Texas	0·9	
(9) U.S.A.	Rogers, Kansas	1·28	
(10) U.S.A.	Pearson, Oklahoma	0·63	
(11) U.S.A.	Amarillo, Texas	1·75	13600000
(12) Canada	Calgary, Alberta	0·3	

The most important helium producer is the Amarillo source (11); the production is as follows:

Year	1954	1955	1956	1957	1958	1959	1960
$ft^3 \times 10^6$	191	221	244	291	334	477	475
$m^3 \times 10^6$	5·45	6·60	6·97	8·31	9·57	13·6	13·5
tons	98	119	125	150	172	245	244

Neon

This element is probably much more important cosmically than it is terrestrially. Virtually the only source is the atmosphere of which the neon content has the low value of 0·00125% by volume, equivalent to a terrestrial abundance of less than 0·001 g/ton. On the other hand the cosmic abundance of neon is such that the Ne : H ratio is about 1 : 2000 and in some B-type stars as much as 1 : 1400. Three stable isotopes of the element occur in Nature and on the Earth the relative amounts are

$$^{20}Ne = 90{\cdot}22\%$$
$$^{21}Ne = 0{\cdot}25\%$$
$$^{22}Ne = 8{\cdot}82\%$$

^{20}Ne is probably synthesized in the α-process by the reaction

$$^{16}O + {}^{4}He \longrightarrow {}^{20}Ne + \gamma$$

^{22}Ne by the reaction

$$^{18}O + {}^{4}He \longrightarrow {}^{22}Ne + \gamma$$

In the ‘ hydrogen burning ’ shells surrounding the inert helium cores of the red giant stars the temperature rises so high that the neon–sodium cycle sets in and ^{21}Ne is produced:

$$^{20}Ne + p \longrightarrow {}^{21}Na + \gamma \qquad {}^{21}Na \longrightarrow {}^{21}Ne + \beta^+ + \nu$$
$$^{21}Ne + p \longrightarrow {}^{22}Na + \gamma \qquad {}^{22}Na \longrightarrow {}^{22}Ne + \beta^+ + \nu$$
$$^{22}Ne + p \longrightarrow {}^{23}Na + \gamma \qquad {}^{23}Na + p \longrightarrow {}^{20}Ne + \alpha$$

Fowler and the Burbidges have suggested that a resonance may favour the formation of ^{21}Ne which may be built up to substantial quantities in the cores of certain stars.

The terrestrial rarity of neon (see Fig. 7.1) seems anomalous and presents a number of problems.

Since in Nature it forms no compounds it cannot be 'fixed' in a non-volatile form so that it may have largely escaped from the primordial atmosphere. This process would, however, require a rather high temperature (see Fig. 2.4). Under such conditions gases like methane, ammonia, and water vapour would be even more readily lost. If on the other hand the Earth originated by the accretion of initially cold solid particles then the presence of any great quantity of such a gas would be unlikely. In that event the small neon content of the atmosphere could have been derived by release (possibly by de-gassing) of some of the neon known to be still retained in the lithosphere.

The very small amount of neon found as occluded gas in rocks varies very little as between different igneous types from dunite to granite. The average is about 0·08 cm^3/ton which is only a very small fraction of the argon content.

Argon

This element is the most abundant of the inert gases in the atmosphere, amounting to about 0·9 % by volume. It is likewise the most abundant of these gases occluded in rocks. The cosmic abundance of argon, however, is third in order of the inert gases which is consistent with the progressive decrease in abundance of elements with increasing mass number.

The isotopic composition of atmospheric argon is

$$^{36}Ar = 0{\cdot}337\,\%$$
$$^{38}Ar = 0{\cdot}063\,\%$$
$$^{40}Ar = 99{\cdot}60\,\%$$

The view of Weizacker (1937) that the high terrestrial abundance of ^{40}Ar is due to the fact that it is formed in the decay of ^{40}K is now generally held.

The most abundant nuclide cosmically is ^{36}Ar and it is probable that it is one of those synthesized in the so-called 'equilibrium process' which is thought to develop during that phase of stellar evolution following the consumption of the helium in the α-process.

The potassium isotope ^{40}K undergoes decay by two processes, by *K*-electron capture to give ^{40}Ar and by β-emission to give ^{40}Ca. The first process accounts for 11·2 % of the decay and various calculations from the known potassium content have been made to deduce the amount of existing terrestrial ^{40}Ar. Not all of this argon is yet present in the atmosphere but it appears that occluded gases continue to be released by weathering and by de-gassing in volcanic regions.

A rigorous quantitative interpretation of this problem is not yet possible but it is certain that at least a major proportion of terrestrial argon is radiogenic and therefore, like helium, is no older than the Earth itself.

Production

Argon is readily available as a by-product gas from fractionation of liquid air and in the ' bleed gas ' in the manufacturing of synthetic ammonia. Only the U.S.A. figures are available.

Year	1956	1957	1958	1959	1960	1961
$ft^3 \times 10^6$	230	397	372	499	564	640
$m^3 \times 10^6$	6·8	11·7	11·5	14·7	16·6	18·8
tons	6120	10430	10350	14230	14940	16920

Krypton

The low terrestrial abundance of krypton places it among the rarer elements. The only source is the atmosphere which contains 0·0003 % of the gas by weight.

Terrestrial krypton is a mixture of six isotopes:

$^{78}Kr = 0{\cdot}354\%$ $^{83}Kr = 11{\cdot}55\%$
$^{80}Kr = 2{\cdot}27\%$ $^{84}Kr = 56{\cdot}90\%$
$^{82}Kr = 11{\cdot}56\%$ $^{86}Kr = 17{\cdot}37\%$

There is the possibility of the natural occurrence of ^{81}Kr which is a long-lived radio-isotope (half-life $2{\cdot}1 \times 10^5$ years). It is known that ^{83}Kr, ^{84}Kr, and ^{86}Kr are among the products of natural fission of uranium, therefore some of the terrestrial krypton has originated in this way but the amount (about 0·30 %) is very small.

Xenon

Like krypton, this element is obtainable only from the atmosphere which contains 0·000036 % by weight. There are nine stable isotopes in Nature of mass numbers 124, 126, 128, 129, 130, 131, 132, 134, and 136.

Some of these may be partly radiogenic being derived by β-decay from ^{129}I, ^{128}Te, and ^{130}Te.

Fissiogenic xenon is believed to be present in natural uranium minerals. On account of these reactions xenon measurements in minerals may be applied to age determination.

CHAPTER 8

Oxygen

General

Oxygen is an abundant element both terrestrially and cosmically. The very high abundance (*circa* 50 %) in the Earth's crust is undoubtedly due to differentiation, but even in the Sun and many stars it stands high in order of relative abundance. By Oddo and Harkins' rule a nuclide of even mass number ($= 4 \times 4$) could be expected to be of high stability and it is not surprising that the two other natural isotopes, ^{17}O and ^{18}O, both occur in very small amounts. The average isotopic composition is

$$^{16}O = 99{\cdot}759\,\%$$
$$^{17}O = 0{\cdot}0374\,\%$$
$$^{18}O = 0{\cdot}2039\,\%$$

Modern theories of nucleogenesis account for the origin of the most important isotope, ^{16}O, by nuclear reactions in the so-called 'helium burning' stars where it is thought that the 'ramming together' of three α-particles to form a carbon atom is possible. From carbon a further addition of an α-particle gives ^{16}O. ^{18}O is thought to be formed by α-particle capture by ^{14}N. The much rarer ^{17}O is supposed to be formed by the reactions

$$^{16}O + {}^{1}H \rightarrow {}^{17}F + \gamma$$
$$^{17}F \rightarrow {}^{17}O + \beta^{+} + \nu$$

but most of the ^{17}O is then destroyed by the reaction

$$^{17}O + {}^{1}H \rightarrow {}^{14}N + \alpha$$

Forms of Occurrence

While it is not surprising that such an active element should exist in many different forms of combination, it is the more remarkable that it should occur massively in the free state. There is spectroscopic evidence for the existence of oxygen atoms in various states of excitation in stellar atmospheres, a few oxygen radicals or ions are detected in cometary atmospheres, and carbon dioxide appears to be a major constituent of the Venusian atmosphere. There does not seem to be any certain evidence of the existence of molecular oxygen anywhere in the Universe apart from the Earth.

Elementary molecular species. Recent work indicates that at different levels the species O, O_2, and O_3 are present in the atmosphere; the relation between them is closely connected with the photochemical action of sunlight. At low levels the ordinary molecular form O_2 is of course inherently stable and the only form perceptibly present (the O_3 content at sea-level is variable at about 0·5 p.p.m.).

Diatomic oxygen is dissociated by radiation in two wavelength regions. There is very strong absorption from 1760 Å, the Schumann–Runge region and from 2420 Å, the Herzberg region. The first band of radiation is virtually all absorbed above 60 miles altitude and in this zone the recombination of free oxygen atoms occurs as a third-body reaction

$$O + O + M \rightarrow O_2 + M$$

Physical measurements with rocket techniques show that whereas at 60 miles the monatomic and diatomic species are about equal in concentration, at greater heights the monatomic form predominates and at 116 miles the O/O_2 ratio may be 10 : 1. The formation and existence of ozone in the so-called 'ozonosphere'—a region at about 12 miles altitude—is determined by the following reactions:

$$\begin{array}{ll} (1) & O + O_2 + M \rightarrow O_3 + M \\ (2) & O_3 + h\nu \rightarrow O_2 + O \\ (3) & O_3 + O \rightarrow 2O_2 \\ (4) & O_2 + h\nu \rightarrow O + O \end{array}$$

From these it is possible to deduce the ratio of concentration of O_2 and O_3 in terms of the radiation intensities effective for (2) and (4), the reaction constants for (1) and (3), and the concentration of the third body M. The actual amount of ozone at its maximum is equivalent to a partial pressure of 0·02 cm Hg, at a height of about 15 miles, corresponding to about 1 % by volume. Variations in the vertical distribution of ozone are likely to be of considerable meteorological significance.

One of the most important functions of the ozonosphere is that it provides an effective blanket against much of the short-wave solar radiation which would otherwise reach the Earth and which would of course be inimical to all forms of life. It may be assumed that in the absence of an oxygen atmosphere the surface of our planet would be subjected to intense radiation and many photochemical reactions would be possible which do not now take place.

Covalent molecular compounds. The most important oxygen compound of this type is clearly water, but also included are carbon dioxide and a large number of organic oxygen compounds. It is usual to consider the silicates as ionic structures but the partial covalent character of the Si—O — should not be overlooked.

Ionic oxygen. The greater part of the oxygen in combination in the solid crust of the Earth is considered to be in ionic form.

The most important types are, (*a*) free metallic oxides, (*b*) complex oxygen ions.

Free metallic oxides. In contrast with the sulphides, most of the metallic oxides are essentially ionic. Simple oxides of a large number of the metals are found as minerals, some are very abundant, but others are mere mineralogical curiosities. Many of them will be found referred to under their appropriate elements. Elements whose oxides do not seem to occur in Nature include the alkali metals, the alkaline earths (except Ca and Mg), silver, gold, and the platinum metals. The following list includes most of the better-known oxide minerals. It should be noted that many compound oxides are known which do not contain complex anions. Doubtful species are marked with an asterisk. Minerals occurring in substantial quantity are in italics.

OXIDE MINERALS

Cuprite	Cu_2O	*Ilmenite*	$FeTiO_3$
Bromellite	BeO	Tenorite	CuO
Lime*	CaO	*Periclase*	MgO
Monteponite*	CdO	Zincite	ZnO
Corundum	Al_2O_3	Montroydite	HgO
Anatase	TiO_2	*Rutile*	TiO_2
Baddeleyite	ZrO_2	Brookite	TiO_2
Cassiterite	SnO_2	Thorianite	ThO_2
Manganosite	MnO	Plattnerite	PbO_2
Pyrolusite	MnO_2	Hausmannite	Mn_3O_4
Magnetite	Fe_3O_4	Wustite	FeO
Bunsenite	NiO	*Hematite*	Fe_2O_3
Spinel	$MgAl_2O_4$	Chrysoberyl	$BeAl_2O_4$
Franklinite	$ZnFe_2O_4$	Gahnite	$ZnAl_2O_4$
Perovskite	$CaTiO_3$	*Chromite*	$FeCr_2O_4$

HYDROXIDE OR HYDRATED OXIDE MINERALS

Psilomelane	$(BaH_2O)_2Mn_5O_{10}$	Boehmite	$AlO(OH)$
Brucite	$Mg(OH)_2$	Manganite	$MnO(OH)$
Diaspore	$HAlO_2$	Goethite	$HFeO_2$
Gibbsite	$Al(OH)_3$		

Complex oxygen ions. These are very numerous in Nature in the form of oxy-salts of various metals and examples only of the numerous types are given in the list below. Many elements are met with only in the form of oxy-salts—these are the characteristically lithophile elements and for this reason the term ' oxyphile ' has been used as an alternative.

Mineral	*Formula*	*Valency in Anion*
Borax	$Na_2B_4O_7.10H_2O$	3
Calcite	$CaCO_3$	4
Enstatite	$MgSiO_3$	4
Xenotime	YPO_4	5
Pucherite	$BiVO_4$	5
Lautarite	$Ca(IO_3)_2$	5
Soda niter	$NaNO_3$	5
Anhydrite	$CaSO_4$	6
Crocoite	$PbCrO_4$	6
Scheelite	$CaWO_4$	6
Wulfenite	$PbMoO_4$	6

Redox potential in Nature

There are numerous examples in Nature of elements existing in several different oxidation states. The following list indicates examples of this kind.

Cu^{+}, cuprite Cu_2O	Cu^{2+}, tenorite CuO
Fe^{2+}, wustite FeO	Fe^{3+}, hematite Fe_2O_3
Mn^{2+}, manganosite MnO	Mn^{4+}, pyrolusite MnO_2
Cr^{3+}, chromite $FeCr_2O_4$	Cr^{6+}, crocoite $PbCrO_4$
Pb^{2+}, massicot PbO	Pb^{4+}, plattnerite PbO_2
I^{-}, sol. iodides	I^{5+}, lautarite $Ca(IO_3)_2$
Cl^{-}, halite $NaCl$	Cl^{7+}, perchlorates (Caliche)

In the case of copper, iron, manganese, and lead the oxidation can take place by the action of atmospheric oxygen under various conditions. The elements iron and manganese alone are sufficiently abundant to have an effect on the geochemical balances. Their significance lies in the inherent possibility that the complete oxidation of all ferrous and manganese compounds in quite a thin layer of the Earth's crust would by ordinary weathering processes suffice to remove all the free oxygen from the atmosphere. There is the possibility that conditions of this kind may well have developed to an advanced stage on the planet Mars. It does not seem that any other of the numerous natural processes which draw upon atmospheric oxygen takes place on such a massive scale as to be of importance in this respect.

The most highly oxidized compounds in Nature appear to be the nitrates, chromates, and iodates.

There are two modes by which nitrogen reaches its highest oxidation state, firstly by means of atmospheric discharges, etc., which give rise to nitrogen oxides and ultimately nitric acid, and secondly by the agency of certain types of chemautotrophic bacteria resident in the soil.

The mechanism of the formation of chromates and iodates and possibly also of per-iodates in saline deposits is obscure, but bacterial action cannot be excluded. It is significant that these types of oxy-salts occur together in the peculiar form of evaporite deposit found in Chile and this site would appear to represent a *locus* of oxidation potential higher than any other terrestrial situation.

It should be added that minute amounts of hydrogen peroxide are found in the atmosphere probably as a result of ultra-violet irradiation, but there is no evidence that it gives rise to any other per-oxidized compounds. It is interesting to note that there are many known oxidation states of elements which never appear to be met with in Nature. Thus manganates and permanganates are absent nor do we find any of the higher valency compounds of the Group VIII metals. In the lower states divalent tin and chromium for example seem to be wanting and the rarer metals of Group IV (except perhaps germanium) seem to be always tetravalent. Thallium and the lanthanides which often occur in trace quantities in minerals may, however, be found in any of their oxidation states.

The lowest redox conditions in Nature tend to be in the lowest zones of the Earth where oxygen is lacking. In practice the metallic meteorites provide samples which may be analogous to the deep interior of the Earth and they show evidence of very strongly reducing conditions by the presence of free metals and carbon, carbides, sulphides, and phosphides.

Just as very high redox potential may result from biological action of certain bacteria, so also very low potentials may be developed as in various forms of anaerobic bacterial decomposition leading to hydrides of carbon, sulphur, nitrogen, and possibly phosphorus.

Distribution in the Earth's Crust

The lithosphere is believed to be largely ionic in structure and because of the large size of the oxygen ion (radius 1·40 Å) it is oxygen which makes up more than 90% of the volume. Experience shows that the oxygen content of rocks actually decreases in depth and the problem has been discussed by Barth on a thermodynamic basis.

The lithosphere is believed to be made up of four layers:

(i) Sediments (which are not continuous).
(ii) 'Sial', extending to about 25 km depth and of a composition about equal to the average of igneous rocks.
(iii) 'Sialma', a layer at 20–70 km depth of about the composition of basalt.
(iv) A peridotitic or dunitic layer constituting the 'mantle'.

These zones are characterized by certain structural features in the metallic silicates of which they are mainly composed, ranging from the very open 'tektosilicates' (feldspars, etc.) of (i) and (ii) through the more compact phyllo- and ino-silicates to the most compact of all, the nesosilicates, as typified by the mineral olivine of which the mantle is probably mainly composed.

These changes in chemical and crystallographic composition are of course consistent with the increase of pressure and it is clear that the structures of the close-packed type containing the smaller ions will be more stable at great depth.

Barth considered that the large oxygen ions would be squeezed out of a structure under sufficient pressure and would tend to migrate upward.

In the relatively simple case of iron oxides it was postulated that the stable form at the surface was Fe_2O_3 with Fe_3O_4 at moderate depth and FeO at greater depth.

More recent work on the physics of the Earth suggests that the problem is much more complicated and that the extent to which pressure increase gives rise to chemical, as distinct from crystallographic, changes has not yet been fully established.

Presence in the Atmosphere

The existence of oxygen in the atmosphere has been the subject of much discussion. The evidence is on the whole in favour of the view that the bulk of the atmospheric oxygen is of secondary origin, produced by the photosynthetic activity of green plants.

This means that the oxygen of the air is derived from the oxygen of water and not from carbon dioxide. It is not certain whether free oxygen was initially present in the early stages of the Earth's pre-geological evolution. In spite of the high abundance of oxygen in the upper layers of the Earth, it is clear that, in the Earth as a whole, the quantity of readily oxidizable material is much more than sufficient to combine with the whole of the existing oxygen, unless the temperature was so high that compounds could not exist. In this latter case, however, the oxygen, along with other light gases would have been lost to outer space.

The persistent and widespread evolution of carbon dioxide and steam as volcanic gases suggests that they may be the main carriers of oxygen from the deeper layers of the Earth's crust. The initial formation of free oxygen in the atmosphere may well have been due to the photo-dissociation of water vapour in the upper atmosphere with the escape into outer space of the hydrogen also formed.

It is certain that the possession of an atmosphere containing a large proportion of free oxygen is unique in the solar system and it suggests a combination of causes which have not been reproduced outside the Earth. It is remarkable that such a highly reactive element could have remained in the uncombined state and it may well be that the present position is not a permanent one. The matter is further discussed below. The oxygen content of the atmosphere may well be subject to slow change during geological time but present evidence is inadequate to enable us to decide how it may have changed during the last few hundred million years.

Natural Inorganic Cycles

Goldschmidt pointed out that the amount of combined oxygen in the lithosphere is such as to leave some of the polyvalent cations unsaturated. This applies particularly in the case of iron which is the most abundant of such elements. According to Barth, the equivalent mass of ferrous oxide in the 'ten-mile crust' is 7000 Gg, but the total mass of atmospheric oxygen is only 11·5 Gg which is clearly grossly inadequate to oxidize all the ferrous iron to the ferric state. Even allowing for some of the Fe^{2+} compounds being, like Fe_3O_4, resistant to oxidation, still, all the atmospheric oxygen could be absorbed by sub-aerial (lateritic) weathering of a thickness of only 240 m of igneous rock over the whole Earth. This process, which would virtually 'fossilize' all the oxygen, would require geologically stable conditions for a fairly long period. Thus there seems to be a possibility that this process might eventually eliminate the free oxygen and leave the Earth under 'Martian' conditions. However, Barth points out that major orogenic movements can cause down-folding of the upper oxidized layers of the crust and at great depths they may suffer de-oxidation with the eventual return of the oxygen to the surface again in the form of water.

In such case a complete cycle may exist as shown in Fig. 8.1.

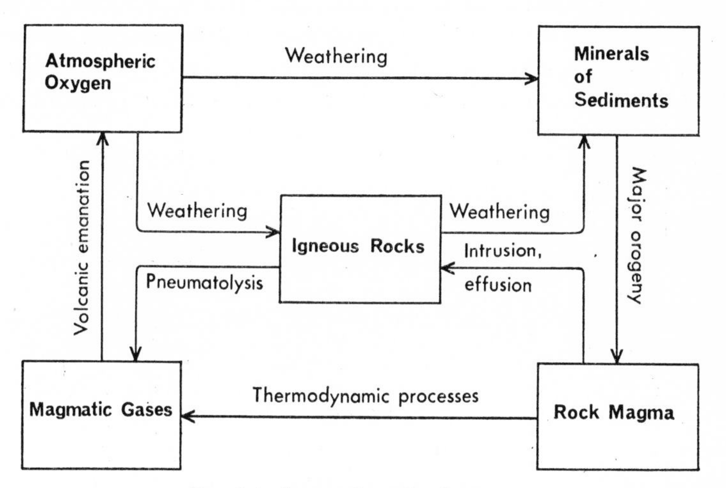

FIG. 8.1. Inorganic cycle of oxygen.

Organic Cycle

Oxygen is indissolubly linked with biological processes and all living organisms require it. It is taken directly as O_2 by animals and to some extent by higher plants. In the main, higher plants obtain their oxygen as H_2O and CO_2. In this case an amount of free oxygen is liberated equivalent to the CO_2 absorbed in the synthesis, this oxygen coming from the water and not from the CO_2.

The vital process of elaborating organic molecules is essentially the reduction of carbon dioxide with water as the reducing agent. But for some organisms, mainly bacteria, other chemical mechanisms have become established and carbon dioxide can be reduced in other ways.

The use of oxidation reactions as a source of energy is characteristic of practically all living things and where, as in the case of the anaerobic bacteria, free oxygen is not used, various oxygen compounds are used instead. It is true that life could be maintained in the absence of an oxygen atmosphere but no highly active and complex organisms such as animals could exist under such circumstances.

The oxygen produced by photosynthesis can be related to the amount of carbon in ' fossil fuel ', now in sedimentary rocks, and Goldschmidt found that a reasonable balance exists. In other words the total oxidizable carbon on the Earth corresponds to the total free oxygen and the photosynthetic origin of the oxygen of the atmosphere therefore seems to be certain.

Oxygen in solution in natural waters is of great importance in several ways. On the inorganic side it tends to cause at least surface waters of rivers, lakes, and the ocean to be regions where oxidative reactions are

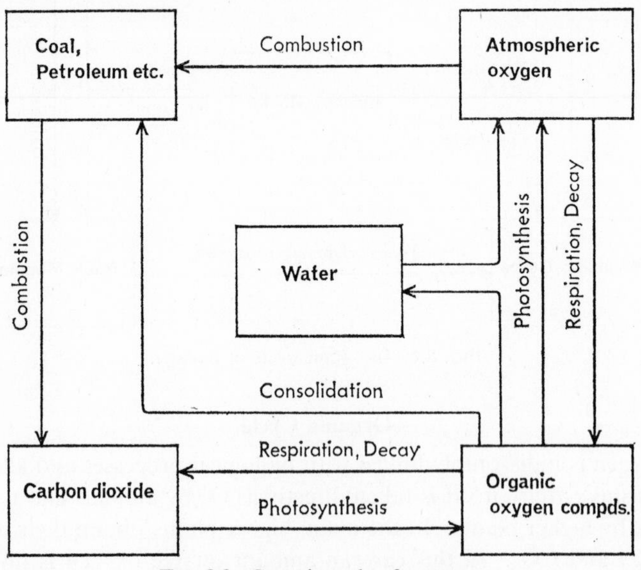

FIG. 8.2. Organic cycle of oxygen.

favoured—*e.g.*, Fe^{2+} to Fe^{3+}. Biologically, dissolved oxygen is essential to many aquatic organisms such as fish.

It seems probable that most of the dissolved oxygen in natural waters is provided by photosynthesis and the presence of green plants is therefore essential. There are diurnal variations in the dissolved oxygen content of rivers, etc., which are closely related to the amount of light and it may well happen that under favourable conditions supersaturation in oxygen is reached for a few hours a day.

On the other hand in the deeper waters of enclosed seas there is an oxygen deficiency, anaerobic bacteria are active, and the zone is one of reducing reactions.

Oxygen and Human Activities

Large quantities of free oxygen are utilized for the various operations of civilized Man, but its use in combustion reactions is by far the most important. It is probable that about 10^{11} tons of carbonaceous fuels have so far been consumed by human activity, requiring for its combustion 3×10^{11} tons of oxygen $= 3 \times 10^{17}$ g. To this should be added the uncertain amount due to forest fires and other similar natural occurrences. This great amount is, however, not more than 0·005 Gg and as the atmosphere contains 11·5 Gg, it is clear that the consumption of oxygen is not a serious matter as compared with the rate at which fuel reserves are being depleted. The estimated rate of photosynthesis by Riley, as measured in terms of organic carbon, can be represented as an equivalent amount of oxygen. This should be 0·003–0·004 Gg annually, so at this rate the present oxygen content of the atmosphere could have been formed in 2000–3000 years. It is evident that human operations in regard to oxygen are still on a comparatively puny scale.

Production

The technical production of oxygen as gas or liquid is a major industrial process and output tends to increase for several reasons, particularly the use of oxygen instead of air in steel-making and other metallurgical processes. The principal process of extraction which has been in use for many years is purely physical and depends on fractionation of liquid air. Figures of output are not readily available but it is believed that the present production (1961) of oxygen in the United Kingdom is 2000–3000 tons per day of which about 1500 tons is 'tonnage oxygen' used by the steel industry. The estimated consumption for the mid-1960's in the U.K. is 5000 tons per day for the steel industry. An estimate of world production related to steel output would therefore be perhaps one-twentieth of the steel tonnage. This might be of the order of 15 million tons of oxygen per annum. There is also a small production of electrolytic oxygen—*i.e.*, oxygen from water —which is probably mainly incidental to other processes such as the production of pure hydrogen for the food industries.

Ultimate Forms

It will have been seen from the above that on the macro-scale there is evidently a biochemical equilibrium for free oxygen in the Earth. From the more restricted point of view of human operations, however, it would seem that free oxygen tends on the whole to pass into the form of stable oxides. There is little doubt that the most important, and the only ones which need to be noted, are carbon dioxide, water, and ferric oxide. The basic processes involved in the utilization of the element are clearly respiration, combustion, corrosion, and weathering.

CHAPTER 9

The Alkali Metals

General

The elements of this family, so often cited in elementary texts as examples of close chemical similarity, are actually characterized by well marked differences in their natural chemistry. It is probable that, with one or two very special exceptions, these metals always occur in Nature in the form of their stable univalent cations and their geochemistry is very closely associated with the relative sizes of these ions.

Abundance

There is not a great deal of variation in the relative abundances cosmically and terrestrially, but in respect of the three heavier elements —potassium, rubidium, and caesium—the terrestrial abundance is relatively higher. This probably reflects the fact that they are essentially elements of the upper regions of the crust. The most striking feature of this family of elements is the marked scarcity of the lightest member, lithium, and this is undoubtedly related to its instability in thermonuclear reactions.

Lithium

Two stable isotopes of this element exist in Nature:

$$^{6}Li = 7{\cdot}52\%$$
$$^{7}Li = 92{\cdot}48\%$$

The lithium nucleus is easily destroyed in thermonuclear reactions, and so it cannot have been created in stellar interiors. Thus lithium is about three hundred times rarer in the Sun than on the Earth. It may have originated at an early stage in the formation of the solar system in a low-density region where large magnetic fields could accelerate particles to high energy. These particles are presumed to have given rise to spallation reactions on heavier nuclei such as C, N, and O.

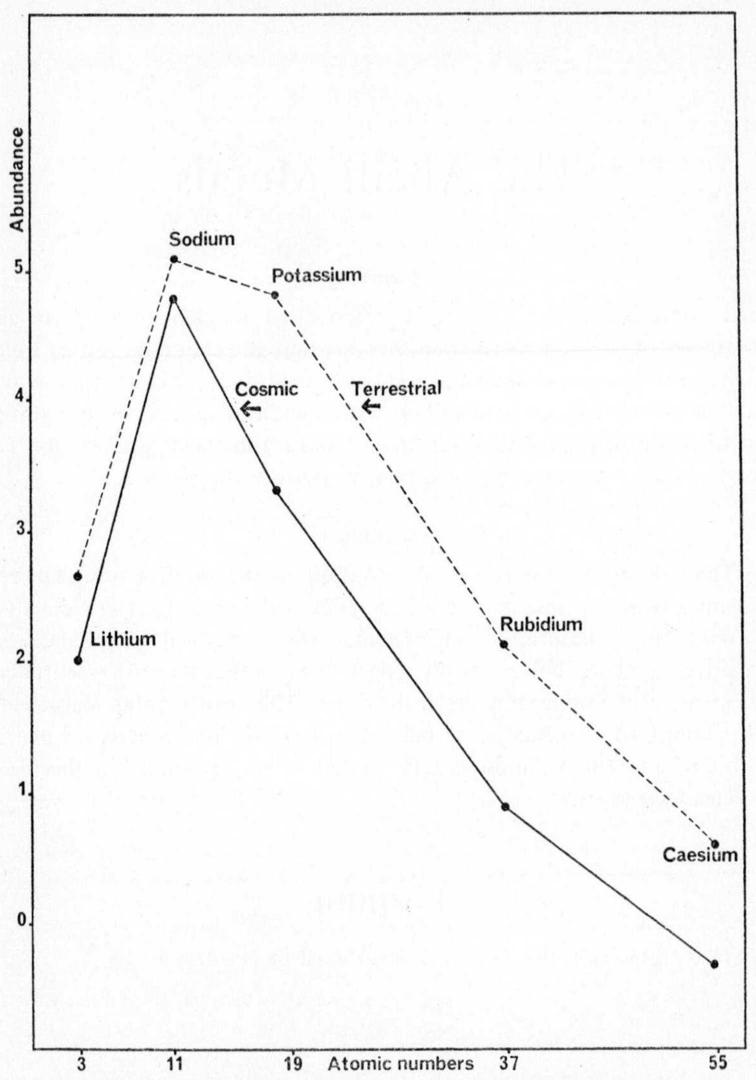

FIG. 9.1. Relative abundance of the alkali metals (logarithmic).

Terrestrial Occurrence

Lithium is essentially an element of the upper layers of the crust and is invariably in the ionic state. It is very typically lithophile. Owing to its small size (about 30 % less than the sodium ion), the lithium ion is not much associated with those of the other alkali metals. It is not very readily accepted into the crystal lattices of the common silicate rock minerals and therefore tends to concentrate in the residual crystalline products—*i.e.*, pegmatites associated with an acidic (granitic) magma. The estimated abundance for the Earth's crust is 60 g/ton.

The presence of minute amounts of lithium in the upper atmosphere, and one of the causes of the twilight glow, has been recently demonstrated. The total Li content of the atmosphere cannot be greater than about 500 g. Its origin is at present obscure.

LITHIUM MINERALS

(Commercial sources marked *)

Name	*Formula*	*Percentage of Li* (Approx.)
Cryolithionite	$Li_3Na_3Al_2F_{12}$	
Petalite*	$Li_2Al_4Si_{14}O_{35}$	1–2
Spodumene*	$LiAlSi_2O_6$	2–4
Eucryptite	$LiAlSiO_4$	
Cookeite	$LiAl_4(Si,Al)_4O_{10}(OH)_8$	
Lepidolite*	$K(Li,Al)_3(SiAl)_4O_{10}(F,OH)_2$	1–2
Irvingite	$(Na,K)_3(Li,Al)_5(Si,Al)_8(O,OH,F)_{24}$	
Zinnwaldite*	$K_2(Li,Fe,Al)_6(Si,Al)_8O_{20}(F,OH)_4$	
Manandonite	$Li_4Al_{14}B_4Si_6O_{29}(OH)_{24}$	
Lithiophylite*	$LiMnPO_4$	
Triphylite*	$LiFePO_4$	1–3
Sicklerite	$(LiMnFe)PO_4$	
Lithiophorite	$(Al,Li)MnO_2(OH)_2$	
Tainiolite	$KLiMg_2Si_4O_{10}F_2$	
Montebrasite	$LiAlPO_4OH$	
Palermoite	$(Li,Na)_4SrAl_9(PO_4)_8(OH)_6$	

Perceptible amounts of lithium (up to 3000 g/ton) may occur in minerals of acid igneous rocks, particularly those of the mica group. In addition an important source of lithium salts is provided by the brines obtained in connexion with the recovery of sodium salts from Searle's Lake, California. In this instance the lithium is recovered as Li_2NaPO_4.

Production

The production of lithium compounds began to rise rapidly from 1952 and may well have continued to do so. However, presumably for defence reasons, several countries now withhold figures of production and it is not possible to draw any useful conclusions from the limited information now available. It may be noted that up to 1957 the largest producer was Southern Rhodesia, mainly of the silicates lepidolite and petalite.

Geochemistry

On account of the high solubility of lithium salts the existing solid minerals are complex fluorides, aluminosilicates, or phosphates. In the process of weathering of igneous rocks the lithium passes into solution and would presumably follow a similar course to sodium and magnesium. However, the lithium content of sea water is low and in relation to magnesium it is much below the ratio for igneous rocks. In some sedimentary iron ores of marine origin the lithium content at 100–120 g/ton is about double that for the average of igneous rocks, showing that lithium tends to be adsorbed by hydrolysate sediments of this kind.

It is not known that lithium is of any biological significance although

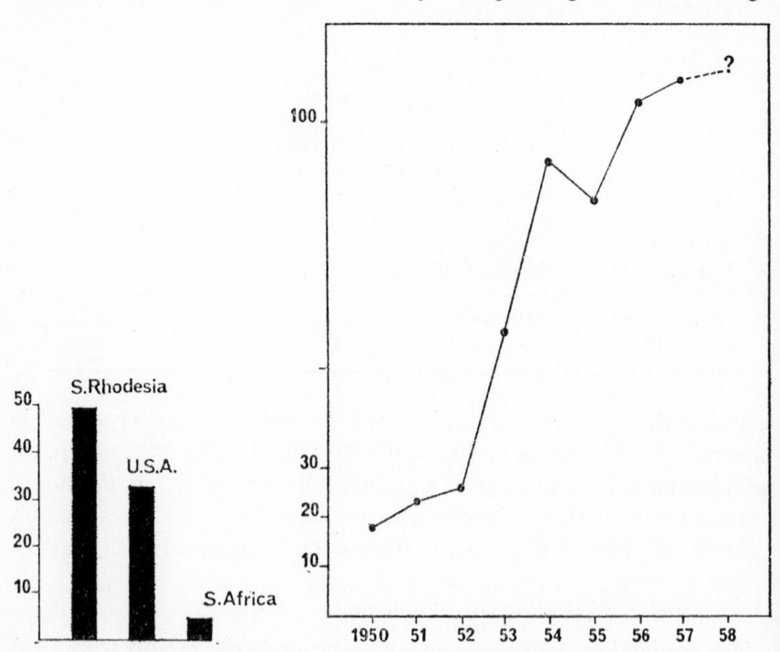

FIG. 9.2. Chief producers of lithium minerals (estimated 1957).

FIG. 9.3. Lithium mineral production in thousands of tons.

minute amounts are found quite widely in plant and animal material. Some plant ash—*e.g.*, tobacco, is enriched in lithium up to 0·44%. Soils may occur enriched in the element up to as much as 5000 g/ton.

Sodium

Only one stable isotope of sodium, ^{23}Na, is present in Nature. It is thought that sodium atoms are one of the products of the thermonuclear reactions in ' helium burning ' stars. In the ' hydrogen burning ' shells of the ' red giant ' stars ^{23}Na could be formed by a succession of proton capture stages from ^{20}Ne.

At temperatures of $1{\cdot}3 \times 10^9\,°K$ in the later stages of helium burning stars the reaction

$$^{12}C + {}^{12}C = {}^{23}Na + {}^{1}H$$

may be the chief means by which the element is synthesized in Nature.

Another mechanism involves neutron capture on a slow time-scale by ^{22}Ne.

Terrestrial Occurrence

Sodium stands high among the elements with an abundance of 28 300 g/ton. It is present in all outer spheres of the Earth.

Atmospheric sodium is one of the minor constituents of the upper atmosphere and is apparently present in the form of neutral sodium atoms to the extent of 10^4 atoms/cm^3 at a height of 70 km. It is believed that excitation of sodium atoms to the ^2P level is brought about by interactions involving atomic oxygen and is one of the causes of twilight and night luminescence of the sky.

In the hydrosphere sodium is the most abundant cation at 10 556 g/ton.

There are three main modes of occurrence of sodium compounds in the solid crust of the Earth:

(*a*) In a limited number of complex aluminohalide minerals.

(*b*) In a very large number of complex silicate minerals, typically associated with igneous rocks. No simple silicate of sodium exists in Nature.

(*c*) In a considerable variety of soluble salts, simple and complex, which are found mainly in deposits of the evaporite type.

SODIUM MINERALS

Name	*Formula*
Villiaumite	NaF
Halite*	$NaCl$
Hydrohalite	$NaCl.2H_2O$
Cryolite*	Na_3AlF_6
Chiolite	$Na_5Al_3F_{14}$
Malladrite	Na_2SiF_6
Ferruccite	$NaBF_4$
Borax*	$Na_2B_4O_7.10H_2O$
Natron*	$Na_2CO_3.10H_2O$
Trona*	$Na_3H(CO_3)_2.2H_2O$
Gay-Lussite*	$Na_2Ca(CO_3)_2.5H_2O$
Burkeite	$Na_6SO_4(CO_3)_2$
Nitratine*	$NaNO_3$
Darapskite	$Na_3NO_3SO_4.H_2O$
Pectolite	$NaCa_2Si_3O_8OH$
Soda-tremolite	$Na_2CaMg_5Si_8O_{22}(OH)_2$
Riebeckite	$Na_2Fe_3^{\cdot\cdot}Fe_2^{\cdot\cdot\cdot}Si_8O_{22}(OH)_2$
Aegirine	$NaFe^{\cdot\cdot\cdot}Si_2O_6$
Iron-albite	$NaFe^{\cdot\cdot\cdot}Si_3O_8$
Nepheline	$NaAlSiO_4$
Albite	$NaAlSi_3O_8$
Jadeite	$NaAlSi_2O_6$
Natrolite	$Na_2Al_2Si_3O_{10}.2H_2O$
Analcime	$NaAlSi_2O_6.H_2O$
Anorthoclase	$(Na,K)AlSi_3O_8$
Stilbite	$NaCa_2Al_5Si_{13}O_{36}.14H_2O$
Stercorite	$NH_4NaHPO_4.H_2O$
Matteucite	$NaHSO_4.H_2O$
Thenardite*	Na_2SO_4
Mirabilite*	$Na_2SO_4.10H_2O$
Aphthitalite	$NaKSO_4$
Krönkite	$Na_2Cu(SO_4)_2.2H_2O$
Vanthoffite	$Na_6Mg(SO_4)_4$
Loeweite	$Na_2Mg(SO_4)_2.2\frac{1}{2}H_2O$
Natroalunite	$NaAl_3(SO_4)_2(OH)_6$
Schairerite	$Na_3SO_4(F,Cl)$
Sulphohalite	$Na_6(SO_4)_2ClF$

The above list is representative of the wide variety of minerals which contain sodium ions. The total number is of course much larger. Many of these substances are of technical interest although not necessarily as sources of sodium compounds.

Geochemistry

The main features of processes involving sodium have been indicated in Chapter 2. The sodium ion is rather widely separated in size from those of lithium (radius 0·68 Å) and potassium (radius 1·33 Å). In consequence sodium ions are less associated with those of potassium than a pure chemist would perhaps expect. The sodium ion (radius 0·98 Å) is much nearer to that of calcium (radius 0·99 Å) and is very frequently associated with it in the important silicate minerals of igneous rocks.

Furthermore the two elements are of nearly equal abundance in the Earth and are often met with in series of solid solutions of isomorphous minerals. The most important example is provided by the feldspars—characteristic and often dominant minerals of the igneous rocks of the intermediate and upper zones of the crust. The three simple types are orthoclase $KAlSi_3O_8$, albite $NaAlSi_3O_8$, and anorthite $CaAl_2Si_2O_8$.

The plagioclase feldspars are an isomorphous series between albite and anorthite and are very common particularly in basic and intermediate rock types.

As complete solid solution between soda and potash feldspars only exists above 660 °C, the mineral anorthoclase $(Na,K)AlSi_3O_8$ is relatively rare and is often replaced by a micro-intergrowth of albite and orthoclase called perthite.

Another solid solution series exists between diopside $CaMgSi_2O_6$ and aegirine $NaFeSi_2O_6$, and there are also variable amounts of sodium in the hornblendes $NaCa_2(Mg, Fe, Al)_5(Si, Al)_8O_{22}(OH)_2$. The so-called alkaline rocks contain the feldspathoid minerals such as nepheline $NaAlSiO_4$ and sodalite $Na_8(AlSiO_4)_6Cl_2$, and also account for large amounts of sodium in the Earth's crust. There are large tracts of alkaline rocks—*e.g.*, in South Norway where sodium predominates over potassium. Among minerals of secondary origin which contain sodium the most important are the zeolites. As these minerals possess cationic exchange properties it is not uncommon to find sodium replacing calcium in several of the species which are not primarily sodium compounds. It is of interest that sodium is on the whole absent from some groups of silicates such as the micas, which, however, contain potassium and sometimes lithium.

The typical minerals of sedimentary rocks are devoid of sodium because in weathering the sodium passes into solution and there is no natural process by which it can be rendered insoluble and immobilized. Thus Goldschmidt was able to show that the estimated amount of sodium extracted by weathering from the primary rocks could be largely accounted for in the sodium content of ocean water. There is not much tendency for sodium ions to be adsorbed by colloidal sedimentary material and the small Na^+ content of marine muds and shales is mainly due to minute grains of unaltered albite.

In addition to sea water the technically important deposits of sodium compounds are all of the evaporite type.

Sodium chloride is the most abundant, but other salts which are extracted include the carbonate and sulphate. In addition the nitrate, iodate, and borates of sodium are important sources of the particular anions they contain.

The nature of solid saline deposits is determined by the nature of the original solution, sea water, lake, or spring water, etc., and the conditions under which it has been evaporated.

For oceanic salt deposits—*e.g.*, those of Stassfurt, etc., the physical chemistry of deposition has been fully worked out. Very broadly speaking, the process, which involves quite complex phase relations, gives deposits of calcium salts followed by sodium and then by those of potassium and magnesium. Approximately one-seventh of the total salt production is derived from the evaporation of sea water.

Producing Countries

In the United Kingdom most of the salt is produced from brines obtained from boreholes put down into the bedded deposits in Keuper Marl in Cheshire. Other deposits are found in Staffordshire, Worcestershire, Somerset, Lancashire, and County Durham. Other producers are shown on Fig. 9.4.

Biogeochemistry

In the vegetable kingdom sodium is rather common but it is not known to be an essential element. Halophyte plants are relatively rich in sodium but it is claimed that if these plants are grown in an environment of low sodium content they absorb much less of the element and contain about the same amount of sodium as average types of land plants. In animals sodium is of considerable importance because it is the chief cation in the body fluids and among other things it helps to maintain ionic balance.

Sodium chloride is the principal form in which sodium ions are

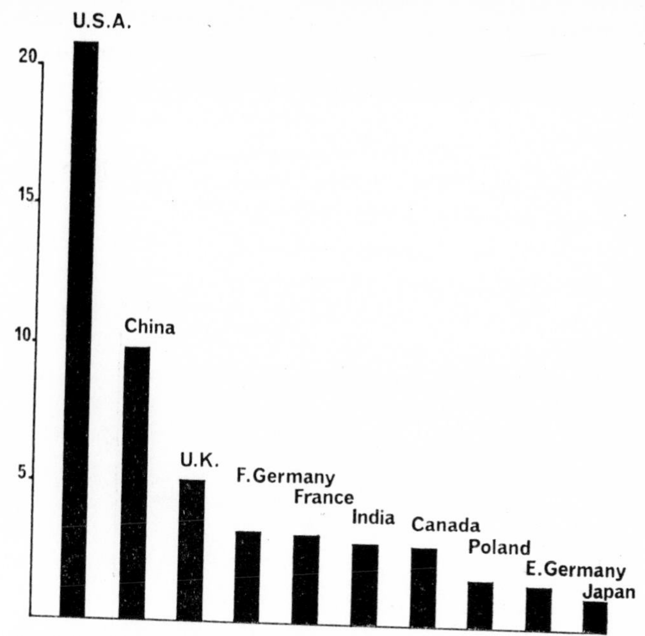

FIG. 9.4. Chief producers of common salt in 1959 in millions of tons.

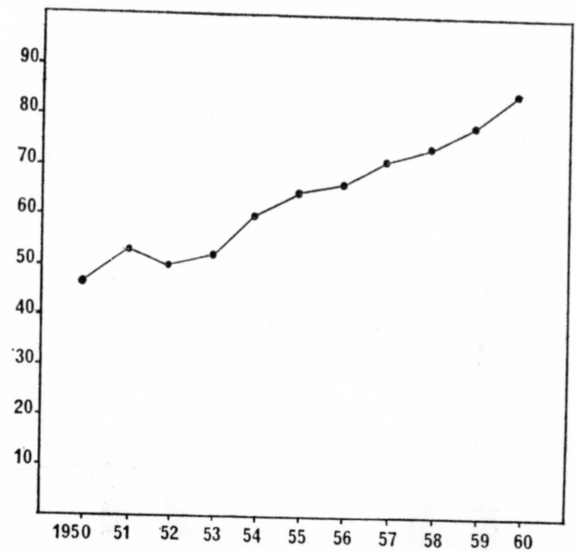

FIG. 9.5. World production of common salt in millions of tons.

ingested and also the form in which they are largely excreted (perspiration, tears, urine). Thus biological sodium will tend in the long run to be returned to the ocean.

Ultimate Forms

In respect of human activities a high proportion of sodium salts used must ultimately find its way into drainage systems and the sea. Only

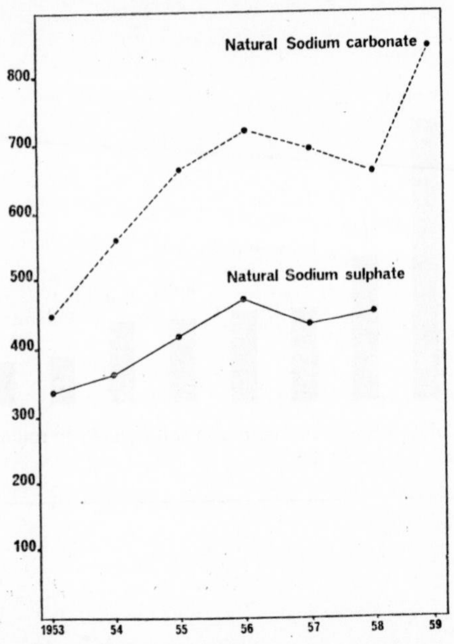

FIG. 9.6. World production of sodium sulphate and sodium carbonate in thousands of tons.

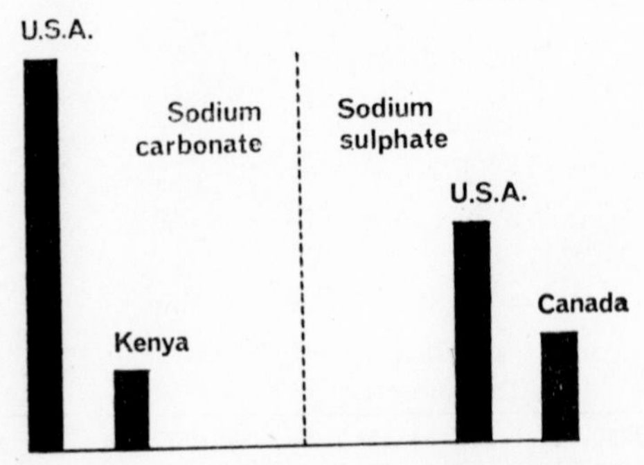

FIG. 9.7. Chief producers of sodium sulphate and sodium carbonate in 1958.

a few sodium compounds are relatively immobile. Glass probably represents the most abundant sodium compound which is relatively resistant to solution and is likely to retain the element. As soda glass is the commonest kind, the total amount of sodium fixed for some considerable time in this form must be quite large but difficult to estimate.

Potassium

There are three natural isotopes of potassium, ^{39}K, ^{40}K, and ^{41}K. The first and last are stable and ^{39}K is the most abundant at 93·08 %.

Potassium atoms may have been synthesized in the later stages of the α-process in the 'helium burning' stage of stellar evolution. The s-process of slow neutron capture could also have contributed to their formation.

The nuclide ^{40}K is radioactive and present in natural potassium to the extent of 0·0119 %.

Radioactivity

^{40}K breaks down in two ways, 88·8 % of it decays by β-emission to produce ^{40}Ca and the balance of 11·2 % by *K*-electron capture to give ^{40}Ar as noted in Chapter 8. The half-life is $1{\cdot}27 \times 10^9$ years. Hence with a terrestrial age of about 4×10^9 years the present ^{40}K content of the Earth is now only a fraction of the original. The significance of ^{40}K in Nature may be noted under three headings:

(1) As the source of the present atmospheric argon ^{40}Ar (see Chapter 8).

(2) As a considerable factor in the production of radiogenic heat in the Earth's crust.

(3) As a convenient system for application to geochronology. The value of ^{40}K lies in its ready availability in igneous rocks so that dating methods can be applied to a wide variety of specimens and in its rather long half-life which makes it valuable for dating the older ($> 10^8$ year) geological formations.

Terrestrial Abundance

Whilst cosmically potassium is of about one-thirtieth of the abundance of sodium, in the Earth's crust the two are almost equal. The modes of occurrence of potassium compounds in the Earth are rather similar to those of sodium but the two elements tend to be differentiated from each other for several reasons. These are largely related to ionic radius which determines mode of crystallization in igneous rocks, relative ease of hydration in solution, and degree of adsorption on colloids, and to some extent the relative solubility of the different salts.

Potassium minerals are numerous and like those of sodium fall into three main classes of complex halides (often of volcanic origin), silicates (mainly rock minerals), and soluble salts in variety (mainly in evaporite deposits).

POTASSIUM MINERALS

Name	*Formula*
Sylvine	KCl
Erythrosiderite	$K_2FeCl_5.H_2O$
Hieratite	K_2SiF_6
Fairchildite	$K_2Ca(CO_3)_2$
Nitre	KNO_3
Ferri-orthoclase	$KFeSi_3O_8$
Kalsilite	$KAlSiO_4$
Leucite	$KAlSi_2O_6$
Orthoclase Microcline	$KAlSi_3O_8$
Muscovite	$KAl_3Si_3O_{10}(OH)_2$
Phlogopite	$KMg_3AlSi_3O_{10}(OH)_2$
Glauconite	$K_{1\frac{1}{2}}(Fe,Al,Mg,Fe)_4(Si,Al)_8O_{20}(OH)_4$
Lepidomelane	$K_2(Fe,Fe,Mg)_{4-6}(Si,Al,Fe)_8O_{20}(OH)_4$
Biotite	$K_2(Mg,Fe,Al,Fe)_{4-6}(Si,Al)_8O_{20}(OH)_4$
Mercalite	$KHSO_4$
Kainite	$KMgSO_4Cl.3H_2O$

Production

The following are the main producing countries at present.

Germany. Werra-Fulda district, Sud-Harz district.

France. Alsace.

U.S.A. New Mexico, Texas.

Canada. Saskatchewan, Manitoba, Alberta.

Spain. Navarre, Catalonia.

Poland. Kalusz.

U.S.S.R. Solikamsk, Urals.

United Kingdom. There are no deposits worked at present but considerable beds are known in Eastern counties of England and have been proved by boring. They are generally of similar type to the European ones and have a succession of saline layers.

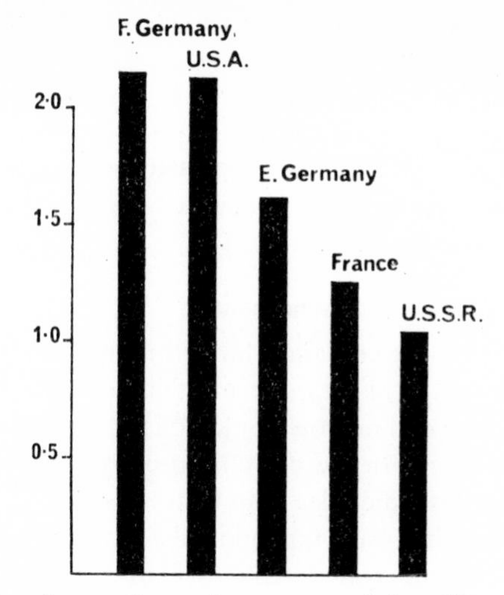

FIG. 9.8. Chief producers of potassium compounds in millions of tons in 1959.

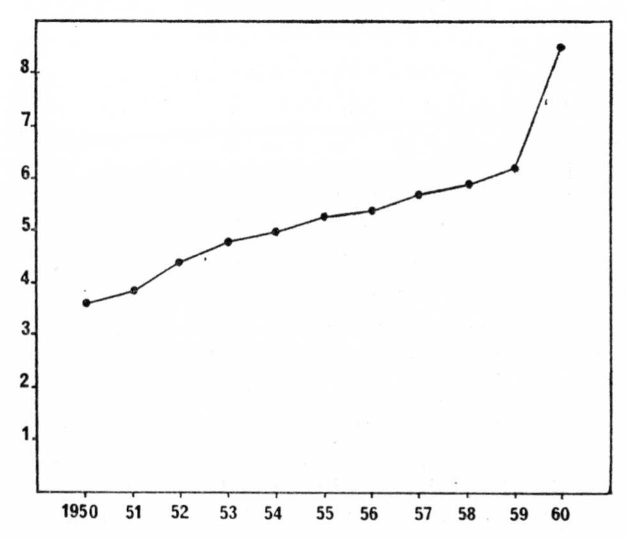

FIG. 9.9. World production of potassium compounds in millions of tons.

Saline waters. Potassium salts are obtained from a few of these, the most important being the Dead Sea and Searle's Lake, California.

Small pilot plants for the extraction of potassium salts from sea water have been operated in Norway and recently in Holland. The process involves the precipitation of the sparingly soluble potassium salt of hexanitrodiphenylamine.

Use

Over 90 % of the world's output of potassium salts is used as fertilizers and the very steady rise in production clearly reflects the corresponding development of agriculture.

Geochemistry

Primary igneous rocks contain a limited number of potassium compounds in the form of silicates, the most important being the potash feldspars (orthoclase and microcline) $KAlSi_3O_8$, and the micas (mainly muscovite and biotite). In the crystallization of a magma, if water is present then the hydroxidic micas can be formed and they may then be the principal potassium-carrying minerals in the rock. If water is deficient or absent conditions for the formation of feldspars are more favourable. On account of the occurrence of biotite in fairly deep-seated basic rocks, as well as its association with muscovite and potash feldspars in granite, potassium is rather widespread in its occurrence in igneous rocks. In certain volcanic rocks the feldspathoid leucite is abundant—*e.g.*, Vesuvian lavas. Much of the 'potash' feldspar of granites and pegmatites is not very pure but consists partly of perthite and partly of anorthoclase. Feldspar sources for technical uses therefore require careful selection in order to get a sample which contains somewhere near the theoretical potassium content. In the weathering of igneous rocks potassium passes initially into solution and the relative ease of attack depends on the particular mineral. Biotite and leucite are rather easily decomposed and much of the potash in soils is derived from biotite. Leucite from the volcanic material around Vesuvius is one of the causes of the great fertility of the soils in this area. Potash feldspars are more resistant and are found to remain even after the soda-feldspars have weathered completely.

The potassium ion, being larger than that of sodium, is less readily hydrated and is much more readily adsorbed on to colloids. Thus it is that potassium ions can be retained in appreciable quantity in soils and sediments of various kinds.

It also appears that some of the mica-like minerals such as chlorite and sericite can take up potassium to form what are in effect re-constituted micas. Thus micas containing potassium are by no means uncommon in sedimentary and many metamorphic rocks. Of the potassium ions which ultimately reach the sea, some are taken up in the formation of the mineral glauconite and some are adsorbed by oxidate products such as hydrated manganese dioxide. For these reasons it will be seen why the potassium content of sea water is very much lower than the sodium content.

Biogeochemistry

Potassium has been noted (Chapter 2) as being an essential element in plants and animals. Although the exact function of potassium is obscure it is known to be taken by the roots of land plants and by the whole surface of aquatic forms and its preferential absorption against sodium is very marked.

The ash of plants contains potassium which can be extracted as carbonate but the amount is variable and may be quite small, particularly if the burning of the plant material has not been been properly carried out. The organic decay of vegetable material would of course liberate most of the potassium into the soil again.

In animals, potassium is the principal cation of the intracellular fluid. It is also in the extracellular fluid where it influences muscle activity. Within the cell it influences acid–base balance and osmotic pressure, including water retention. The total potassium content of higher animals is not very different from that of sodium.

Ultimate Forms

It is probable that much of the potassium from human activities returns to the solid earth in the form of adsorbed ions. Potassic wastes discharged into drainage systems will tend to be adsorbed on sludge and recovered in some processes of sewage treatment. A fair amount of potassium compounds is volatilized at high temperatures as in some furnace operations and if not specially recovered will wash down in rain and again return to the soil.

The general tendency of these processes is to a high degree of dispersion of potassium from the comparatively localized deposits from which a large part of it was originally obtained.

Rubidium

Two isotopes of this element occur in Nature. ^{85}Rb is stable and constitutes 72·15% of natural rubidium. ^{87}Rb is radioactive. It decays by β-emission and has a long half-life of about 6×10^{10} years, the decay product being ^{87}Sr. This process is employed as one of the more recent methods of geochronology and depends essentially on the determination of the Rb/Sr ratio in selected materials. On account of the long half-life of ^{87}Rb the method is of most value for material of high geological age.

Terrestrial Occurrence

Rubidium at 120 g/ton is about 200 times scarcer than potassium in the Earth's crust. The rubidium ion (radius 1·47 Å) is not much more than 10% larger than the potassium ion and consequently it can be accommodated in the same crystal lattices, particularly in the later stages of crystallization. For this reason there are virtually no distinct rubidium minerals. As the pegmatites represent the later stages of consolidation of a granitic magma they are favourable sources of rubidium. Particularly high figures, up to 25000 p.p.m. Rb, have been reported from potassic minerals, notably microcline feldspars from such formations. Lepidolite mica is also an important carrier of rubidium and it may contain up to 1·5% of Rb. Lepidolite is characteristic of rocks called greisen which are an end product of cooling of granitic magma and therefore represent a late stage of crystallization.

Production

Rubidium salts are, or have been, extracted from mother liquors remaining after separation of lithium from lepidolite and from the residues from electrolysis of artificial carnallite at Solikamsk, U.S.S.R.

Caesium

Natural caesium is a simple element consisting only of the nuclide ^{133}Cs.

It is known that several caesium isotopes are produced in nuclear processes but although the presence of ^{132}Cs has been postulated, its existence, or that of its decay product ^{132}Ba, have not been conclusively established in the most likely natural sources.

Terrestrial Occurrence

Although caesium is so much rarer than rubidium (Cs : Rb = 1 : 100) yet on account of its ion being so much larger (radius 1·65 Å) it tends to remain to the last stages of magmatic crystallization and is able to form the distinct mineral pollucite $(Cs,Na)AlSi_2O_6.nH_2O$ and may possibly also be present in some quantity in avogadrite $(K,Cs)BF_4$.

The chief production of caesium is from two African localities, Joost mine, Karibib, South West Africa, and Bikita, Southern Rhodesia. In both cases the mineral is pollucite. The former locality is also the world's largest producer of rubidium.

The mineral pollucite was first found in the Island of Elba but other localities include S. Dakota, U.S.A., Maine, U.S.A., and Varutraesk, Northern Sweden. It may well be less rare than formerly supposed.

Caesium salts are also recoverable from mother liquors which contain rubidium.

Biological Aspects

Both rubidium and caesium are reported in traces from biological material where they presumably accompany potassium. It is claimed that whilst rubidium is concentrated by some species of plants it is toxic to animals and caesium is said to be toxic to both plants and animals.

CHAPTER 10

Group Ib: Copper, Silver, and Gold

General

These three elements, which occupy a special place in human activity, have long been associated with each other and exhibit relationships in their natural chemistry which are in the main related to their chemistry as observed in the laboratory. As a consequence of lower chemical reactivity their modes of occurrence are entirely different from those observed for the metals of Group Ia. A particular feature they exhibit, and which is of course quite unknown to the alkali metals, is their not infrequent existence in the free metallic state in Nature. As far as their general geochemistry is concerned they are typically chalcophile or sulphophile—occurring in comparatively concentrated ore bodies and other deposits and not being in general elements of importance in contributing to the structure of rocks. If we compare them with the three corresponding elements of Group Ia, potassium, rubidium, and caesium, it is seen that abundance declines much more steeply with rising atomic number and that in any case they are very much less abundant than the metals of the alkalis.

Thus in respect of terrestrial abundance:

$$\text{Cu} : \text{Ag} = 450 : 1 \qquad \text{K} : \text{Cu} = 370 : 1$$
$$\text{Ag} : \text{Au} = 20 : 1 \qquad \text{Rb} : \text{Ag} = 3100 : 1$$
$$\text{Cu} : \text{Au} = 9000 : 1 \qquad \text{Cs} : \text{Au} = 14000 : 1$$

Nevertheless on account of their tendency to occur in concentrated deposits the actual availability of these elements is high and this is shown both in production figures and prices.

The history of these three metals is longer than that of any others and they are very closely connected with the development of human civilization.

Copper

Two stable isotopes of copper exist:

$$^{63}Cu = 69{\cdot}1\%$$
$$^{65}Cu = 30{\cdot}9\%$$

There is no great significance in their behaviour in Nature as at present ascertained.

Terrestrial Abundance

The average figure is 45 g/ton for the Earth's crust and the more acidic rocks (granitic) are much poorer in the metal at about 15 g/ton. The average for basic rocks is as much as 149 g/ton or about ten times that in the acidic rocks. Some sediments such as limestones are low in copper but argillaceous sediments are often considerably enriched in it.

Copper does not contribute appreciably to rock-forming silicate minerals and seems to occur in rocks mainly in the form of sulphides.

Copper Minerals

Copper is characteristically thiophile and the largest concentrations of the element are in the form of various sulphur compounds but it also forms stable compounds of many types in Nature. There is a marked propensity for copper to form basic salts and very many of the natural oxy-salts of the metal are of this type.

Upwards of 150 copper minerals have been described of which a limited number actually constitute important ores of the metal. These are marked below with an asterisk.

From the point of view of mode of occurrence these very numerous minerals can be classed into two broad groups. Native copper together with its various sulphides and compounds with the metalloids are the chief constituents of ' primary ' deposits. The oxy- and halide salts are in general encountered in the upper oxidized zones of cupriferous deposits. Owing to the characteristic green or blue colour of so many of these compounds, mere traces of copper are often readily observed, but they are frequently too poor or restricted in size to be of commercial interest.

Geochemistry

The primary copper minerals which are either simple or complex sulphides are unlike the silicates and other oxy-salts in being

COPPER MINERALS

Name	*Formula*
Native copper*	Cu
Domeykite	Cu_3As
Chalcocite*	Cu_2S
Covelline*	CuS
Berzelianite	Cu_2Se
Bornite*	Cu_5FeS_4
Chalcopyrite*	$CuFeS_2$
Chalcopyrrhotine*	$CuFe_4S_5$
Cubanite*	$CuFe_2S_3$
Valerite	$Cu_2Fe_4S_7$
Tennantite*	Cu_3AsS_3
Tetrahedrite*	Cu_3SbS_3
Cuprite*	Cu_2O
Tenorite*	CuO
Namaqualite	$Cu_2Al(OH)_7.2H_2O$
Nantokite	$CuCl$
Eriochalcite	$CuCl_2.2H_2O$
Atacamite*	$Cu_2Cl(OH)_3$
Malachite*	$Cu_2CO_3(OH)_2$
Azurite	$Cu_3(CO_3)_2(OH)_2$
Dioptase	$CuSiO_2(OH)_2$
Chrysocolla	$CuSiO_3.2H_2O$
Pseudomalachite	$Cu_5(PO_4)_2(OH)_4.H_2O$
Turquoise	$CuAl_6(PO_4)_4(OH)_8.5H_2O$
Olivenite	$Cu_2AsO_4(OH)$
Uzbekite	$Cu_3(VO_4)_2.3H_2O$
Chalcantite	$CuSO_4.5H_2O$
Antlerite	$Cu_3SO_4(OH)_4$
Brochantite	$Cu_4SO_4(OH)_6$
Linarite	$(Pb,Cu)_2SO_4(OH)_2$
Lindgrenite	$Cu_3(MoO_4)_2(OH)_2$
Cuprotungstite	$Cu_2WO_4(OH)_2$
Gerhardtite	$Cu_2NO_3(OH)_3$

predominantly covalent in character and in many cases they exhibit metallic or sub-metallic properties.

In spite of the variety of compounds the disulphide $CuFeS_2$ chalcopyrite is the most abundant ore mineral. However, owing to the very widespread nature and great variety of workable copper deposits, the metal is extracted from a larger number of different minerals than perhaps any other metal.

Mineralogically the ores can be divided into four groups:

(*a*) Native copper.
(*b*) Sulphides.
(*c*) Oxidized ores.
(*d*) Complex ores.

The most valuable ores in general are the sulphides. The complex ores, however, are often valuable as they may also be sources of silver, gold, etc. Although the actual copper content from rock analyses indicates that basic (gabbroic) types of rocks are the most cupriferous, yet most of the localized copper ores are associated with rather acidic intrusive rocks such as quartz–monzonites.

The economic utility of copper ores depends on a number of factors including ease of extraction, accessibility, and market price of the metal.

Ores as low as 0·6% may be workable and on the average ores carrying 4% or more of metal may be smelted directly. Low-grade oxidized ores can often be extracted economically by leaching processes and the copper directly precipitated from the solution.

As previously indicated, where copper is present in appreciable quantity in a rock magma it tends to associate with sulphur and iron and not with oxygen and silicon, so that the relatively few deposits of magmatic type are mainly of sulphides, particularly the several complex sulphides of iron and copper. It is probable that most of the other kinds of copper minerals are of the hydrothermal type—*i.e.*, have been deposited from aqueous solutions within the temperature range 500°–50 °C and also of course in some cases under considerable pressure. Recent Russian work indicates that solubility of sulphides under such conditions is greater than was once thought.

Oxidation. The primary minerals of copper being essentially sulphides, when the ore body in course of geological changes is brought near to the surface it may well happen that a considerable amount of it comes above the water table. In this zone, within which water, oxygen, and carbon dioxide are available, most of the oxidation products are formed. As most ores contain pyrite, FeS_2, it is suggested that oxidation in accordance with the reaction

$$FeS_2 + 7O + H_2O = FeSO_4 + H_2SO_4$$

provides a source of sulphuric acid. Then the reaction

$$6FeSO_4 + 3O + 3H_2O = 2Fe_2(SO_4)_3 + Fe(OH)_3$$

provides ferric sulphate which can react with copper sulphide as follows:

$$CuFeS_2 + 2Fe_2S(O_4)_3 = CuSO_4 + 5FeSO_4 + 2S$$
$$Cu_2S + Fe_2(SO_4)_3 = CuSO_4 + FeSO_4 + CuS$$
$$CuS + Fe_2(SO_4)_3 = CuSO_4 + 2FeSO_4 + CuS$$

Cupric sulphate in solution, not being susceptible to oxidation and being less readily hydrolysed than the iron salts, is fairly mobile and could travel both laterally and vertically. On reaching an environment of higher pH it would form various basic sulphates and thus be immobilized. The presence of phosphates in solution derived from the action of sulphuric acid on the mineral apatite often present in igneous rocks

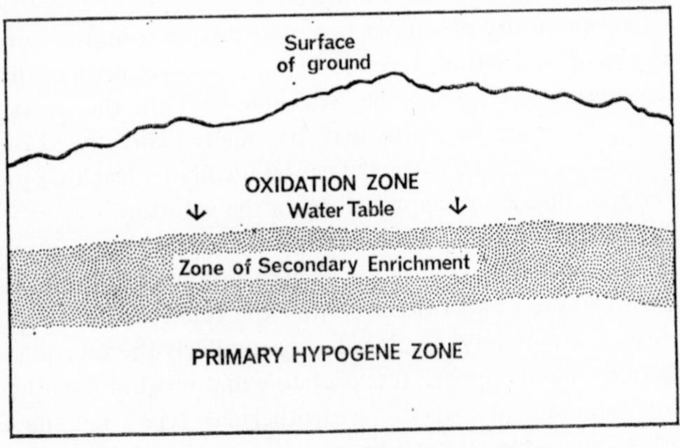

FIG. 10.1. Alteration zones of an ore body.

could provide the means of formation of copper phosphate minerals. The carbonates would form when the pH was high enough and when carbonic acid became available either from the atmosphere or more often from limestone rocks.

On account of the enormous variety in detail of copper deposits and local peculiarities of geology, etc., it is not possible to generalize further on the mode of origin of the very numerous copper oxy-salts.

Supergene enrichment. This is a rather important process which has provided some valuable copper deposits. Metals in solution percolate downward from the oxidation zone to where there is no free oxygen and there undergo deposition as secondary sulphides. The metals removed from above are thus added to those below and a zone of secondary enrichment is thus established. (See Fig. 10.1.)

Many great copper-mining areas owe their success in part to this effect, examples being Utah Copper, U.S.A. and Cananea, Mexico. The sulphides in these deposits appear to have formed from sulphate solutions at the expense of other sulphides.

The following reactions have been proposed:

$$PbS + CuSO_4 = CuS + PbSO_4$$
$$ZnS + CuSO_4 = CuS + ZnSO_4$$
$$5FeS_2 + 14CuSO_4 + 12H_2O = 7Cu_2S + 5FeSO_4 + 12H_2SO_4$$
$$4FeS_2 + 7CuSO_4 + 4H_2O = 7CuS + 4FeSO_4 + 4H_2SO_4$$
$$CuFeS_2 + CuSO_4 = 2CuS + FeSO_4$$
$$8CuFeS_2 + 11CuSO_4 + 8H_2O = 8Cu_2S + 8FeSO_4 + 8H_2O$$
$$Cu_5FeSO_4 + CuSO_4 = 2Cu_2S + 2CuS + FeSO_4$$
$$Cu_5FeS_4 + 11CuSO_4 + 8H_2O = 18Cu_2S + 5FeSO_4 + 8H_2SO_4$$
$$5CuS + 3CuSO_4 + 4H_2O = 4Cu_2S + 4H_2SO_4$$

It will be seen that the characteristic copper minerals in the supergene enrichment zone will therefore be chalcocite Cu_2S and covellite CuS. The full development of these zones is unlikely in countries which have been subject to severe glacial erosion in recent geological time.

The remarkable large-scale occurrence of native copper in the Lake Superior region, U.S.A., is thought to be due to upward-flowing copper solutions which would normally have deposited sulphides but due to reaction with superincumbent hematite lost their sulphur by oxidation and copper was precipitated as metal. The exact mechanism of this process is, however, not very clear.

The many minor deposits of copper in the United Kingdom have long ceased to be worked but are often of great mineralogical interest. In 1800 Great Britain was the world's largest producer of copper. The present output is merely nominal at less than 20 tons per annum. At Allihies in S.W. Ireland are copper deposits which may still be of economic interest.

Biological Aspects

Copper is probably an essential trace element for plants but the amount is small and the average soil content of around 10 p.p.m. is adequate to supply it.

It is well known that fairly large quantities of the element are inimical to organisms but more especially to fungi and algae, hence the use of copper compounds in agriculture. The continued use of such preparations as Bordeaux mixture on vines in parts of Southern Europe has led apparently to the ground beneath them being quite green with copper

EXAMPLES OF IMPORTANT COPPER DEPOSITS

Locality	*Type of Deposit*	*Chief Minerals*
Utah Copper, U.S.A.	Replacement in quartz porphyry	Pyrite, bornite, chalcopyrite
Butte, Montana, U.S.A.	Fissure veins in granodiorite	Chacocite, bornite, chalcopyrite, covellite, tennantite, etc.
Lake Superior, Michigan, U.S.A.	Conglomerate and amygdaloidal lodes and fissure veins in conglomerates and sandstones	Native copper
Bisbee, Arizona, U.S.A.	Hydrothermal replacement in limestone and porphyry	Pyrite, bornite, chalcopyrite and oxidized ores
Sudbury, Ontario, Canada	Probably magmatic concentration	Pyrrothite, chalcopyrite, pentlandite, cubanite
Novanda, Quebec, Canada	Replacement deposits in a complex of intrusive and effusive rocks	Pyrrhotite, pyrite
Chiquicamata, Chile	Replacement in porphyry and extensive oxidized zone	Pyrite, enargite, tetrahedrite, chalcopyrite, bornite and many oxidized minerals—antlerite, atacamite, etc.
Roan Antelope, Northern Rhodesia	Disseminated ore in feldspathic sandstones—bedded deposit	Chalcocite, bornite, chalcopyrite
Union Minière, Katanga, Congo	Oxidized ores disseminated in dolomitic beds	Malachite, chrysocolla
Rio Tinto, Huelva, Spain	Massive replacement deposits in porphyry and slate	Pyrite and chalcopyrite
Djekasgan, U.S.S.R., Kazakstan	Impregnations in sandstones	Pyrite, chalcopyrite
Novilsky, Kola Penin., U.S.S.R.	Magmatic ores	Pyrrhotite, chalcopyrite, pentlandite
Mansfeld, Germany	Bedded deposit in shale impregnated by mineralizing solutions	Bornite, chalcoite
Cornwall, England	Fissure veins in slate associated with granite	Chalcopyrite
Lake District, England	Fissure veins in slates and volcanic tuffs	Chalcopyrite

salts but without any deleterious effect. It must be observed, however, that soluble copper salts are very readily adsorbed by organic matter and immobilized. Copper-deficient soils have been recognized in some coastal regions such as the 'polders' of Holland, the 'everglades' of Florida, and certain parts of West and South Australia. In these places

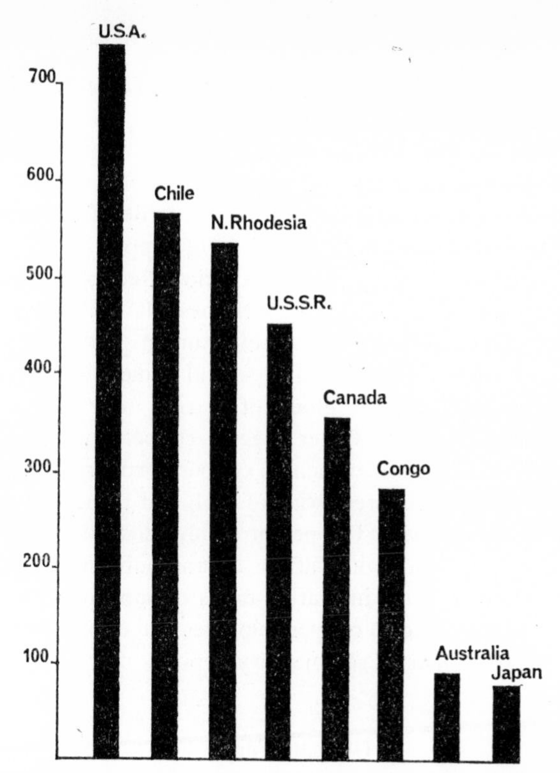

FIG. 10.2. Chief producers of copper in 1959 in thousands of tons (Cu content).

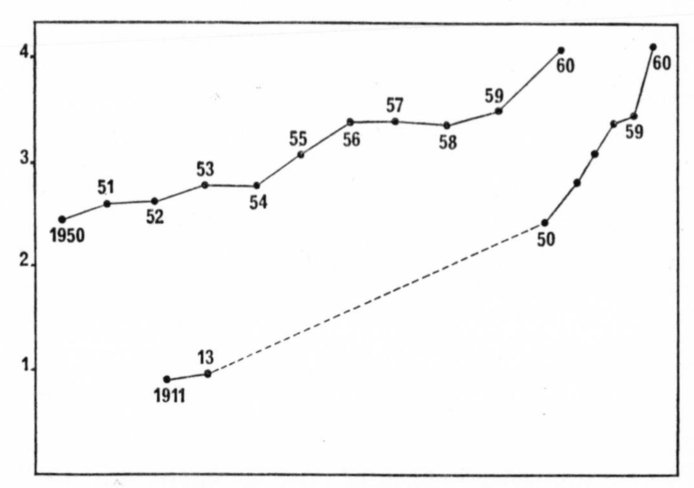

FIG. 10.3. World copper production in millions of tons.

plant diseases develop, including poor cropping of cereals and chlorosis in fruit trees.

Herbivorous animals seem to react perceptibly to copper deficiency and several diseases of sheep and cattle have been reported. Only in exceptional cases, as for example near the outcrop of oxidized copper ores, does the soil contain toxic quantities of copper.

In vertebrate blood copper exists as a series of copper proteins. It is also a constituent of a number of enzymes. It is important in the synthesis of haemoglobin. The adult human body may contain 100–150 mg of copper. Copper is an essential element in the respiratory pigment haemocyanin in the blood of marine invertebrates such as molluscs and crustaceans. Other specific copper compounds which have been described include turacin, a copper porphyrin, in feathers of some birds, and haemocuprein which is believed to be a factor in the synthesis of haemoglobin. Copper probably tends to form complex co-ordination compounds with various organic substances and may be precipitated in such forms in marine muds or sapropels. Subsequent bacterial action may lead to copper sulphides and even metallic copper and some of the copper of sedimentary deposits may have originated in this way.

Ultimate Forms

A high proportion of the copper employed in human operations is in the metallic form and it is of course one of the metals very susceptible to corrosion. Much of the copper produced and used during the last 6000 years or so must now be in the form of corrosion products. These are various basic salts—the basic carbonate verdigris which is the same as malachite, the basic sulphate, and to some extent the basic chloride are all found as common corrosion products. Some of these may of course ultimately pass into solution. The cupreous solutions may ultimately reach the sea where to a large extent the copper will be precipitated by organic matter to settle in the bottom muds. Otherwise it will be retained in a relatively insoluble form in soils. In the long run it would appear that copper will ultimately revert to a sulphide mineral collected in submarine deposits.

Silver

There are two stable isotopes of silver in Nature; they are ^{107}Ag (51·35 %) and ^{109}Ag (48·65 %). As a result of the decay of the natural fission product ^{107}Pd it has been inferred that some of the natural ^{107}Ag would have been formed in this way. This process has not so far been conclusively established as occurring in Nature. There does not appear to be any significant variation in the proportion of the two isotopes in natural processes.

Terrestrial Abundance

Silver is a comparatively rare element at 0·1 g/ton for the Earth's crust, but on account of its mode of occurrence it is rather easily accessible and has of course been known to civilized man from very early times.

Silver is a distinctly chalcophile or sulphophile element and there is little evidence that it ever enters into the composition of ordinary rocks. There does not seem to be any natural silicate of silver and the known minerals are less numerous than those of copper. In part this may be due to the absence of basic salts. The sparingly soluble halides all occur as rare minerals and are ionic in structure.

A few complex sulphate minerals which are also basic salts are encountered but apart from these it is probable that the various compounds with sulphur and metalloids are to be regarded as essentially covalent. The free element and a number of natural alloys are also of some importance.

SILVER MINERALS

Name	*Formula*
Native silver	Ag
Amalgam	Ag,Hg
Allargentum	(Ag,Sb)
Acantite, Argentite	Ag_2S
Stromeyerite	$AgCuS$
Argentopyrite	$AgFe_2S_3$
Smithite	$AgAsS_2$
Proustite	Ag_3AsS_3
Miargyrite	$AgSbS_2$
Pyrargyrite	Ag_3SbS_3
Polybasite	$(Ag,Cu)_{16}Sb_2S_{11}$
Chloroargyrite	$AgCl$
Bromoargyrite	$AgBr$
Embolite	$Ag(Cl,Br)$
Iodoargyrite	AgI
Miersite	$(Ag,Cu)I$
Argentojarosite	$AgFe_3(SO_4)_2(OH)_6$

Geochemistry

The occurrence of silver falls into two categories. The first is the 'straight' silver deposits in which there occur various discrete ores of silver itself and other metals are quite incidental. Perhaps one-quarter of the present output of the metal is from sources of this kind. The second form of occurrence is the existence of relatively small amounts of silver compounds associated with the ores of other metals, particularly those of copper and lead. Most silver deposits are hydrothermal in origin and have resulted from cavity fillings and replacements.

About 70% of the present output of silver is from the American continent but the occurrence of silver is in fact very widespread and considerable amounts are obtained from very many parts of the world. Some of the more important silver deposits are indicated below.

IMPORTANT SILVER DEPOSITS

Locality	*Type*	*Chief Minerals*
Polaris Mine, Sunshine Mine, Idaho, U.S.A.	Quartz veins in quartzite	Argentiferous tetrahedrite, galena
Tintic, Utah, U.S.A.	Replacements in limestone	Siliceous ores, argentiferous lead, and pyritic ores
Comstock Lode, Nevada, U.S.A.	Mineralized fault fissure	Complex minerals with silver, argentite, polybasite, electrum
Cobalt, Ontario, Canada	Mineralized fault fissure veins	Native silver, argentite, polybasite, stephanite, ruby silver, etc.
El Potosi, Chihuahua, Mexico	Fissure veins and irregular replacements in limestone	Base metal sulphides with minor silver minerals
Potosi, Bolivia	Veins in volcanic rocks, etc.	Cerargyrite, silver, argentite, various argentiferous base metal sulphides

A major proportion of silver production is from ores of the second type. These are primarily sulphides of base metals, particularly lead, which act as 'hosts' for silver ions or atoms. According to Goldschmidt this mode of occurrence is due to a process which involves the capture of a cation between anions in a mineral which is semi-metallic and involves an alternation between the charged and uncharged states of the atoms. In galena, the tetrahedral interstices between four nearest sulphur particles (and four lead particles) may capture silver ions. The calculated interionic distance of 2·43 Å is just a little smaller than 2·57 Å, the available distance for Ag—S in the interstices. This mechanism is comparable with the capture of small atoms of boron, carbon, and nitrogen by elements of the iron type.

In the case of galena which is very rich in silver it has been shown that the metal actually occurs as segregations in the baser mineral.

The information available about the general geochemistry of silver is rather scanty. One of the most significant facts is that there is a perceptible concentration of silver in the black bottom muds forming in stagnant waters. These ultimately produce shales which may carry 1 p.p.m. Ag, an amount about ten times that in average igneous rocks.

There is a certain amount of evidence that silver may be concentrated by marine animals, including fish, and some plants have also shown some degree of enrichment. There is no definite information of the function of silver in organisms.

Silver in Britain. The output of silver in Great Britain is now small and is related to the falling lead output. Many of the lead ores worked in the past were quite rich in silver—*e.g.*, some of the Lake District ores carried 20 oz/ton but the actual output was never very important.

Ultimate Forms

Silver is used mainly in the form of metal, a considerable portion of which is involved in bullion and in monetary systems. There seems little doubt that a large and unknown amount of the metal is virtually hoarded and as this process has probably gone on for many centuries much of the metal is ultimately lost although probably still largely in the metallic form. Many of the Eastern countries have long been large

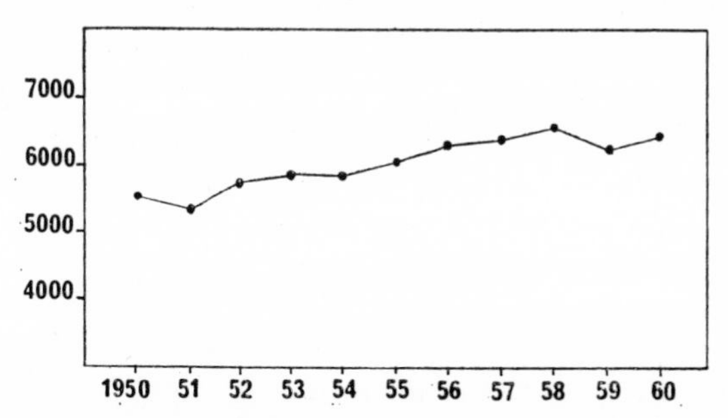

FIG. 10.4. World production of silver in tons.

importers but very small producers of the metal. The conversion of silver into its various salts is mainly of concern to the photographic industry which uses increasing amounts but this is still only a small proportion of the total.

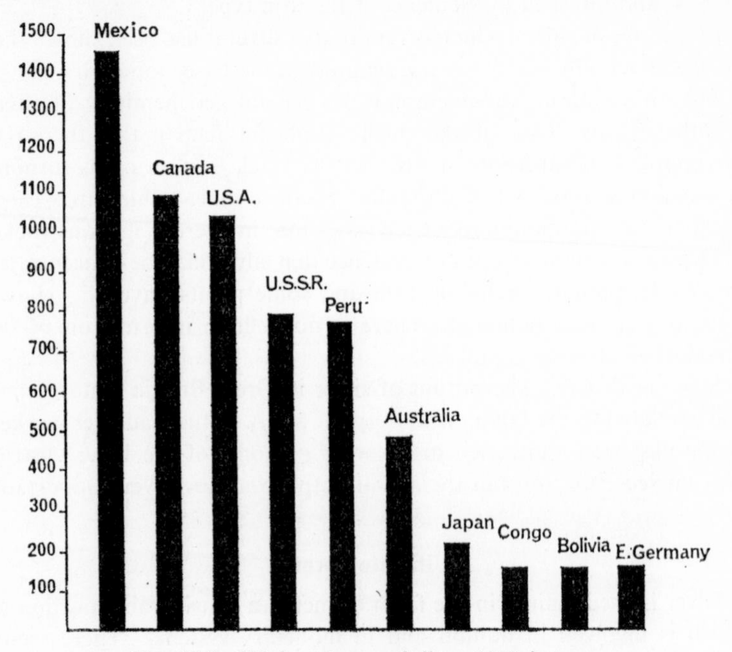

FIG. 10.5. Chief producers of silver, 1959, in tons.

Gold

This metal, both relatively and absolutely one of the rarer elements, has been an object of human cupidity for the whole course of civilization and has therefore attracted what is perhaps a somewhat excessive amount of attention. There is no special interest in its isotopy as only one stable nuclide, ^{197}Au, is known to exist in Nature. In spite of its scarcity the availability of gold is surprisingly high and this is partly due to its inert character and tendency to collect in mechanical concentrations and also in no small part to techniques such as cyaniding and recovery from electrolytic slimes which enable low-grade ores to become commercial sources of the metal.

Gold occurs in as wide a variety of deposits as the other two metals

GOLD MINERALS

Name	*Formula*
Native gold	Au
Cuproauride	Au,Cu
Electrum	Au,Ag
Maldonite	Au_2Bi
Calaverite	$AuTe_2$
Montbrayite	Au_2Te_3
Petzite	$(Au,Ag)_2Te$
Sylvanite	$AuAgTe_4$

of the group but it shows some rather special features. In the first place it shows a marked tendency to occur in the free state and owing to its chemical resistance it tends to remain so. This leads to the very characteristic occurrence of the metal in various kinds of mechanical concentrations, particularly the well-known 'placer' deposits.

In the second place gold, unlike silver, is not much associated with sulphur but in the few well-characterized mineral compounds which it forms it is predominantly in combination with tellurium. Much of the gold of commerce has been won from various base-metal sulphide ores such as chalcopyrite. These minerals often contain it in what are really heavy trace quantities.

Geochemistry

The fact was established by Goldschmidt that gold is an essentially siderophile element and is therefore to be expected in its highest concentrations in the metallic core of the Earth, and it is only incidentally associated with chalcophile elements like silver and copper.

In its crystal chemistry gold presents peculiar problems. It seems that most of its few minerals are covalent structures and in the case of the gold traces carried by sulphide ores it may well be that gold atoms or ions can be captured in mineral lattices containing tellurium, antimony, and bismuth in somewhat the same way as postulated for silver in galena (*q.v.*).

The rather common association of gold with pyrite and arsenopyrite is not very easy to explain but Goldschmidt suggested possible thermal changes in the complex particles in these minerals which would then enable them to accommodate ions or atoms of gold.

The evidence of numerous hydrothermal deposits containing gold seems to suggest that the metal must have been conveyed in solution at some stage. This is a rather difficult problem because of the known intractibility of gold to most reagents. Ferric salt solutions have been postulated as one of the possible means by which gold could be dissolved. It is known on the other hand that gold is frequently left

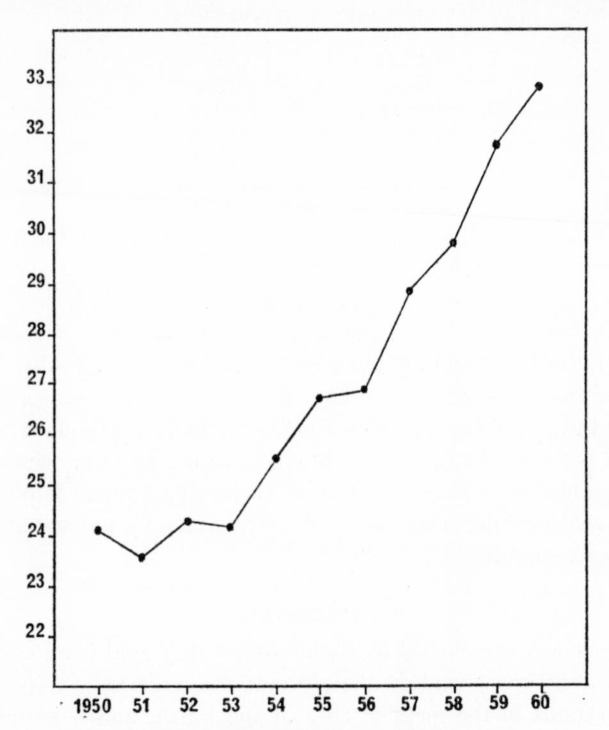

FIG. 10.6. World production of gold in millions of troy ounces.

unchanged in the upper oxidation zone of many deposits but small particles can of course be carried down mechanically. There is a possibility that some, at least, of the gold may be in the form of a collosol but the means of dispersion are not very clear. The minute amount of gold in sea water (around 0·000006 p.p.m.) might be present in colloidal dispersion.

Production

For comparative purposes the figures are quoted in tons. It may be observed that the total amount of the metal extracted by the human race in the course of history is estimated at 50000 tons. The record world output was 41 million oz (1271 tons) in 1941.

Gold in Britain. Very small amounts of gold have been won in the United Kingdom, principal localities being in central Wales (near Dolgelley) where it occurs disseminated in quartz veins in Lower Palaeozoic rocks, and in Scotland where placers are found at Kildonan (Sutherland) and at Leadhills (Lanarkshire). The latter locality is said to have produced the metal for the Scottish crown jewels and although not now regularly worked it produced a nugget of about 1 oz about 1940. A little gold has also been obtained from localities in Ireland—*e.g.*, the Wicklow Mountains.

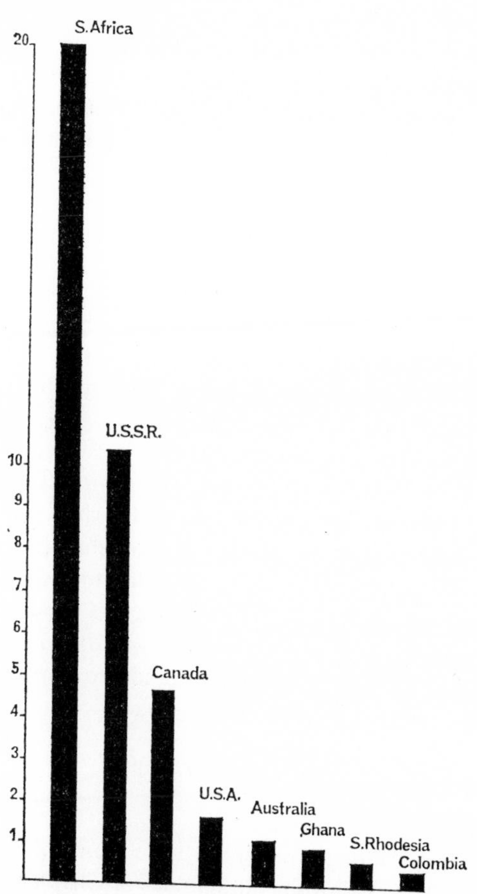

FIG. 10.7. Chief producers of gold, 1959, in millions of troy ounces.

CHAPTER 11

The Elements of Group II

General

In their natural chemistry these elements show much variation. They fall broadly into the chalcophile and lithophile types, the b-group, consisting of zinc, cadmium, and mercury, being of the first type. Beryllium and magnesium will be considered together and then the well-marked family of the ' alkaline earths ', calcium, strontium, and barium. The interesting features of their relative abundances are mainly in the rarity of beryllium and the comparative abundance of barium.

Beryllium

Only one stable nuclide, ^{9}Be, of this element exists in Nature. It is a rare element both cosmically and terrestrially, and its scarcity, as in the case of lithium, is related to the fact that it is unstable in the zone of thermonuclear reactions in stellar interiors.

The problem of the origin of beryllium in Nature is therefore a difficult one. It has been proposed that, along with lithium and boron, beryllium could be synthesized in the expanding hydrogen envelope of a supernova, when, under an intense neutron flux, atoms of carbon, nitrogen, and oxygen could undergo spallation to yield Be, Li, and B. ^{10}Be may be formed in Nature by spallation of O and N in the upper atmosphere by high-energy cosmic ray neutrons. Beryllium in terrestrial minerals, when associated with uranium and thorium as it sometimes is, may sustain α-particle bombardment, when the reaction

$$^{9}Be + \alpha \longrightarrow {}^{12}C + n$$

could act as a natural neutron source.

A number of interesting possibilities arise from this fact. They are discussed in connexion with the lanthanides in Chapter 24.

Terrestrial Occurrence

Beryllium forms a simple cation of small radius (0·34 Å) by reason of which it has little tendency to enter the lattices of common rock-forming silicates. It tends to take up tetrahedral co-ordination with four oxygen or four fluorine ions and is in fact able to form a variety of independent silicate minerals. It is very markedly lithophile or oxyphile in character and is concentrated in the upper parts of the Earth's crust.

Beryllium compounds tend to separate in the last stages of magmatic crystallization and are therefore mainly found in pegmatite veins and dykes associated with granitic rocks. Some beryllium minerals are found in hydrothermal veins also and, having separated at relatively low temperatures, often form very transparent crystals, a matter of some significance in relation to those beryllium minerals which are used as gem stones.

BERYLLIUM MINERALS

Name	*Formula*
Bromellite	BeO
Chrysoberyl	$BeAl_2O_4$
Hambergite	Be_2BO_3OH
Phenacite	Be_2SiO_4
Bertrandite	$Be_4Si_2O_7(OH)_2$
Eudidymite	$NaBeSi_3O_7OH$
Chkalovite	$Na_2BeSi_2O_6$
Harstigite	$Be_2Ca_3Si_3O_{11}$
Barylite	$BaBe_2Si_2O_7$
Beryl	$Be_3Al_2Si_6O_{18}$
Euclase	$BeAlSiO_4OH$
Bavenite	$Ca_4(Be,Al)_4Si_9(O,OH)_{28}$
Beryllium vesuvianite	$Ca_{10}(Mg,Be,Fe^{\cdot\cdot}Fe^{\cdot\cdot\cdot})_2Al_4Si_9O_{34}(OH)_4$
Helvine	$(Mn,Fe,Zn)_8Be_6Si_6O_{24}S_2$
Beryllonite	$NaBePO_4$
Herderite	$CaBePO_4(OH,F)$
Swedenborgite	$NaBe_4SbO_7$

In magmatic crystallization beryllium often separates in the form of complex anions in conjunction with silicon and 4-co-ordinated aluminium. When this is the case the ' admitted ' beryllium causes a weakening of the bonds. For this reason the element only enters to a small extent into the late stages of magmatic minerals.

In weathering processes beryllium behaves much like aluminium in consequence of the two elements having closely similar ionic potentials (Be = 2/0·37, Al = 3/0·57). In soils and sediments generally it is therefore found that the ratio Be/Al is very near to the value for the igneous rocks from which the material has been derived.

Very little is known of the biogeochemical significance of beryllium. Some plants have been reported to show a small degree of concentration of the element. The highly toxic nature of beryllium and its compounds to humans is well known.

Sources of Beryllium Compounds

The localities which produce beryllium minerals are in the main those where granite or nepheline syenite pegmatites occur. More rarely beryl is found in hydro-thermal veins. There is considerable metallurgical interest in the metal but the supplies of ore are limited and output in recent years has fluctuated considerably and the decline noticeable for several of the older producers suggests that the more accessible deposits are approaching exhaustion. At present the ore mined is beryl which contains about 5% of the metal and is a somewhat intractable mineral. In practice a yield of about 70 lb of beryllium is obtained per ton of ore. Chrysoberyl, helvite, and phenacite may be possible ores in the future.

Several beryllium minerals have long been prized as gem stones. Beryl, in the form of deep green and clear crystals, constitutes the emerald and may at times be the most valuable of gems. It has been known for many centuries and was obtained by the Egyptians and Greeks from mines in the Nubian desert and from this source many Eastern potentates were later supplied. Many fine emeralds were brought by the Spaniards from Peru and Colombia. The pale blue beryl called aquamarine was obtained in Brazil. Emeralds were discovered in the Urals in 1831.

Chrysoberyl, euclase, and phenacite have all been used as gem stones and at one time Ceylon was an important producer. According to Goldschmidt the rarity of emerald is due to two causes—the conditions for formation of clear crystals which are produced under hydrothermal conditions and also the necessary presence of either chromium or vanadium to impart the colour.

British occurrence. Small amounts of beryl have been recorded from various parts of Great Britain and it could be expected in localities where granite pegmatites are found. These are, however, poorly developed in Great Britain. A small deposit associated with micas and other pegmatitic minerals was found in the Knoydart district of Argyllshire. The material was generally of the opaque quality and in quantity probably insufficient to be of commercial interest.

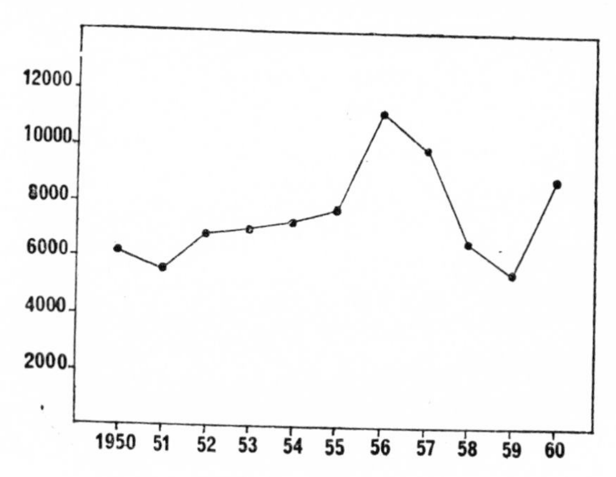

FIG. 11.1. World beryl production in tons.

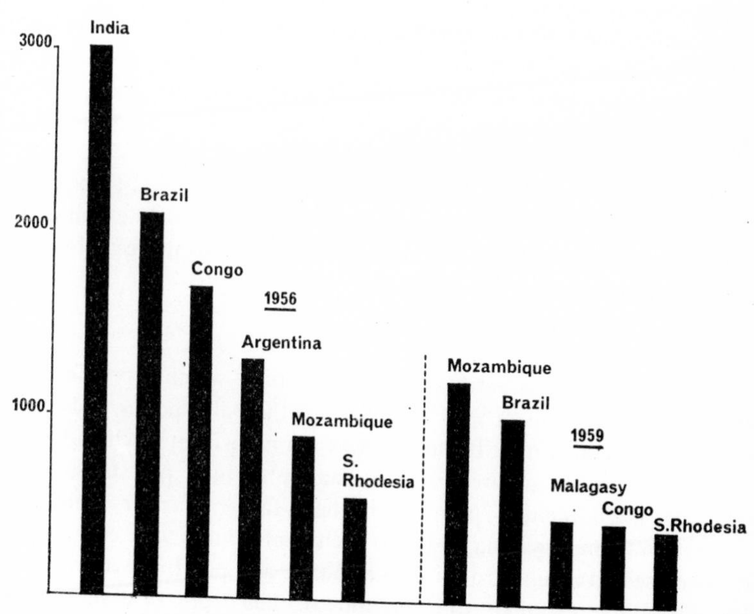

FIG. 11.2. Chief producers of beryl, in tons.

Magnesium

There are three stable natural isotopes of this element:

$$^{24}Mg = 78{\cdot}60\,\%$$
$$^{25}Mg = 10{\cdot}11\,\%$$
$$^{26}Mg = 11{\cdot}29\,\%$$

There is little or no information at present on the possible variation of these proportions by natural processes.

It is probable that magnesium atoms are synthesized in ' helium burning ' stars by such reactions as

$$^{20}Ne + \alpha = {}^{24}Mg + \gamma$$
$$^{12}C + {}^{16}O = {}^{24}Mg + {}^{4}He$$
$$^{21}Ne + \alpha = {}^{24}Mg + n$$

Abundance

Magnesium is an abundant element in Nature, being eighth in order both cosmically and terrestrially. The figure for the Earth's crust places magnesium below the two common alkali metals but in fact the estimates for the Earth as a whole place magnesium fourth in order at 17 %. This estimate (Mason) takes due account of the large mass of the Earth's mantle which consists mainly of magnesium silicates.

Magnesium Minerals

These are very numerous and the list below is intended to be representative only. It will be seen that, apart from a few halides, all of the minerals are oxygen compounds, thus emphasizing the oxyphile and lithophile character of the element.

Geochemistry

Magnesium is characteristically the cation which goes into the orthosilicate minerals of which forsterite is the pure species and olivine the more general form. These minerals are nesosilicates which possess a very compact structure, great resistance to high pressures, and a relatively high melting point. All these factors favour their early separation from a magma and they are therefore met with especially in the deep-seated igneous rocks such as dunite and peridotite of which the Earth's mantle is thought to be composed. The ordinary basic rocks of basaltic and gabbroic type which characterize the deeper zones of the crust also contain magnesian silicates, usually associated with ferrous

MAGNESIUM MINERALS

Name	*Formula*	*Occurrence*
Periclase	MgO	Metamorphic minerals
Brucite	$Mg(OH)_2$	Metamorphic minerals
Spinel	$MgAl_2O_4$	Metamorphic minerals
Sellaite	MgF_2	Volcanic, etc.
Bischoffite	$MgCl_2.6H_2O$	Evaporite deposits
Carnallite	$KMgCl_3.6H_2O$	Evaporite deposits
Suanite	$Mg_2B_2O_5$	Evaporite deposits
Inderite	$Mg_2B_6O_{11}.15H_2O$	Evaporite deposits
Fluorborite	$Mg_3BO_3(F,OH)_3$	Evaporite deposits
Boracite	$Mg_6B_{14}O_{26}Cl_2$	Evaporite deposits
Magnesite	$MgCO_3$	Altered rocks, minerals, etc.
Nesquehonite	$MgCO_3.3H_2O$	Altered rocks, minerals, etc.
Lansfordite	$MgCO_3.5H_2O$	Altered rocks, minerals, etc.
Hydromagnesite	$Mg_5(CO_3)_4(OH)_2.4H_2O$	Altered rocks, minerals, etc.
Dolomite	$CaMg(CO_3)_2$	Altered rocks, minerals, etc.
Nitromagnesite	$Mg(NO_3)_2.6H_2O$	Evaporite deposits
Forsterite	Mg_2SiO_4	Rock minerals
Enstatite	$MgSiO_3$	Rock minerals
Chrysotile	$Mg_3Si_2O_5(OH)_4$	Rock minerals
Talc	$Mg_3Si_4O_{10}(OH)_2$	Rock minerals
Ackermanite	$MgCa_2Si_2O_7$	Rock minerals
Diopside	$MgCaSi_2O_6$	Rock minerals
Tremolite	$Ca_2Mg_5Si_8O_{22}(OH)_2$	Rock minerals
Olivine	$(Mg,Fe)_2SiO_4$	Rock minerals
Hypersthene	$(Mg,Fe)SiO_3$	Rock minerals
Pyrope	$Mg_3Al_2Si_3O_{12}$	Rock minerals
Phlogopite	$KMg_3AlSi_3O_{10}(OH)_2$	Rock minerals
Bobierrite	$Mg_3(PO_4)_2.8H_2O$	Guano deposits
Struvite	$NH_4MgPO_4.6H_2O$	Guano deposits
Kieserite	$MgSO_4.H_2O$	Evaporite deposits
Epsomite	$MgSO_4.7H_2O$	Evaporite deposits

iron. Examples are the pyroxenes and amphiboles, both of which classes are inosilicates, tending to form chain structures, less compact then the orthosilicates. The magnesian minerals of the upper crust are generally of the phyllosilicate class, tending to extended two-dimensional structures, good examples being talc, serpentine, and phlogopite mica. The garnets are especially connected with metamorphic rocks; they are orthosilicates well adapted in structure to high-pressure conditions.

For the above reasons it is clear that a high proportion of the magnesium in the Earth will have separated at an early stage and the associated minerals of the magnesium-rich peridotitic rocks are usually refractory oxides like magnetite and chromite.

Apart from igneous rocks proper there are several other types which may contain considerable amounts of magnesium:

(1) Serpentinous and talcose rocks presumably derived from the weathering or sub-surface alteration by hydrothermal solutions of olivine rocks.

(2) Extreme alteration of such rocks by carbonated solutions giving rise to massive magnesite deposits which may be valuable as source minerals for magnesium.

(3) Replacement deposits where limestone rocks have been altered by action of magnesian solutions, possibly by the reaction

$$MgCl_2 + CaCO_3 = MgCO_3 + CaCl_2$$

This commonly results in the formation of dolomite $CaMg(CO_3)_2$.

This mineral, of which whole mountain masses are made, is not a solid solution series between calcite and magnesite. It is formed by the regular substitution of alternate ions of calcium and magnesium in the crystal lattice of calcite. The reason for this is the large difference in ionic radii between Mg (0·66 Å) and Ca (0·99 Å) which is unfavourable to diadochy. A great many limestones have suffered some degree of dolomitization which thereby reduces their value as sources of lime.

(4) Evaporite deposits. In the course of weathering of rocks magnesium ions may ultimately pass into solution and reach the ocean. In this environment, where the most abundant anion is Cl^-, magnesium may separate as $MgCl_2$ or a related salt if evaporation takes place and proceeds far enough to enable most of the calcium and sodium salts to deposit. There is thus a group of magnesian minerals found in various saline deposits. Carnallite $KMgCl_3.6H_2O$, is probably the commonest of these. Only a small proportion of the total magnesium ions is in fact removed by evaporation in this way so that sea water contains approximately 0·13 % Mg ions. This is equivalent to about 18 Gg for the whole of the oceans, evidently an adequate reserve for human requirements.

Some magnesium compounds are also found in hydrothermal veins. Dolomite is a fairly common gangue mineral in such veins.

In aqueous media magnesium ions tend to acquire six water molecules and the comparative paucity of the element in soils is no doubt due to the lack of adsorbed ions—an effect also observed for the equally readily hydrated ions of sodium. The initially small cation becomes relatively large by hydration whereas the initially larger potassium ion generally remains non-hydrated and is therefore much more readily adsorbed on soil colloids.

Biogeochemistry

Magnesium is essential in the formation of chlorophyll and is therefore necessary for all green plants. Although the magnesium content of soils is usually small the normal requirements of plants are adequately met and magnesium deficiency is on the whole rather uncommon.

In marine organisms calcareous material is partly replaced by magnesite and this tendency is greater in warm waters than in cold. Magnesium is an essential element in animals in skeletal structure, nerve tissue, and muscle. It is also important in carbohydrate metabolism. Excreted magnesium salts are met with in guano deposits as phosphates.

Sources of Magnesium Compounds

Only a limited number of magnesium minerals are of interest as sources of the metal but several others have important applications for various technical purposes. As good-quality magnesite deposits are limited and will ultimately be exhausted it is probable that sea water will become increasingly important as a source of magnesium and its compounds. There is considerable fluctuation of output from the various countries and the figures quoted are incomplete—they do not include U.S.S.R. production or the British output from sea water. A large proportion of the U.S.A. production is from sea water. There are sea water extraction plants at Wilmington, N. Carolina, on the Gulf Coast, U.S.A., and at Hartlepool, England.

British occurrences. There are no commercially interesting deposits of magnesite in Great Britain but dolomite is abundant, particularly in the form of the magnesian limestones of the Permian rocks of North-East England and parts of the Midlands. This provides a source of magnesium compounds to some extent. It also has application as a refractory. It is less satisfactory as a building stone in the air of towns—its use for the Parliament buildings at Westminster is an example.

Serpentine, mainly of interest as an easily carved ornamental stone, occurs in the extreme south-west of England near the Lizard and in the extreme north of Shetland in the Island of Unst. Serpentine also constitutes in part the Connemara ‘ marble ’ of the west of Ireland.

Talc is also produced in Great Britain to the extent of about 5000 tons per annum in a world production of about $1\frac{1}{2}$ million tons.

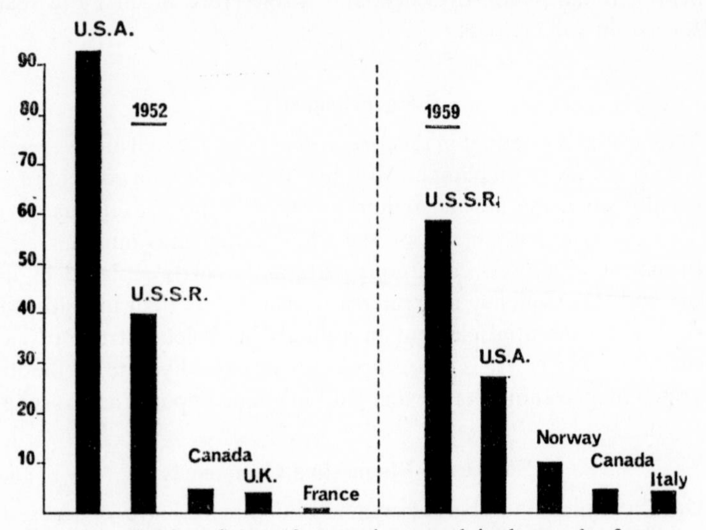

FIG. 11.3. Chief producers of magnesium metal, in thousands of tons.

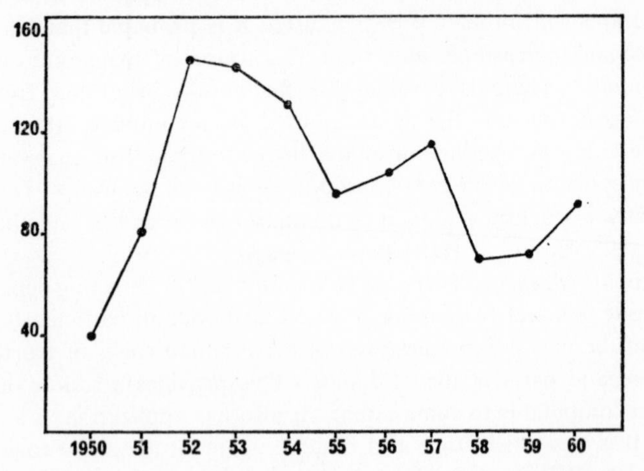

FIG. 11.4. World production of magnesium metal, in thousands of tons.

Calcium

There are six stable isotopes of the element in Nature:

$^{40}Ca = 96{\cdot}97\%$	$^{44}Ca = 2{\cdot}06\%$
$^{42}Ca = 0{\cdot}64\%$	$^{46}Ca = 0{\cdot}0033\%$
$^{43}Ca = 0{\cdot}145\%$	$^{48}Ca = 0{\cdot}185\%$

No precise mechanism for the synthesis of calcium atoms appears to have been proposed but the presence of strong calcium lines in the spectra of certain 'white dwarf' stars suggests that it might be one of the products of the later stages of the α-process.

It is known that some of the terrestrial ^{40}Ca is the stable end-product of the decay by β-emission of ^{40}K. In spite of the difficulties which arise from the contamination of radiogenic ^{40}Ca with non-radiogenic material, the use of the $^{40}Ca/^{40}K$ ratio as a means of age determination has been developed. It is especially applicable to the older formations.

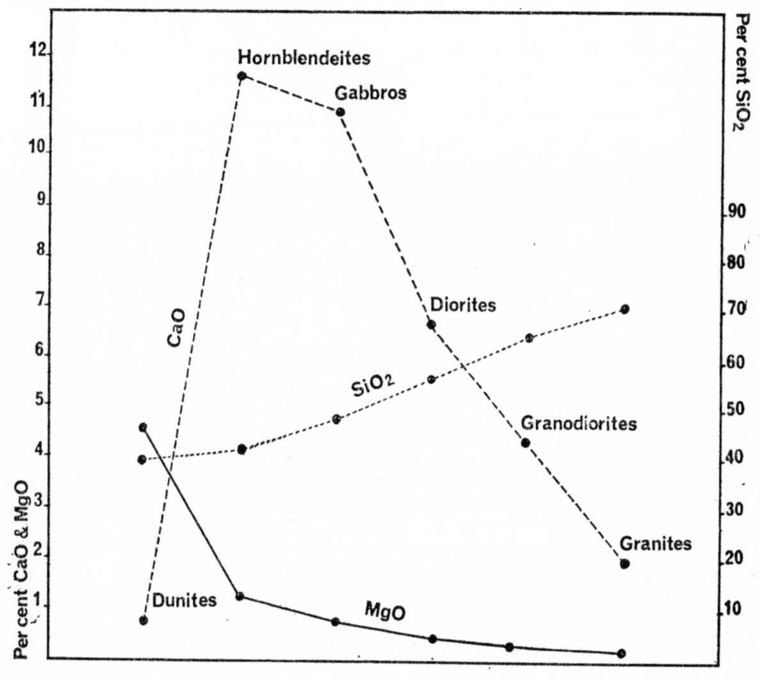

FIG. 11.5. Variation in composition of igneous rocks.

Abundance

Calcium is of comparatively high abundance cosmically and on the Earth it appears to be more abundant than magnesium. This is true only for the crust for in the Earth as a whole the value of 0·61 % places it well below magnesium (17 %). The variation in the calcium content of igneous rocks is shown in Fig. 11.5.

As with magnesium, the number of calcium-bearing minerals is so large that only a representative list is presented here.

CALCIUM MINERALS

Name	*Formula*	*Occurrence*
Oldhamite	CaS	Meteorites
Portlandite	$Ca(OH)_2$	Alteration product
Fluorite	CaF_2	H.T. veins, rocks
Chlorocalcite	$KCaCl_3$	Volcanic sublimates
Colemanite	$Ca_2B_6O_{11}.5H_2O$	Evaporites
Calcite	$CaCO_3$	Sedimentary rocks, mineral veins
Aragonite	$CaCO_3$	
Nitrocalcite	$Ca(NO_3)_2.nH_2O$	Efflorescence in caves
Wollastonite	$CaSiO_3$	Igneous rocks
Rankinite	$Ca_3Si_2O_7$	Secondary minerals of igneous and metamorphic rocks
Larnite	$CaSi_2O_4$	
Tobermorite	$Ca_5Si_6O_{16}(OH)_2.4H_2O$	
Pectolite	$NaCa_2Si_3O_8OH$	
Anorthite	$CaAl_2Si_2O_8$	Igneous rocks
Laumontite	$CaAl_2Si_4O_{12}.4H_2O$	Zeolitic
Datolite	$CaBSiO_4OH$	
Whitlockite	$Ca_3(PO_4)_2$	Phosphate deposits
Hydroxyapatite	$Ca_5(PO_4)_3OH$	
Apatite	$Ca_5(PO_4)_3F$	Rock minerals
Chloroapatite	$Ca_5(PO_4)_3Cl$	
Anhydrite	$CaSO_4$	Evaporites
Gypsum	$CaSO_4.2H_2O$	Evaporites
Scheelite	$CaWO_4$	H.T. veins
Lautarite	$Ca(IO_3)_2$	Evaporites

Geochemistry

It will be seen from Fig. 11.5 that calcium as a rock-forming element tends to be concentrated in the basic and intermediate types of igneous rocks and there are several reasons for this:

(1) Calcium forms a feldspar, anorthite $CaAl_2Si_2O_8$, which readily forms a series of isomorphous mixtures with the pure soda feldspar albite $NaAlSi_3O_8$—these constitute the plagioclase series. They are

the typical feldspars of basic and intermediate igneous rocks and extend even to some extent to the granites.

The pure lime feldspar melts at 1550 °C whereas soda feldspar melts at 1118 °C so that calcic feldspars will tend to crystallize first from a magma and the outer layers of large feldspar crystals will always tend to be richer in albite.

(2) Calcium ions readily enter certain ino-silicates together with magnesium and ferrous iron—these are the pyroxenic and amphibolic minerals which are very characteristic of basic and intermediate rocks such as gabbro, basalt, dolerite, andesite, and diorite.

(3) Residual calcium in igneous magmas can often be taken into various accessory minerals, some of which may be fairly abundant. They include apatites, sphene $CaTiSiO_5$, perovskite $CaTiO_3$, cancrinite $3NaAlSiO_4.CaCO_3$, and even calcite itself.

The ubiquity of calcium ions in igneous rocks is no doubt due largely to their radius (0·99 Å) which is such that a large number of other cations can deputize for calcium and vice versa. The following are examples:

$$Na = 0{\cdot}98\,Å, \quad Mn^{2+} = 0{\cdot}91\,Å, \quad Sm^{2+} = 1{\cdot}26\,Å$$
$$Sr = 1{\cdot}27\,Å, \text{ and the lanthanides—}e.g., Lu = 0{\cdot}99\,Å$$

In practice calcium minerals largely act as hosts for rarer cations.

It has recently been shown that one of the causes of the twilight glow is the presence of excited metal atoms in the upper atmosphere and one of these is calcium. It is possible that this calcium is of meteoric origin. The total amount for the whole atmosphere is about 200 kg.

Calcium in Sedimentary Rocks

The precipitation of calcium carbonate in the sea is the principal cause of formation of limestones. These rocks form an important part of the sedimentary formations and they may in some cases be amongst the purest chemical materials occurring in bulk in the Earth's crust.

The solubility of $CaCO_3$ in sea water is governed by a complex of variables as indicated below.

SOLUBILITY OF $CaCO_3$ IN SEA WATER

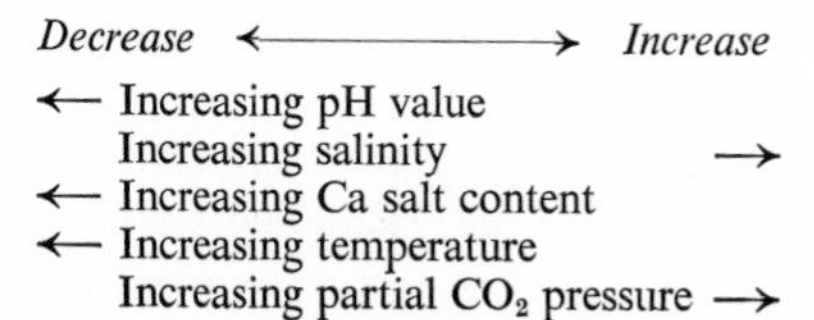

These factors are determined by various causes and are not entirely independent.

Purely inorganic deposition may follow from supersaturation in $CaCO_3$, mainly in warm and rather shallow waters. It probably does not account for very large amounts of limestone.

There are three forms of calcium carbonate in Nature:

(*a*) Calcite—hexagonal-scalenohedral.
(*b*) Aragonite—orthorhombic.
(*c*) Vaterite—hexagonal.

Some opinion favours the biogenic origin of most limestones. This may be directly due to the organisms abstracting $CaCO_3$ from the water to build up their skeletal material, etc., or indirectly in the case of plants by abstracting HCO_3^- ions and so raising the pH. The decomposition of protein by liberating ammonia may also help to raise the pH of the water. The first form of calcium carbonate deposited in shells is probably vaterite which transforms into aragonite in the case of freshwater shells and into aragonite or calcite in the case of marine forms.

There are important types of limestones in which the calcium carbonate is largely concretionary—having been deposited in concentric layers around a small grain of silica or other solid particle. The oolitic limestones of England are of this type. They have long been famed as building stones. Many limestones (including the above) contain varying amounts of calcareous remains of marine organisms. These may be foraminifera as in the English chalk, corals and encrinites as in much of the carboniferous limestone, and sometimes molluscan shells and the remains of calcareous algae. There are considerable deposits of limestone of pre-Cambrian age which have been claimed as of inorganic origin but it is now agreed that life must have existed much earlier than 5×10^8 years ago, although not of a type complex enough to leave positive fossil evidence in the rocks.

Large quantities of crystalline calcite and aragonite are found as gangue in metalliferous veins, particularly when these occur in limestone. Good examples are met with in the lead veins of the North of England and in the iron ore deposits of Cumberland. The famous crystals of Iceland Spar—some of great size—are from cavities in basalt and probably represent a product of alteration of calcium silicate minerals from the volcanic rock. Calcium carbonate deposited from mineral waters constitutes calcareous tufa or travertine. This is obtained in quantity, for example, in the vicinity of Rome and is widely used as an ornamental stone.

Marbles, which are generally described as crystalline limestones, vary in composition. In many cases they have been considerably altered chemically by such reactions as

$$CaCO_3 + SiO_2 = CaSiO_3 + CO_2$$

and if dolomite is present the reaction

$$2CaMg(CO_3)_2 + SiO_2 = Mg_2SiO_4 + CaCO_3 + 2CO_2$$

Some of the limestones of the Scottish Highlands are forsterite marbles of this kind.

In addition to calcium carbonate rocks we have also to note the existence of considerable amounts of the sulphate in evaporite deposits. Anhydrite $CaSO_4$ is formed at rather low temperatures whereas gypsum $CaSO_4.2H_2O$ tends to deposit at appreciably higher temperatures. Most evaporite deposits contain both of these salts. Polyhalite $K_2Ca_2Mg(SO_4)_4.2H_2O$ may also sometimes be present.

Anhydrite is an important raw material as a source of sulphate and will be referred to again in Chapter 22. The phosphate deposits, which are largely calcium salts, will likewise be dealt with in Chapter 17.

Biogeochemistry

The importance of calcium to plants was alluded to in Chapter 4. Calcium is of importance physiologically and to a lesser extent structurally. The calcareous algae which bear a superficial resemblance to corals are a good example of calcium salts entering into plant structure. Calcium oxalate is found stored in the solid state in some plants.

Calcium carbonate in soils regulates the pH, and the working of soils in agriculture, by improving drainage, tends to encourage the leaching out of the calcium as bicarbonate. Heavy manuring both by natural and artificial fertilizers tends to build up soil acidity so that intensive agriculture requires in most cases the periodic addition of lime in some form to maintain the pH at a value not far removed from the neutral point which is favourable to the growth of most food crops. Lime also tends to flocculate soil colloids and therefore improves porosity.

Calcium in animals as skeletal material occurs as carbonate, phosphate, and to some extent as fluoride. Most of this calcium is absorbed directly as soluble salts and not to any extent indirectly as vegetable food. The abundance and low or non-toxicity of these calcium salts favours their use as the basis of structures. The skeletal material of invertebrates is mainly based on calcium carbonate whereas that of vertebrates is mainly calcium phosphate.

Calcium is actually the fifth element in order of abundance in both plant and animal material (on dry weight). In the vertebrates, calcium is present in the body in larger amounts than any other cation.

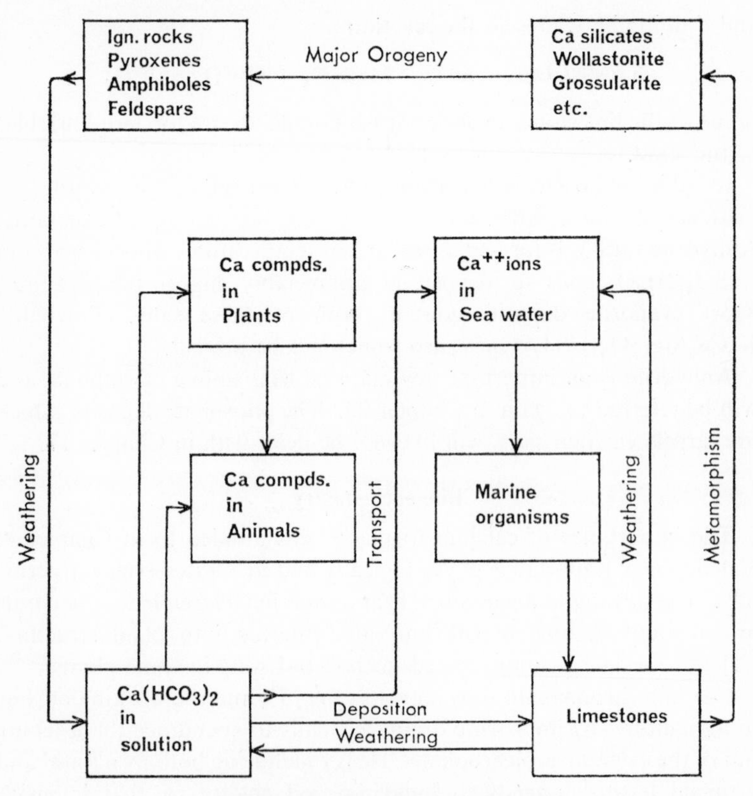

FIG. 11.6. Natural cycle of calcium salts.

Production of Calcium Compounds

Calcium compounds are used in a wide variety of ways which involve some sort of chemical operation, the more important being the following:

(*a*) Agriculture—soil conditioning.
(*b*) Building—cements and mortars.

(*c*) Metallurgy—fluxes.
(*d*) General chemical industry.

In the majority of cases the source is calcium carbonate in the form of limestone, chalk, and even sea shell where it is available. For category (*d*) the raw material must be of good quality and largely free from other metallic carbonates such as those of magnesium and iron.

In the U.K. such material may be obtained in some localities in the Chalk and from the Carboniferous limestones in some parts of the Pennine area as at Settle, Yorkshire and Buxton, Derbyshire.

The world production of calcareous material cannot be accurately evaluated but it must be many hundreds of millions of tons. The output of metallic calcium is small, being of the order of 50 tons per annum, mainly from Canada.

Strontium

There are four stable isotopes of strontium in Nature:

$$^{84}Sr = 0{\cdot}56\% \qquad ^{86}Sr = 9{\cdot}86\%$$
$$^{87}Sr = 7{\cdot}02\% \qquad ^{88}Sr = 82{\cdot}5\%$$

As mentioned in Chapter 9, ^{87}Sr is the stable decay product of ^{87}Rb. Not all ^{87}Sr is radiogenic, however, and it is evident that some variation in ^{87}Sr content of strontium minerals from different localities is to be expected.

Abundance

For the Earth's crust the strontium content is estimated at 450 g/ton whence it is eighty times as scarce as calcium but only 12% more

STRONTIUM MINERALS

Name	*Formula*	*Occurrence*
Veatchite	$Sr_3B_{16}O_{27}.5H_2O$	Veins in limestone
Strontianite	$SrCO_3$	L.T. hydrothermal veins
Strontium–heulandite	$(Na,Ca,Sr)_4Al_6(Al,Si)_4Si_{26}O_{72}.24H_2O$	Zeolitic
Brewsterite	$(Sr,Ba)Al_2Si_6O_{16}.5H_2O$	Zeolitic
Goyazite	$SrAl_3P_2O_7(OH)_7$	Pegmatites
Svanbergite	$SrAl_3PO_4SO_4(OH)_6$	Metamorphic minerals
Celestine	$SrSO_4$	Bedded deposits with gypsum

abundant than barium. The number of independent strontium minerals is relatively small.

The peculiarities of the geochemistry of strontium are very largely related to the size of its ion compared with those of calcium, barium, and to a lesser extent potassium:

$$Ca^{2+} = 0{\cdot}99 Å, \quad Sr^{2+} = 1{\cdot}12 Å, \quad K^{+} = 1{\cdot}33 Å, \quad Ba^{2+} = 1{\cdot}34 Å$$

On account of the rather low abundance of strontium it tends to be a guest ion in both magmatic and aqueous conditions.

The strontium ion is 13% larger than the calcium ion and 16% smaller than the barium or potassium ions. There is therefore about equal tendency for strontium ions to enter calcium minerals or barium minerals but, on account of the greater abundance of the former, strontium is more often found in small amounts in calcium minerals.

As will be seen later barium can replace potassium to a greater extent than strontium, with its smaller ion, is able to do. The general position is that strontium is frequently found in the calcium minerals of the alkalic types of igneous rocks and also in the basic types but to a much smaller extent in the silicates of granitic rocks.

There is a marked deficiency of strontium in the complex silicates which contain calcium with magnesium and ferrous iron, and only rarely is it found in potash feldspars. There is a small tendency for strontium to accumulate in magmatic residua and therefore discrete deposits of strontium minerals are sometimes found in hydrothermal veins. The mineral celestine, however, is mainly found in bedded deposits with gypsum and sometimes sulphur.

Biogeochemistry

Strontium has been described as the ‘ fellow traveller ’ of calcium in respect of biological material. It is not known to perform any specific function but may occur in any of the circumstances in which calcium is normally found in an organism. It has been stated that radiolarians have skeletons which may be largely composed of strontium sulphate. In the case of the protozoon *Acanthometra* celestine $SrSO_4$ makes up about 65% of the skeleton.

Production of Strontium Minerals

Strontium is of special interest to British chemists in that it is probably the only one of the elements for which England is the principal producer and it is a strange fact that of the group of elements named after localities it is the only one which carries a name of British derivation. The original locality of Strontian in Argyllshire, Scotland only produced the carbonate in small quantity.

About two-thirds of the world output of strontium minerals comes from the district around Yate in Gloucestershire, England. The deposits take the form of bedded celestine associated with gypsum in

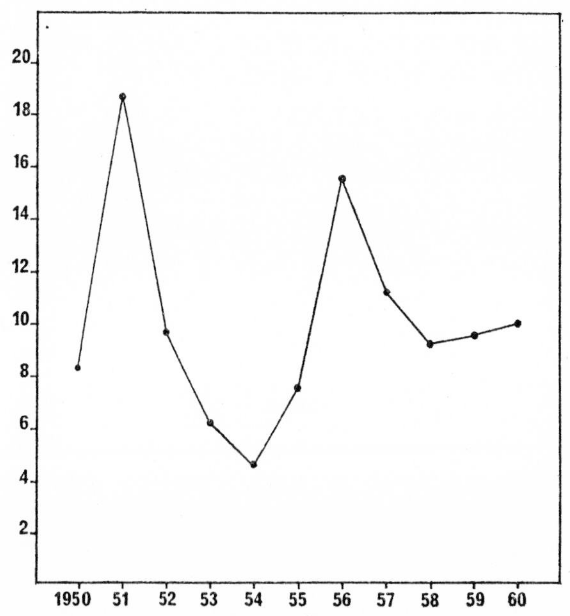

FIG. 11.7. World production of strontium minerals, in thousands of tons.

Triassic marls. Other localities of mineralogical interest include Caltanisetta in Sicily, Wassy, Haute-Marne, France, where it is in Cretaceous marls. In Eastern Russia it occurs in Permian dolomites. Celestine is found associated with zeolites in basalts at St. Andrews, Scotland. There are considerable quantities of strontianite as a vein mineral in Westphalia, Germany, and the original locality at Strontian, Argyllshire, Scotland also produced the carbonate as a gangue mineral in lead veins. This source is now virtually exhausted.

Barium

There are seven stable isotopes of barium in Nature:

$^{130}Ba = 0{\cdot}101\%$	$^{136}Ba = 7{\cdot}81\%$
$^{132}Ba = 0{\cdot}097\%$	$^{137}Ba = 11{\cdot}32\%$
$^{134}Ba = 2{\cdot}42\%$	$^{138}Ba = 71{\cdot}66\%$
$^{135}Ba = 6{\cdot}59\%$	

There do not appear to be any significant aspects of natural nuclear chemistry as far as barium is concerned.

The various barium isotopes lie close to one of the peaks of the curve for neutron capture and it seems probable that barium atoms are products of the r- and s-processes which are supposed to take place in the more advanced stages of stellar evolution.

The conditions are such that some isotopes are produced by one of these processes and some by the other, and some could be produced by both processes. Other isotopes could be fission products of very heavy atoms and yet others could be formed by decay of unstable fissiogenic nuclides.

Abundance

Barium is rather abundant for a heavy element and this can be ascribed to the somewhat favourable conditions for its synthesis as indicated above. Cosmically it is about ten times as abundant as caesium which precedes it and has almost the same abundance as the commoner lanthanides which succeed it in order of atomic number.

In the Earth's crust the relative abundance of barium is even more marked—it is 400 times as abundant as caesium and nine times as abundant as cerium. Barium, like the other alkaline earth metals, is predominantly lithophile in character.

Geochemistry

As noted in the section on strontium, the ionic radius of barium (1·34 Å) is such that it can with facility replace potassium and to some extent lead. The difference in ionic radius between calcium and barium is such that barium does not commonly enter into calcium minerals and is even less likely to enter into magnesium minerals. There are rather more independent barium minerals species than in the case of strontium. There are several barium silicates which form rather rare rock minerals, including the barium feldspar celsian. There are also barium members of the zeolite group.

BARIUM MINERALS

Name	*Formula*	*Occurrence*
Psilomelane	$(Ba,Mn)Mn_4^{IV}O_8(OH)_2$	Hydrothermal veins
Witherite	$BaCO_3$	
Bromlite	$BaCa(CO_3)_2$	
Sanbornite	$BaSi_2O_5$	Rare rock minerals, etc.
Gillespite	$BaFeSi_4O_{12}$	
Celsian	$BaAl_2Si_2O_8$	
Cymrite	$BaAlSi_3O_8OH$	
Harmotome	$BaAl_2Si_6O_{16}.6H_2O$	
Banalsite	$Na_2BaAl_4Si_4O_{16}$	
Hyalophane	$(K,Si,Ba,Al)AlSi_2O_8$	
Leucosphenite	$Na_3CaBaBTi_3Si_9O_{29}$	
Gorceixite	$BaAl_3P_2O_7(OH)_7$	
Baryte	$BaSO_4$	Hydrothermal veins
Barytoanglesite	$(Ba,Pb)SO_4$	
Nitrobarite	$Ba(NO_3)_2$	

The only appreciable concentration of barium compounds occurs in hydrothermal deposits and the two most important minerals are witherite and barite (barytes). The latter is by far the most abundant barium mineral.

The very local minerals bromlite and barytocalcite are examples of double salts between barium and calcium. They are not members of isomorphous series. The barium ion extends its resemblance to potassium also in its ability to be adsorbed by colloids. Hence barium ions are fairly readily retained by soils or are precipitated with hydrolysates such as hydrated manganese oxides—*e.g.*, as in psilomelane. For this reason the barium content of sea water is lower than that of strontium.

Biological Aspects

Barium is toxic to higher animals and there is no indication that it performs any useful function in any kind of organism. Marine organisms which show some enrichment in strontium do not likewise absorb barium. Some plants growing near to barium deposits accumulate a certain amount of the element, probably adventitiously, and this has been proposed as the basis of a possible method of biogeochemical prospecting.

Production of Barium Compounds

Workable deposits, mainly of the sulphate, are far more common than in the case of strontium and they are mainly of the hydrothermal type. Barytes is a common gangue mineral in some metalliferous districts. The carbonate is much more local and is only mined in useful

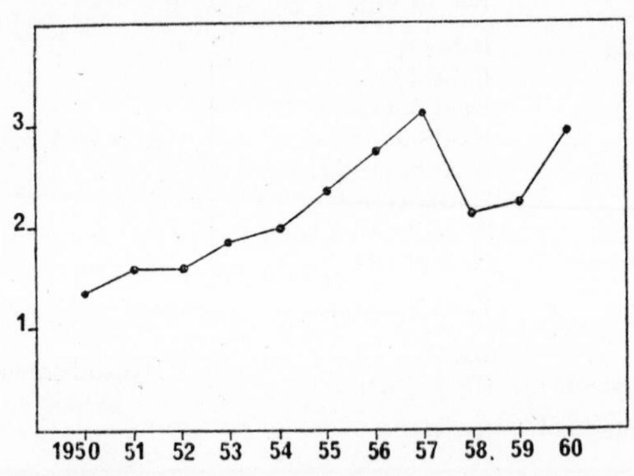

FIG. 11.8. World production of barium minerals, in millions of tons.

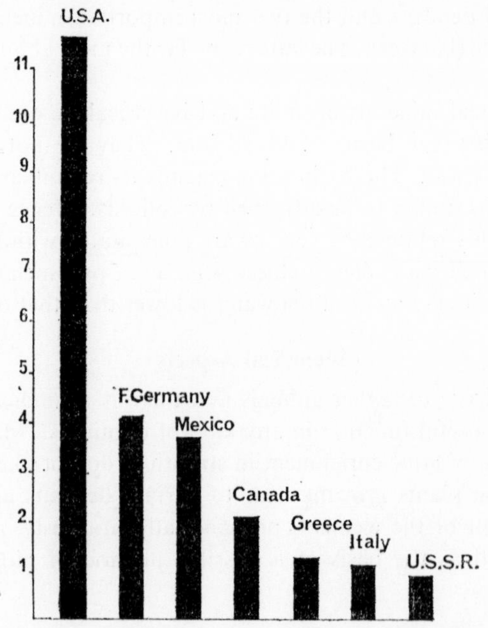

FIG. 11.9. Chief producers of barium minerals, 1957.

quantities on the fringe of the Alston Moor region in the North of England, particularly at Settlingstones which has produced over 500000 tons of the mineral.

Waters draining from certain coal mines in the vicinity of Backworth, in Northumberland, England, contain perceptible amounts of barium chloride in solution and the amount is sufficient to justify its recovery by precipitation as barium sulphate to the extent of several thousands of tons per annum.

Zinc

There are five stable isotopes of zinc in Nature:

$^{64}Zn = 48{\cdot}89\%$ $^{68}Zn = 18{\cdot}56\%$
$^{66}Zn = 27{\cdot}81\%$ $^{70}Zn = 0{\cdot}62\%$
$^{67}Zn = 4{\cdot}11\%$

There do not appear to be any topics of special interest in regard to these isotopes.

Abundance

The value of 65 g/ton for the abundance in the Earth's crust means that zinc is a relatively rare element but it is rather more abundant than copper. The way in which zinc is distributed is related to its markedly thiophile character which leads to it being concentrated in the sulphide phase of magmas where these exist.

In the upper parts of the crust it shows some lithophile character and with an ionic radius of 0·74 Å it can enter lattices in place of Fe^{2+} (0·74 Å), Mg (0·66 Å) and several other similar ions. It therefore behaves as a trace element in some silicate minerals. Zinc is generally more abundant in basic and intermediate igneous rocks than in those of the acid type. It is readily involved in hydrothermal processes and therefore it is mainly found in concentrations in this type of deposit.

Zinc Minerals

Zinc forms only one sulphide mineral of any consequence, but oxygen compounds, chiefly oxy-salts, are fairly numerous. Most of these are mainly of mineralogical interest, occurring in the oxidation zone of zinc deposits.

ZINC MINERALS

Name	*Formula*	*Occurrence*
Blende, sphalerite	ZnS	Hydrothermal veins
Heterolite	$ZnMn_2O_4$	Metamorphic deposits
Gahnite	$ZnAl_2O_4$	
Franklinite	$(Zn,Mn,Fe^{2+})(Fe^{3+},Mn^{3+})_2O_4$	
Roweite	$Ca(Mn,Mg,Zn)B_2O_5.H_2O$	Oxidized zones of mineral veins
Smithsonite	$ZnCO_3$	
Hydrozincite	$Zn_5(CO_3)_2(OH)_6$	
Aurichalcite	$(Zn,Cu)_5(CO_3)_2(OH)_6$	
Willemite	Zn_2SiO_4	
Hemimorphite	$Zn_4Si_2O_7(OH)_2.H_2O$	
Hardystonite	$Ca_2ZnSi_2O_7$	
Clinohedrite	$CaZnSiO_3(OH)_2$	
Larsenite	$PbZnSiO_4$	
Hopeite	$Zn_3(PO_4)_2.4H_2O$	
Scholzite	$Ca_3Zn(PO_4)_2(OH)_2.H_2O$	
Köttigite	$Zn_3(AsO_4)_2.8H_2O$	
Goslarite	$ZnSO_4.7H_2O$	
Zincaluminite	$Zn_3Al_3SO_4(OH)_{13}.2\frac{1}{2}H_2O$	

Geochemistry

As noted above, zinc is predominantly a thiophile element and its occurrence in the silicate shells of the Earth is in a highly dispersed form. On the other hand zinc ions are readily extracted by aqueous media and hydrothermal deposits very commonly contain the element. The primary mineral is invariably the sulphide sphalerite and it is interesting to note that in spite of marked difference between the solubility products of ZnS and PbS, the sulphides of zinc and lead are very commonly found in close association in Nature. In fact, nearly all important lead-producing regions also produce zinc and there is often quite intimate admixture of the two sulphides. Formerly such mixed ores were looked on with disfavour, but with the development of such methods as the blast-furnace process for smelting, it is now possible to produce both metals simultaneously from the same low-grade ore. The conditions under which the two sulphides have been deposited in Nature probably differ considerably from those employed in chemical laboratory techniques, where the controlling factor is largely that of pH value.

Although zinc blende or sphalerite is virtually the only sulphide ore of zinc, it nearly always contains iron and the amount may be quite considerable. Zinc is only a very minor constituent of other sulphide ores. In the variety of its oxidation products zinc shows some resemblance to copper, but the number of species is smaller.

In the weathering of rocks the zinc would pass into solution and would subsequently be precipitated in various forms such as carbonates, silicates, phosphates, etc. Most of these compounds are, however, amenable to acid attack, for example, in soils. What is called ‘ available zinc ’ in soils averages 6–10 p.p.m.

It is known that zinc accumulates in bottom muds or sapropels in the form of sulphide and a few of the commercial ore deposits have apparently originated in this way.

Biological Aspects

It has been known for many years that zinc is an essential trace element for plants. Effects of zinc deficiency are recognized by such diseases as ‘ pecan rosette ’, ‘ little leaf ’, and ‘ mottle leaf ’ in fruit trees. It is known that zinc is present in the granum protein of chloroplasts. Zinc is also present in the prosthetic group of several enzymes. Zinc seems to be necessary for at least some marine animals including coelenterates and actinians. It is also important for the growth and breeding of oysters.

Zinc is a functional component of the enzyme carboxypeptidase and is also present in the hormone insulin. It occurs in blood leukocytes. Intracellular zinc is bound to protein. The highest known concentration of zinc in living matter is the melanin-pigmental tissue of the eyes of certain fresh-water fish.

Production

Workable deposits of zinc ores are very widespread and are to be found all over the world and are mainly of hydrothermal origin. Very frequently they occur along with lead ores and the two are mined together. A selection of some of the well-known deposits is given below.

Joplin, Missouri, U.S.A. Hydrothermal cavity and replacement deposits in limestone from unknown magmatic source. Blende with subordinate base-metal sulphides.

Franklin Furnace, New Jersey, U.S.A. Possibly an oxidized hydrothermal deposit subsequently metamorphosed. They are mainly oxide ores.

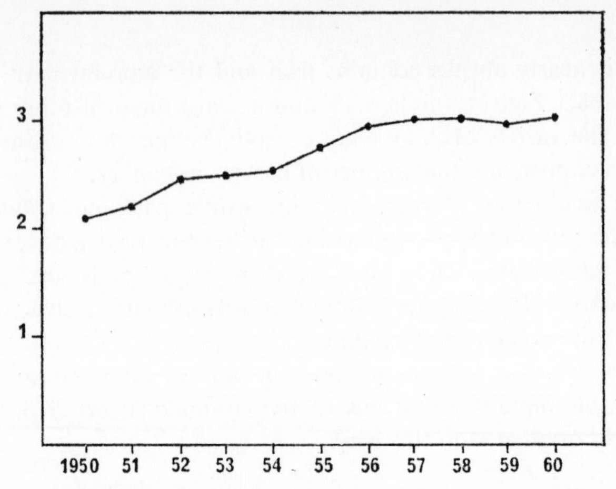

FIG. 11.10. World production of zinc (metal content of ore), in millions of tons.

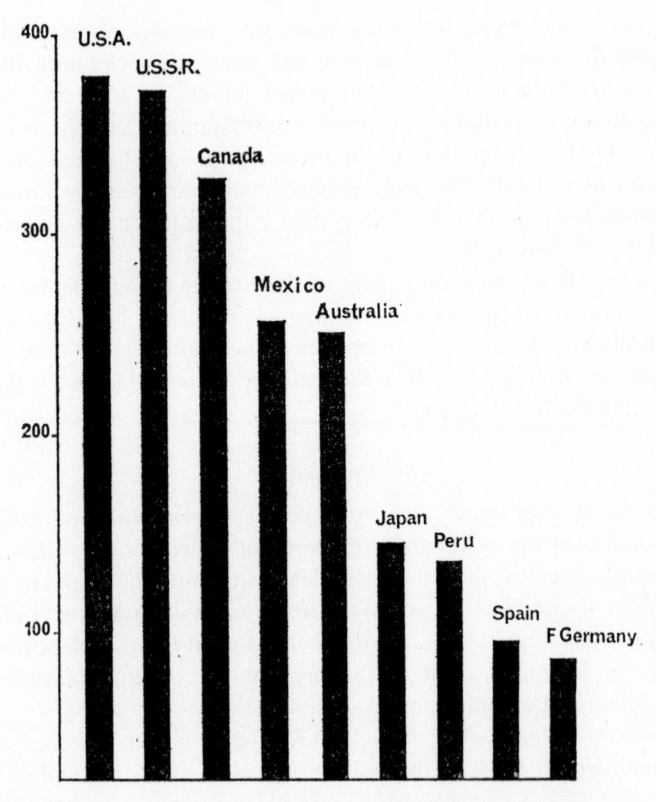

FIG. 11.11. Chief producers of zinc in 1959, in thousands of tons.

Sullivan Mine, Kimberly, B.C., Canada. High-temperature hydrothermal replacement in pre-Cambrian quartzite. Multiple sulphides include blende, galena, and pyrite.

Mexico. Mostly irregular replacements and fissure veins in limestone.

Broken Hill, N.S.W., Australia. High-temperature replacement deposits with metamorphism in pre-Cambrian schists. Lead–zinc sulphides with zinc spinel, magnetite, fluorite, etc.

Mount Isa, Queensland, Australia. Bedded sulphide deposits in black shales derived from bottom muds. Zinc and other base-metal sulphides.

Silesia, Poland, East Germany. Bedded deposits in Permian dolomites. Mainly sulphides. Replacement and cavity fillings by low-temperature hydrothermal solutions.

Sardinia, Italy. High-temperature veins in schists with granite. Blende with gangue of barytes, etc.

Alston Moor, Cumberland, England. A former considerable producer and the deposits are probably not exhausted. Zinc and lead sulphides with gangue of barytes, calcite, fluor, etc., from veins and flats in Carboniferous limestone.

Cadmium

Eight stable isotopes of cadmium exist in Nature:

^{106}Cd = 1·215%	^{112}Cd = 24·07%
^{108}Cd = 0·875%	^{113}Cd = 12·26%
^{110}Cd = 12·39%	^{114}Cd = 28·86%
^{111}Cd = 12·75%	^{116}Cd = 7·58%

These values are probably connected with the rather flat portion of the curve for neutron capture in nucleosynthesis between mass numbers 110 and 120, which indicates that if the various isotopes have originated in this way their relative abundances would be within the range 20 : 1. The actual ratio between the highest and the lowest is in fact 33 : 1. The most interesting nuclear-chemical aspect of cadmium is the very high thermal neutron cross-section of ^{113}Cd which means that it would be possible for this isotope to absorb neutrons from the natural fission of uranium.

Geochemistry

With an estimated abundance of 0·2g/ton for the Earth's crust, cadmium may be considered one of the rarer elements. It is, however,

fairly accessible, mainly as a by-product. Independent cadmium minerals are little more than mineralogical curiosities. The most important are

Příbramite	(Zn,Cd)S
Greenockite	CdS
Cadmium hausmannite	$CdMn_2O_4$
Otavite	$CdCO_3$

Cadmium is mainly distributed in the Earth as a ' guest ' ion or atom. The tetrahedral covalent radius of cadmium is only 13 % greater than that of zinc and it can therefore enter the sphalerite structures in which zinc sulphide crystallizes. On the other hand, the cadmium ion being 31 % larger than the zinc ion, in oxidized compounds it is unlikely that cadmium would be associated with zinc. It is more probable that it would be found in calcium minerals, but the degree of dispersal is very high in such cases.

There is no definite indication that cadmium has any biological significance. It is known to possess toxic properties.

Cadmium is produced practically as a by-product of zinc smelting. The cadmium content of ores ranges from 0·10 % in some North American ores to as much as 1·5 % in concentrates from South West

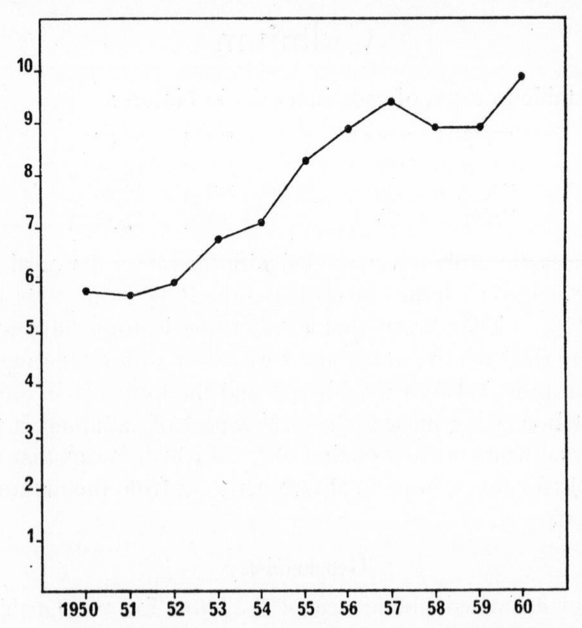

FIG. 11.12. World production of cadmium, in thousands of tons.

Africa. The recovery of cadmium ranges from 4·4 to 18 lb/ton of zinc, depending on the source of the ore.

Mercury

Seven stable isotopes of mercury exist in Nature:

^{196}Hg = 0·146%	^{201}Hg = 13·22%
^{198}Hg = 10·02%	^{202}Hg = 29·80%
^{199}Hg = 16·84%	^{204}Hg = 6·85%
^{200}Hg = 23·13%	

There is no very certain information as to isotopic variations under natural conditions but with such a volatile element it is possible that fractionation by diffusion or distillation might well take place under suitable circumstances in the Earth.

Abundance and Geochemistry

Mercury is relatively abundant for a heavy element but it must nevertheless be considered among the scarcer of the elements, the estimated value for the Earth's crust of about 0·5 g/ton being about the same as that of silver. Mercury shows a much higher abundance in the sedimentary layers of the crust than it does in igneous rocks generally, and it seems certain that mercury has reached the surface layers by processes other than weathering. It would seem that the element is particularly mobile under magmatic and hydrothermal conditions. According to Goldschmidt, mercury compounds in magmas would in general be reduced to the metal by ferrous ions. The highly volatile metal would then pass readily upwards where at a later stage it would be mainly fixed by combination with sulphur.

Concentrations of mercury in Nature are extremely few but they may nevertheless be very rich. It seems probable that very special conditions are necessary in order to immobilize mercury in the form of its usual sulphide ore. It has been suggested that silica may act as a precipitant on alkaline solutions of sulpho-salts of mercury and the presence of bituminous material has also been postulated as a reducing and sulphurizing agent.

The actual formation of mercury minerals from thermal waters in volcanic regions of California has been observed. It is also probable that a mercury deposit once formed may be rather easily removed again by hydrothermal and weathering processes.

There is little or no indication that mercury is of any importance in biological processes. Like most of the heavier elements it is of a toxic character.

MERCURY MINERALS

Name	*Formula*
Mercury	Hg
Amalgam	Ag,Hg
Cinnabar	HgS
Tiemannite	HgSe
Coloradoite	HgTe
Livingstonite	$HgSb_4S_7$
Montroydite	HgO
Calomel	Hg_2Cl_2

Production

In general, mercury deposits have been formed under low-temperature hydrothermal conditions. They occur in any kind of rock which has been fractured and are particularly associated with late Tertiary vulcanism. Cinnabar is the most important ore with much smaller amounts of the native metal, the other minerals being relatively unimportant.

Some of the more important producing localities are indicated below.

Almaden, Spain. These famous deposits, worked since ancient times, are replacements in Silurian quartzites. They contain cinnabar and native mercury.

Monte Amiata, Tuscany, Italy. Hydrothermal fillings in late Pliocene sediments and volcanics, contain cinnabar with some stibnite.

New Idria, California, U.S.A. Hydrothermal veins in fault fissures. Cinnabar with pyrite and quartz.

The two most important producers are Almaden and Monte Amiata, both of which have a history extending back to Roman times. The low-grade deposits at Terlingua, Texas, U.S.A., are interesting for the variety of mercury compounds they contain.

Mercury is not much in evidence in Great Britain. According to Goldschmidt, however, mercury was found as a deposit in condensing units in a British gas works and it may have been derived from the coal used.

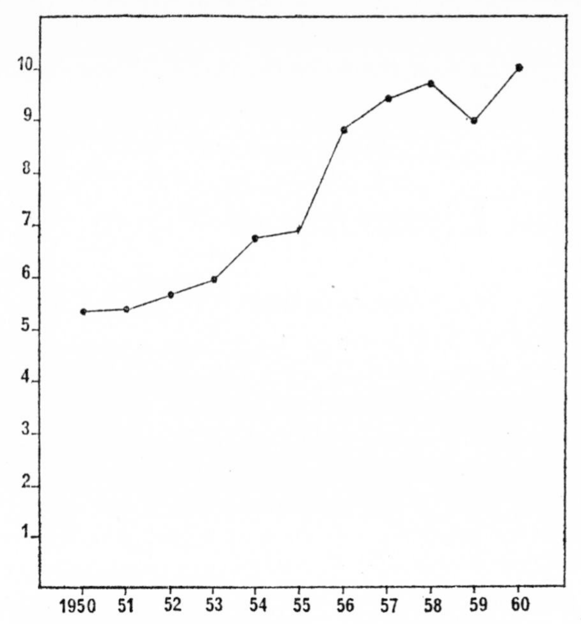

FIG. 11.13. World production of mercury, in thousands of tons.

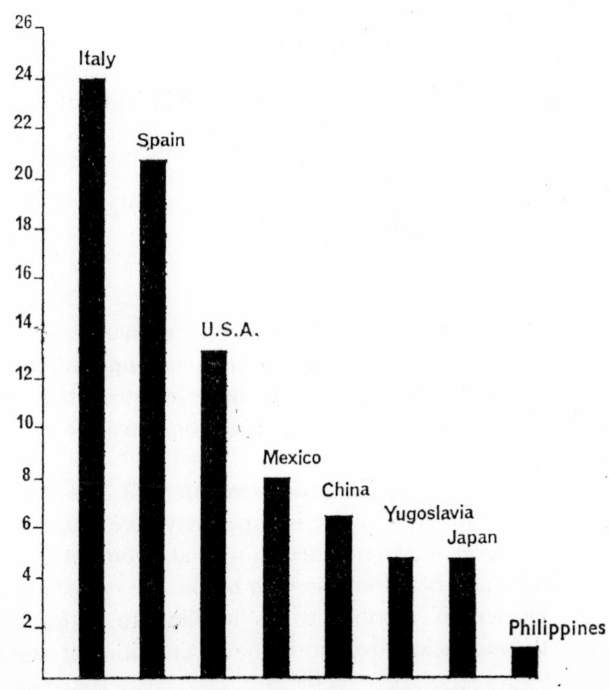

FIG. 11.14. Chief producers of mercury, 1959, in hundreds of tons.

CHAPTER 12

The Elements of Group III

General

In this chapter we shall deal with the elements of Group III with the exception of the so-called ' rare earth ' elements with which we shall include scandium and yttrium. The elements to be considered will therefore be boron, aluminium, gallium, indium, and thallium.

On the whole the geochemical aspects of these elements present less variation than those of Group II—there is a less clearly marked division between lithophile and thiophile types. The relative abundances are indicated opposite. The three rarer metals of the group are found in general in a very high degree of dispersal whereas aluminium is the most abundant metal in the crust of the Earth.

Boron

There are two stable isotopes of this element in Nature:

$$^{10}B = 18{\cdot}98\% \qquad ^{11}B = 81{\cdot}02\%$$

The lighter of the two isotopes can absorb slow neutrons readily and it may be supposed that the reaction

$$^{10}B + n = {}^{7}Li + \alpha$$

can take place to a slight extent where boron compounds occur in the presence of suitable neutron sources such as radium with beryllium. The extent of this reaction in Nature is, however, probably much too small for it to have any appreciable effect on the ratio of the two isotopes.

On account of the relatively large difference in mass, it is possible that some degree of fractionation of the isotopes may take place in Nature. This has not, however, yet been detected with any certainty.

As with lithium and beryllium, so with boron, the rather low abundance of the element is ascribed to its inability to withstand high-temperature thermonuclear reactions in the interiors of stars. The

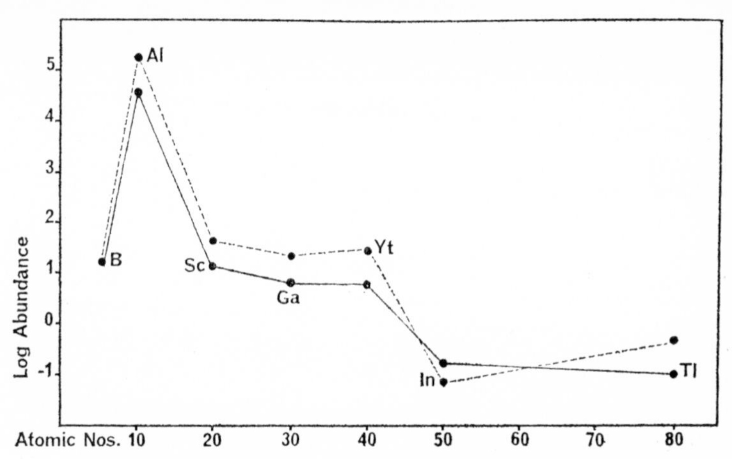

FIG. 12.1. Relative abundance of the elements of Group III, excluding the lanthanides. Dotted line, cosmic. Continuous line, terrestrial.

mode of its origin is thought to be similar to that of the other two—that is by spallation reactions on elements like nitrogen and oxygen in the hydrogen envelopes of supernovae.

Boron Minerals

These are fairly numerous and they are all, except two, compounds of oxygen and the exceptions are fluorine compounds. Boron is clearly a markedly oxyphile and lithophile element.

Geochemistry

There are in the main three aspects of the geochemistry of this element. They can be stated as follows:

(*a*) Occurrence in the late stages of magmatic crystallization.
(*b*) Occurrence in volcanic emanations.
(*c*) Occurrence in evaporite deposits.

(*a*) Boron is an oxyphile and lithophile element and it is characterized by small atomic and ionic dimensions. The radius of the atom for tetrahedral co-ordination is 0·89 Å and the radius of the ion is only 0·20 Å. The corresponding values for silicon are 1·17 and 0·41 Å, respectively and the values are of some significance. Boron has very little tendency

BORON MINERALS

Name	*Formula*	*Occurrence*
Avogadrite	KBF_4	Volcanic deposits
Ferruccite	$NaBF_4$	
Sassolite	$B(OH)_3$	
Kernite	$Na_2B_4O_7.4H_2O$	Evaporites
Borax	$Na_2B_4O_7.10H_2O$	
Kaliborite	$KMg_2B_{11}O_{19}.15H_2O$	
Colemanite	$Ca_2B_6O_{11}.5H_2O$	
Ulexite	$NaCa_5B_5O_9.8H_2O$	
Eremeyerite	$AlBO_3$	Granitic rocks
Sinhalite	$MgAlBO_4$	
Ludwigite	$(Mg,Fe)_2Fe^{3+},BO_5$	Metamorphic rocks
Fluorborite	$Mg_3BO_3(F,OH)_3$	Evaporites
Boracite	$Mg_6B_{14}O_{26}Cl_2$	
Searlesite	$NaBSi_2O_6.H_2O$	
Danburite	$CaB_2Si_2O_8$	Rock minerals
Dumortierite	$(Al,Fe)_7BSi_3O_{18}$	
Leucosphenite	$Na_3CaBaBTi_3Si_9O_{29}$	
Tourmaline	$(Na,Ca)(Li,Mg,Fe,Al)_3(Al,Fe^{3+})_6B_3Si_6O_{27}(O,OH,F)_4$	

to enter the silicate lattices of common rock-forming minerals and therefore it tends to remain until the last stages of magmatic crystallization.

The most abundant and virtually the only important boron mineral in rocks is tourmaline, and as noted above it is a very complex silicate with considerable variation in ionic content. Tourmaline typically crystallizes in granites and granite–pegmatites and as a result of the weathering of these rocks it provides the small but necessary boron content of soils and sediments. It appears that boron can replace silicon in a few silicate minerals apart from tourmaline and in the case of danburite and datolite there are discrete BO_4 tetrahedra in a framework of SiO_4 tetrahedra. There is no doubt that the great part of the boron of the Earth's crust is concealed in silicate minerals but as boron is much less readily accepted into silicate lattices than aluminium it tends to separate in the later stages of magmatic crystallization.

(*b*) The presence of boron in volcanic gases and waters is rather common and it seems that it is closely related to the presence of a volatile boron compound which reacts with water and various rock minerals before it can be actually emitted at the surface of the Earth.

The fact that the commonest boron mineral, tourmaline, contains fluorine and that several fluorborates are found near volcanic vents points strongly to the compound being boron trifluoride BF_3. This is further confirmed by the presence of volatile boric acid in fumarolic gases and in some localities the presence also of hydrogen fluoride—*e.g.*, Valley of Ten Thousand Smokes, Alaska. It is probable that the boron content of sea water, of various saline lakes and of the solid saline deposits may be related to the volcanic emanations of boron compounds.

(*c*) The evaporite deposits of boron compounds are encountered in several parts of the world; they are mostly derived from lake waters which have received volcanic products as noted above.

Although sea water contains a perceptible amount of borate the oceanic salt deposits in general have not provided useful sources of the element at the present time.

Biological Aspects

Boron was early recognized as one of the essential trace elements for the growth of land plants. It is known also that an excess of the element in soils can be harmful. Disorders due to boron deficiency include ' brown heart ' of vegetables and ' dry rot ' of sugar beet.

The functions of boron as a trace element in animals are not understood at present but small amounts have been found in many organisms, particularly those with calcareous skeletons.

Production of Boron Compounds

A very high proportion of the world output of boron minerals comes from North America and almost entirely from localities in California. The deposits are of three kinds:

(*a*) Bedded deposits beneath old playas.
(*b*) Brines of saline lakes and marshes.
(*c*) Encrustations around playas and lakes.

The bedded deposits near Kramer and Searle's Lake in California consist of borax and kernite with subsidiary ulexite and colemanite. Large amounts of borax with other salts of sodium, potassium, and lithium are extracted from the brines of Searle's Lake.

The Italian output which is now small is from fumarolic steam vents which carry boric acid. They occur in the volcanic region of Tuscany.

There are no useful quantities of boron compounds in Great Britain. Tourmaline in granites, particularly in Cornwall, is common but is of no practical value as a source of the element.

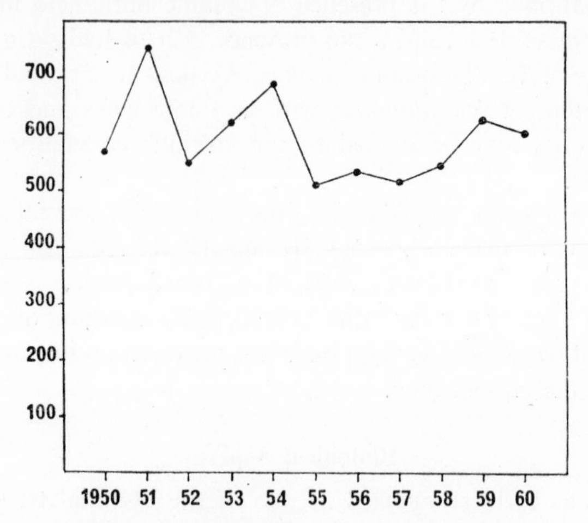

FIG. 12.2. World production of boron minerals, in thousands of tons.

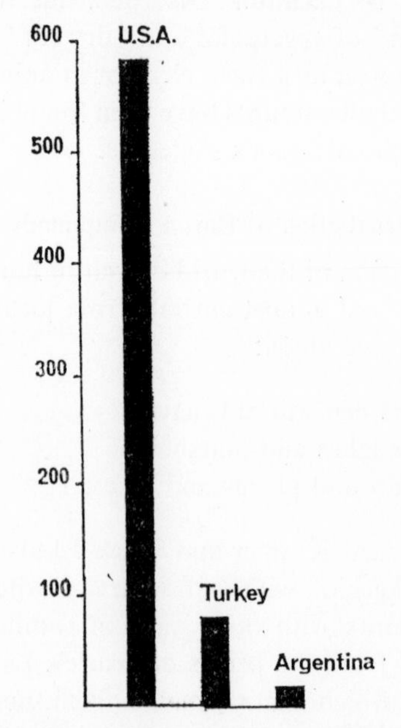

FIG. 12.3. Chief producers of boron minerals, 1959, in thousands of tons.

Aluminium

There is only one stable nuclide, ^{27}Al, existing in Nature. The relative abundance of aluminium in the Earth's crust and in the Cosmos is about 10 : 1 which reflects, no doubt, the high concentration of the element in the outer shells of our planet. The synthesis of the aluminium nucleus may well occur in the later stages of the α-process in ' white dwarf ' stars but a precise mechanism has not yet been proposed.

Abundance

In the Earth's crust aluminium is the dominant metal; in the mantle its place is taken by magnesium. The average aluminium content (by weight) of igneous rocks, excluding ultrabasics, is approximately 8 %, about one-third of the silicon content. The value for the crust as a whole is 81 300 g/ton and the estimated value for the whole Earth is 0·44 % only.

Geochemistry

Aluminium is not only abundant but widespread throughout the crust. This is no doubt partly due to the fact that the aluminium ion can exist, under different circumstances, in both four-fold and six-fold co-ordination. It is therefore able to enter into a great range of silicate structures as well as forming a variety of other oxygen compounds.

There is also an interesting group of fluorine compounds of aluminium. As might be expected for so strongly lithophile an element, aluminium does not form any terrestrial sulphide minerals, nor do they exist in meteorites.

The ionic radius of aluminium is 0·51 Å for six-fold co-ordination which compares with 0·64 for Fe^{3+}, 0·63 for Cr^{3+}, and 0·41 for Si^{4+}. In minerals where aluminium is in this co-ordination it is usually possible for diadochy to take place, particularly with ferric iron and chromium. For tetrahedral four-fold co-ordination the aluminium functions as the centre of an anionic AlO_4 group and in effect is deputizing for silicon. It is thus possible for aluminium to enter both cationic and anionic structures and both of these may sometimes exist in the same mineral. The following examples may be noted:

Corundum	Al_2O_3	Six-fold co-ordination—Al replaceable by Fe^{3+} or Cr^{3+}
Spinel	$MgAl_2O_4$	

ALUMINIUM MINERALS

Name	*Formula*	*Occurrence*
Corundum	Al_2O_3	Metamorphic rocks
Diaspore	$AlOOH$	Hydrolysate sediments
Boehmite	$AlOOH$	
Gibbsite	$Al(OH)_3$	
Spinel	$MgAl_2O_4$	Metamorphic rocks
Fluellite	$3AlF_3.4H_2O$	Volcanic emanations
Cryolite	Na_3AlF_6	Granite pegmatites
Chiolite	$Na_5Al_3F_{14}$	
Elpasolite	K_2NaAlF_6	
Weberite	Na_2MgAlF_7	
Eremeyerite	$AlBO_3$	L.T. hydrothermal veins
Dawsonite	$NaAlCO_3(OH)_2$	
Alumohydrocalcite	$CaAl_2(CO_3)_2(OH)_4.3H_2O$	
Andalusite, Kyanite, Sillimanite	Al_2SiO_5	Metamorphic rocks
Mullite	$Al_9Si_3O_{19\frac{1}{2}}$	
Pyrophyllite	$Al_2Si_4O_{10}(OH)_2$	
Dickite, Kaolinite, Nacrite	$Al_2Si_2O_5(OH)_4$	Clay minerals
Nepheline	$NaAlSiO_4$	Igneous rocks
Albite	$NaAlSi_3O_8$	
Orthoclase	$KAlSi_3O_8$	
Microcline	$KAlSi_3O_8$	
Analcime	$NaAlSi_2O_6.H_2O$	Zeolitic
Muscovite	$KAl_3Si_3O_{10}(OH)_2$	Igneous rocks
Phlogopite	$KMg_3AlSi_3O_{10}(OH)_2$	
Anorthite	$CaAl_2Si_2O_8$	
Almandine	$Fe_3^{3+}Al_2Si_3O_{12}$	Metamorphic rocks
Cordierite	$(MgFe^{2+})_2Al_4Si_5O_{18}$	
Topaz	$Al_2SiO_4(OH,F)_2$	Igneous–pneumatolytic weathered rocks
Berlinite	$AlPO_4$	Metamorphic rocks
Variscite	$AlPO_4.2H_2O$	
Wavellite	$Al_6(PO_4)_4(OH)_6.9H_2O$	Metamorphic rocks
Turquoise	$CuAl_6(PO_4)_4(OH)_8.5H_2O$	Altered aluminous rocks
Alunogen	$Al_2(SO_4)_3.18H_2O$	Altered aluminous rocks
Aluminite	$Al_2(SO_4)(OH)_4.7H_2O$	
Alum	$KAl(SO_4)_2.12H_2O$	
Kalinite	$KAl(SO_4)_2.11H_2O$	
Alunite	$KAl_3(SO_4)_2(OH)_6$	

Orthoclase	$KAlSi_3O_8$	Four-fold co-ordination—alumino-silicate ion
Anorthite	$CaAl_2Si_2O_8$	
Kyanite	Al_2SiO_5	Half the Al ions four-fold, half the Al ions six-fold

Undoubtedly the most important rock minerals are the feldspars and they have structures in which up to one-half of the tetrahedral positions in the framework are occupied by Al and positive ions are present in sufficient number to neutralize the negative charge of the $(Si,Al)O_2$ framework. As the feldspars are the most abundant of the essential components of igneous rocks (excepting the ultrabasic rocks) it will be understood that by far the largest amount of aluminium in the Earth's crust is contained in these minerals.

Two other classes of aluminosilicates should be mentioned, although of less importance than the feldspars. These are the feldspathoids and the zeolites. Both types have the same basic $(SiAl)O_2$ framework as the feldspars but it is more open in structure. The feldspathoids, of which examples are nepheline $NaAlSiO_4$ and leucite $KAlSi_2O_6$, are readily attacked by acids with the removal of the aluminium and the cations, leaving gelatinous silica. They are therefore rather susceptible to weathering processes.

The zeolites are hydrous minerals of very open structure, an example being analcime $NaAlSi_2O_6.H_2O$. They readily lose water when heated and have exchangeable cations. The zeolites are secondary products often found in the cavities of igneous rocks.

In the formation of the crust of the Earth it is clear that aluminium is associated with the main crystallization, but a certain amount remains to the later stages so that feldspars, micas, and some other alumino-silicates are often found well crystallized in pegmatite dykes and veins. Aluminium is, however, conspicuous by its paucity or absence in the pneumatolytic and hydrothermal stages. A special aspect of aluminous pegmatitic minerals is the occurrence of cryolite and related alumino-fluorides at Ivigtut in Greenland and at one or two other localities. It would appear that the conditions for the formation of such compounds in Nature are rather uncommon.

Weathering of Aluminous Minerals

The action of meteoric and hydrothermal waters on the primary minerals of aluminium may take place in several ways. These are indicated below.

(*a*) The mineral is dissolved, the cations leached out and the aluminium and silicon undergo hydrolysis and re-combination to yield aluminium silicate clay minerals of various kinds. This process gives rise to most of the common clays and under special conditions where granite has been pneumatolysed in depth, a massive china clay rock is formed. This process gave rise to the important china clay deposits in Cornwall, England.

(*b*) In tropical and sub-tropical climates of heavy rainfall the rate of weathering of rocks is high and most of the products formed do not remain *in situ*. Within the pH range 5–9 aluminium and ferric hydroxides are less soluble than silica and consequently deposits of the two basic hydrated oxides are formed, largely free of silica. These are known as laterites and if low in iron may actually be valuable deposits of the mixture of hydrated aluminium oxides called bauxite.

There is a marked differentiation in sedimentation so that arenaceous (sandy) and calcareous (limestone) rocks may be almost free of aluminium whereas the argillaceous rocks (clays, shales, slates) may be relatively rich in the element.

The rather uncommon argillaceous limestones are sometimes altered by weathering so as to remove the calcium carbonate. The residue may then become available as a form of bauxite deposit.

Main Products.

Illite	$(H_3O,K)_4Al_8(Si,Al)_{16}O_{40}(OH)_8$
Kaolinite	$Al_2Si_2O_5(OH)_4$
Montmorillonite	approx. $(Al,Mg)_2Si_4O_{10}(OH)_2nH_2O$
Chlorite	$(Mg,Fe,Al)_6Si_4O_{10}(OH)_8$
Vermiculite	$Mg_3Si_4O_{10}(OH)_2.nH_2O$

Metamorphism of Aluminium Minerals

The most interesting and important cases are those involving the thermal alteration of aluminous sediments.

The baking of a clay–slate or shale may often produce a sintered stony product, hornfels, which is analogous to a ceramic material. Further action may lead to the formation of such minerals as andalusite, sillimanite, kyanite, and cordierite. Extreme thermal action may give rise to the mineral mullite and even to silicate glasses. In the absence of silica there may be formed crystalline alumina or corundum. This mineral in transparent coloured varieties yields the gem stones ruby and sapphire. The colour of the former is ascribed to chromium and of the latter to cobalt and titanium. Impure corundum, containing magnetite, etc., constitutes the abrasive called emery.

The metamorphism of aluminous and magnesian minerals may give rise to spinel $MgAl_2O_4$ which may also sometimes constitute a gem stone.

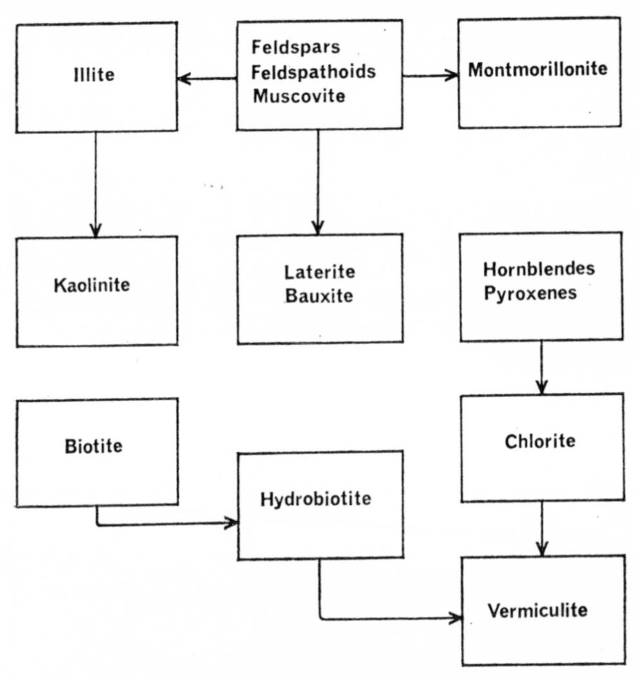

FIG. 12.4. Weathering of aluminous minerals.

The oxidation and weathering under hydrous conditions of aluminous shales containing pyrite may give rise to the formation of sulphuric acid and subsequently of several aluminium sulphate minerals, including alum if potassium ions are also present.

Corundum Minerals

Massive corundum occurs in South Africa in altered pegmatites and also as placer material. In Eastern Ontario, Canada, it is found in the pegmatites related to nepheline syenites and in the U.S.S.R. there are dykes of massive corundum associated with an anorthositic rock called kyschtynnite.

Pegmatites connected with syenites and schists in India yield ruby and sapphire and these gems are also found in placer and sedimentary deposits in Burma, Ceylon, as well as various Indian localities, and in Queensland, Australia.

In Great Britain there are small mineralogical occurrences of blue corundum—sapphire of indifferent quality—in the Isle of Mull and Ardnamurchan, Scotland. They are found in xenoliths embedded in Tertiary igneous rocks.

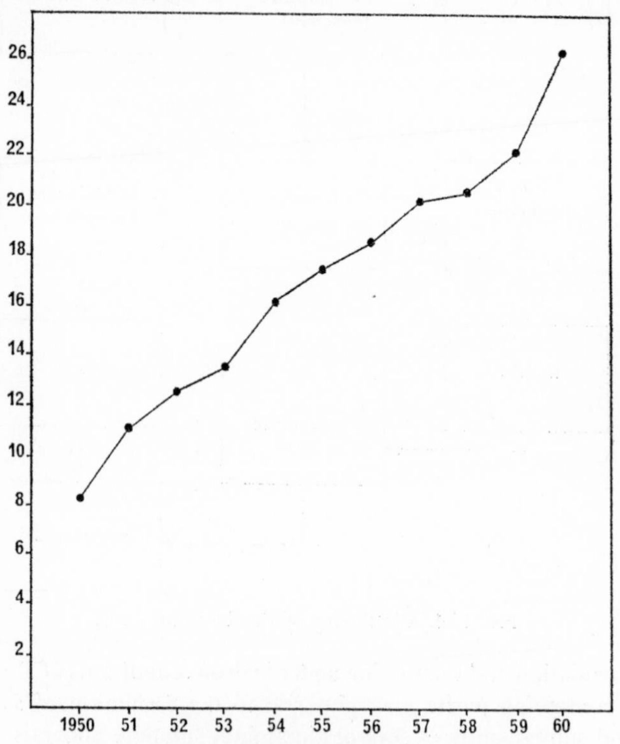

FIG. 12.5. World production of bauxite, in millions of tons.

Aluminium Silicate Minerals

The most important of these are either unaltered feldspars from pegmatites or similar material from nepheline syenites, or occasionally massive andalusite from metamorphic rocks.

The other form is the highly altered weathered product which constitutes china clay or china stone. These materials are mainly of interest, not as sources of aluminium but as raw materials for the ceramic industry.

Production of Aluminium Minerals

The production of natural compounds of aluminium is carried out for such a variety of purposes, and only a relatively small amount of the material is to be regarded as an ore of the metal, that it will be best to consider the topic under several headings.

(1) Aluminium ore. Bauxite is at present the only ore of importance. It is probable that ultimately some of the silicate minerals will have to be used but in the meantime the world bauxite reserves are fairly considerable. The material known under this name is in fact a mixture of

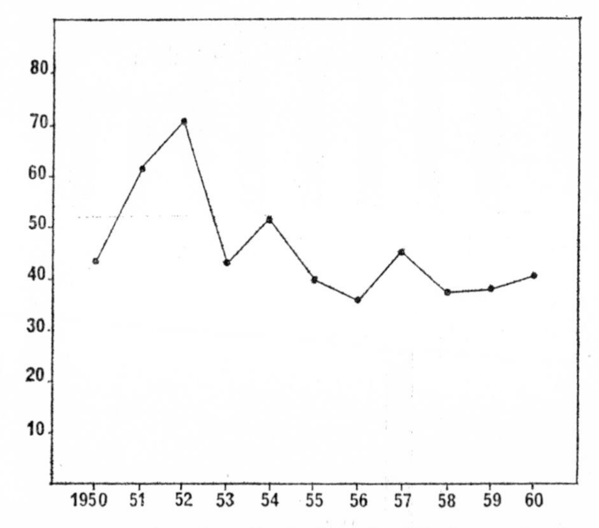

FIG. 12.6. Production of cryolite in Greenland, in thousands of tons.

the three hydrated oxide minerals, gibbsite, boehmite, and diaspore in variable proportions with more or less impurity in the form of clay, iron oxide, and silica, etc. All deposits result from residual weathering and occur as (*a*) blankets at or near the surface, (*b*) interstratified deposits laid down unconformably, (*c*) pockets in limestone, and (*d*) transported deposits.

(2) Corundum minerals. The chief uses of these materials is as abrasives and gem stones. They have been referred to above.

(3) Aluminium sulphate minerals. There is a limited output of these, mainly in the form of the mineral alunite from which alum can be made. The chief producer of a few thousand tons annually is Japan. Production in other countries such as Spain and Italy is sporadic.

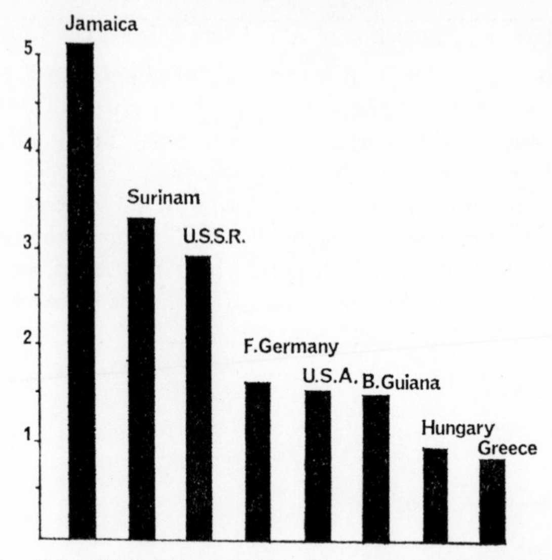

FIG. 12.7. Chief producers of bauxite, 1959, in millions of tons.

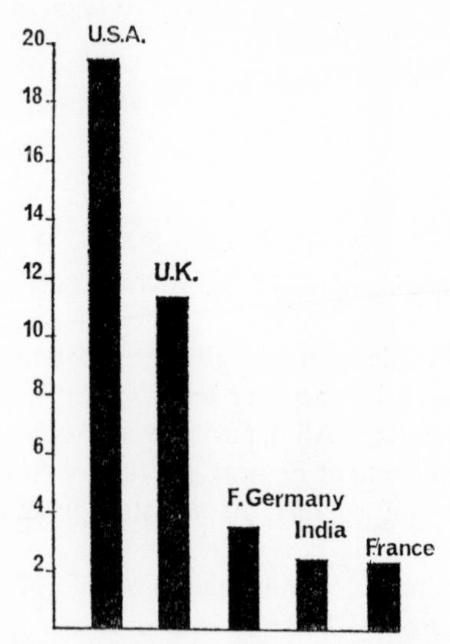

FIG. 12.8. Chief producers of china clay, 1959, in hundreds of thousands of tons.

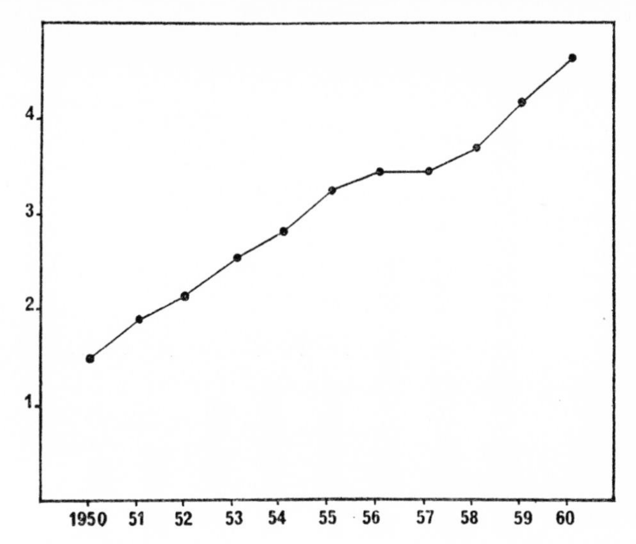

FIG. 12.9. World production of aluminium metal, in millions of tons.

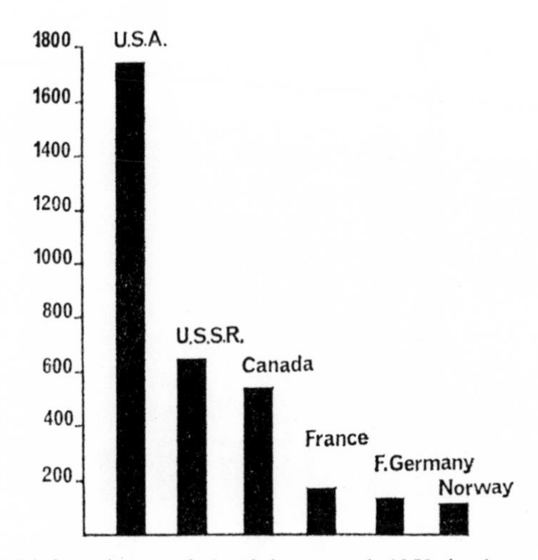

FIG. 12.10. Chief producers of aluminium metal, 1959, in thousands of tons.

Biogeochemistry

Following an exhaustive study by Hutchinson (1943) it appears that aluminium is probably not a generally essential element for plants and animals but there are certain species that accumulate it. It is not clear

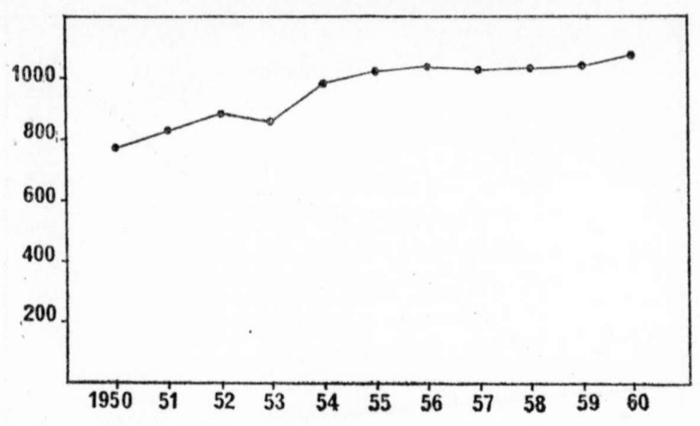

FIG. 12.11. World production of feldspar and china stone, in thousands of tons.

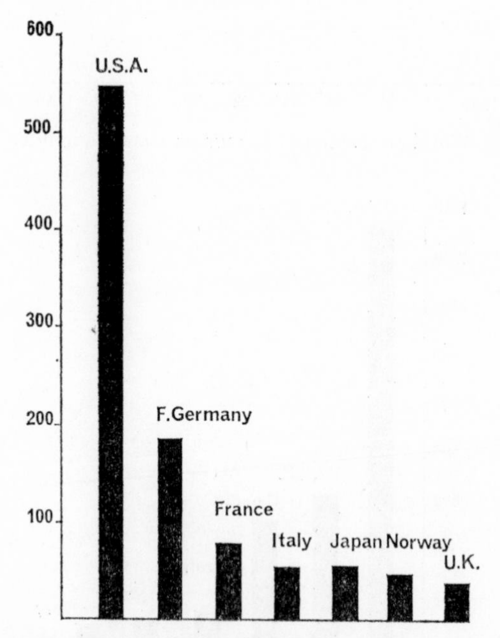

FIG. 12.12. Chief producers of feldspar and china stone, 1959.

that it performs any specific function in such species. It has been stated that certain lichens are able to extract aluminium from the rock surfaces on which they grow. The fact that some of them can behave as natural mordant dyes may be cited as evidence for the presence of aluminium colour-lakes in them. If this is so they would appear to be the natural counterpart of the ‘ metallized ’ dyestuffs widely used in the textile industry.

The aluminium lake of an anthocyanidin is the cause of the blue colour of certain fruits and of hydrangea flowers. The ‘ blueing ’ of hydrangeas, said to have been discovered by a gardener at the Imperial Russian gardens, can often be brought about by making the soil acid around the plant, which is done by addition of alum or ferrous sulphate. Iron is not needed but the soil requires to be within the rather narrow pH range within which Al^{3+} ions are mobile.

Plants known to be aluminium accumulators include *Symplocus*, *Orites*, and *Lycopodium*. The ash of such plants may contain up to 33 % Al_2O_3. The mineral mellite found in brown coal is the hydrated aluminium salt of mellitic acid (benzene hexacarboxylic acid) and may have been derived from organic aluminium in the original coal plants.

Gallium

There are two stable isotopes in Nature:

$$^{69}Ga = 60{\cdot}2\,\%$$
$$^{71}Ga = 39{\cdot}8\,\%$$

There do not appear to be any topics of special interest in connexion with the isotopy or natural nuclear chemistry of this element.

Geochemistry

In common with its neighbours in the Periodic Table, gallium shows siderophile, lithophile, and chalcophile properties. The fact that it is enriched in the metallic phase of meteorites suggests that the first is important; the cosmic abundance may therefore be expected to be greater than the terrestrial. From its position in the third periodic group gallium has an ionic radius and other electronic properties near to those of aluminium and so its lithophile character is the most obvious one. Gallium forms the classic example of a completely dispersed element. Its actual terrestrial abundance of about 15g/ton is by no means insignificant, yet it does not form a single independent mineral. Essentially oxyphile and lithophile, it does, however, show a slight thiophile character in its occurrence in certain sulphide minerals and particularly in the rare species germanite where it may be present to the extent of 1·85 %. The ionic radius of Ga^{3+} is 0·62 Å which is somewhat greater than aluminium (0·51 Å). Gallium is therefore able to enter readily into the crystal lattices of most aluminium minerals, but not apparently into those of the cryolite group of aluminofluorides. Thus gallium is in the main a ' concealed ' element in most aluminium minerals and in particular in the important rock-forming species. In general gallium has the highest content in the most aluminous types of igneous rocks, reaching a maximum (*circa* 23 p.p.m.) in the intermediate rocks of about 57 % SiO_2 content. This point is brought out by the curves (drawn from various sources) of Fig. 12.13.

On account of similarity of ionic radius gallium is also found concealed in minerals containing ferric iron and chromium but its association with aluminium is more important. It is to be expected that the

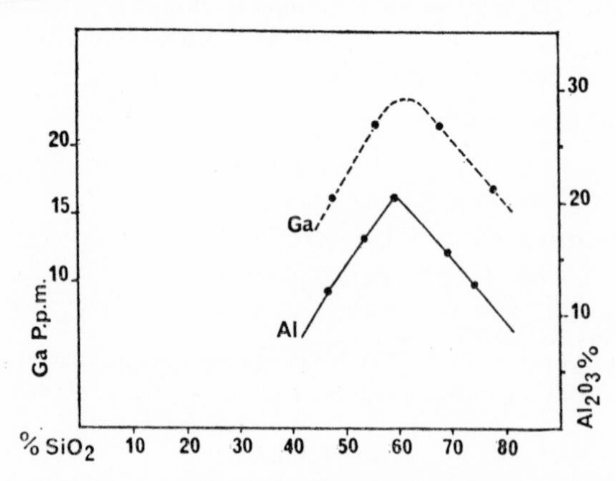

FIG. 12.13. Association of gallium and aluminium in rocks.

slightly larger ion of gallium (which can enter both six-fold and four-fold co-ordinated structures) would tend to be accepted at the later stages of magmatic crystallization of aluminium minerals. This is found to be so to some extent so that the earliest aluminium minerals have a low Ga : Al ratio (say, 1 : 100000) and later stages as in some nepheline syenites may have 1 : 6000 or more.

The pH range for precipitation of $Ga(OH)_3$ is sufficiently different from that of $Al(OH)_3$ to cause some degree of concentration of gallium in sedimentary hydrolysate minerals like bauxite where the Ga : Al ratio may be 1 : 800.

The occurrence of gallium in sulphide minerals led to its discovery and the chief example is provided by zinc blende. A mean value for the Ga content of blende is 45 p.p.m. or about three times that for igneous rocks. It is generally held that gallium is a minor constituent of blende, preferentially concentrated in the low-temperature varieties and evidence generally favours the view that it exists as a solid solution of either GaS or Ga_2S_3 in the mineral.

As previously stated the rare mineral germanite is the only natural material known to contain substantial amounts of gallium. Germanite is formulated $Cu_3(Ga,Ge,Fe,Zn)(As,S)_4$.

Biogeochemistry

There is some indication that gallium may be an essential element for certain plants as for example *Aspergillus niger* and it has been claimed as an element present in some marine organisms.

Sources

There are two main sources at present, these being:

(1) Flue dusts from which germanium is also extracted, the gallium being obtained from the residual liquors.

(2) The residual alkaline liquors from the process of purification of bauxite in aluminium manufacture, a recovery of Ga of as much as 1 oz/ton of bauxite being claimed.

The world production of gallium is uncertain but is unlikely to exceed more than a few tons per annum at present.

Indium

There are two stable isotopes of this element:

$$^{113}In = 4{\cdot}23\,\%$$
$$^{115}In = 95{\cdot}77\,\%$$

Following much inconclusive experimental work it now appears certain (1952) that ^{115}In is feebly radioactive, being a β-emitter with a very long half-life of 6×10^{14} years, decaying to ^{115}Sn. The possible natural radioactivity of ^{113}In is at present unconfirmed.

Geochemistry

Indium is a rare element with a crustal abundance of about 0·1 g/ton and its geochemical character is rather uncertain. On the whole it appears to be thiophile and is probably carried in the sulphide accessory minerals of igneous rocks, mainly chalcopyrite. There is some degree of accumulation of indium in pneumatolytic and hydrothermal minerals, particularly in cassiterite and in some sulphides and sulpho-salts, including sulpho-salts of tin, sphalerite, chalcopyrite, and galena. The reason for the association of indium with tin in cassiterite is obscure.

Indium certainly occurs in perceptible concentrations in zinc sulphide and up to 5000 p.p.m. have been reported. The author found 1000 p.p.m. in a pyromorphite from the oxidized zone of a lead–zinc vein in the northern Lake District, England. There is no definite information concerning the possible biogeochemistry of the element.

Production

At present the principal sources of indium are the flue dusts from lead and zinc smelting and the cadmium-bearing residues left from zinc sulphate purification.

Canada appears to be the chief producer but the returns are incomplete. The output in 1957 was about 12 tons of the metal.

Thallium

There are two stable isotopes of this element:

$$^{203}Tl = 29{\cdot}5\,\%$$
$$^{205}Tl = 70{\cdot}5\,\%$$

In addition several short-lived isotopes occur in Nature, *viz.*,

$$^{206}Tl, \quad ^{210}Tl, \quad ^{207}Tl, \quad \text{and} \quad ^{208}Tl$$

as decay products in the uranium, actinium, and thorium series of radioactive elements.

Geochemistry

Thallium is about ten times as abundant as indium in the Earth's crust at approximately 1 g/ton. Compared with indium, thallium shows a much stronger lithophile character and compared with gallium it shows regularity of distribution in igneous rocks. It appears that in general the Tl^{+} ion occurs in silicates and replaces potassium ($Tl^{+} = 1{\cdot}47$ Å, $K^{+} = 1{\cdot}33$ Å and $Rb^{+} = 1{\cdot}47$ Å). The thallium ion is evidently the same size as the rubidium ion and it is therefore not surprising that late pegmatitic potash feldspar and the mineral pollucite both contain relatively large amounts of the element (about 100 p.p.m.). Thallium also occurs in sulphides, particularly sphalerite from certain regions—*e.g.*, Europe, but not in the great U.S.A. deposits of the

Mississipi valley region, whereas gallium is absent from the European deposits and present in the American ones. Unlike gallium and indium, thallium does actually form a few independent minerals which are, however, rare and local.

THALLIUM MINERALS

Name	*Formula*	*Occurrence*
Crookesite	$(Cu,Tl,Ag)_2Se$	Skrikerum, Sweden, with other selenides
Lorandite	$TlAsS_2$	Allchar, Rozsdan, Salonika, Macedonia, with arsenic sulphides
Vrbaite	$Tl(As,Sb)_3S_5$	
Hutchinsonite	$(Tl,Pb)_2AgAs_5S_{10})$	Langenbach, Valais, Switzerland, with arsenic sulphide minerals

Most of the thallium in the above minerals is evidently in the monovalent state. Oxidation to the Tl^{3+} ion may occur in a suitable environment as with iron and manganese oxidate sediments. The biogeochemistry of thallium is little known. Thallium is known to be toxic to higher animals and may be used as a poison for vermin.

Production

The main production (which could probably be much increased if the demand existed) is as a by-product from furnace fume and flue dust from zinc and lead smelting. Chief producers are believed to be Belgium, France, Germany, U.S.S.R., and U.S.A. The world output may be of the order of 30–35 tons per annum.

CHAPTER 13

Carbon

There are two stable isotopes of this element in Nature:

$$^{12}C = 98{\cdot}892\,\%$$
$$^{13}C = \;\;1{\cdot}108\,\%$$

In addition, radioactive carbon ^{14}C, with a half-life of 5568 years, also occurs terrestrially.

The ratio of $^{12}C/^{13}C$ is in fact variable and the variations are of considerable significance. It ranges from about 94 : 1 for certain terrestrial material to a possible 5 : 1 for some stellar interiors. There are two main causes of these variations, the first of which applies cosmically and the second terrestrially.

(1) Nuclear reactions under different stellar conditions and different stages of evolution. The nuclear chemistry of carbon is closely connected with the origin of the element. The basic reaction for the synthesis of carbon atoms may be, as suggested by Hoyle, the ' ramming together ' of three α-particles to give an excited $^{12*}C$ nucleus which can decay to normal ^{12}C.

This reaction should occur in the hot dense cores of certain types of ' helium burning ' stars. If a star is very massive it is thought that ^{12}C will be the main product but in less massive stars further reactions lead to ^{16}O and ^{20}Ne. The isotope ^{13}C is involved in the so-called ' carbon cycle ' which is postulated as one of the ' hydrogen burning ' reactions. Proton capture reactions can build up to ^{15}N which breaks down:

$$^{15}N + {}^{1}H = {}^{12}C + {}^{4}He$$

^{13}C is formed in this chain of reactions but the extent to which the chain is completed is related to temperature and therefore the ratio of ^{12}C to ^{13}C finally present varies for different types of stars. Evidently in the long run the conditions will not favour the retention of a high proportion of ^{13}C; furthermore it can be lost by such reactions as

$$^{13}C + \alpha = {}^{16}O + n$$

which is thought to be one of the sources of neutrons in stellar nuclear processes.

The isotope ^{14}C does not seem to be permanently present in stars. It is, however, of great interest as a terrestrial nuclide formed by cosmic ray neutron capture in the reaction

$$^{14}N + n = {}^{14}C + {}^{1}H$$

At the same time some of the neutrons react:

$$^{14}N + n = {}^{12}C + {}^{3}H$$

These reactions take place in the atmospheric zone at 9–12 km height where the neutron population is at its maximum.

^{14}C is the second most important source of natural terrestrial radioactivity, being roughly about half of that due to potassium. The total radioactivity of the Earth due to ^{14}C is estimated to be about 3×10^8 c. The annual rate of production of ^{14}C is about 9·8 kg and the total terrestrial content of the isotope is about 70000 kg. This represents a state of equilibrium between the rates of formation and decay and on account of the fairly long half-life it is supposed that there is plenty of time to ensure adequate mixing of ^{14}C uniformly throughout the Earth. On account of the special association of carbon with biological material, the possibility of using radioactivity measurements as a means of deducing the age of such material within the range 1000–30000 years was mooted as early as 1947 by Anderson and others. There are several complications and possible causes of error inherent in the method. It has been used with considerable success in many cases where all necessary precautions have been taken to avoid vitiation of the results, but further refinement of the method appears to be required before archaeologists can fully accept it for pre-historic dating purposes.

(2) Isotope exchange reactions between ^{12}C and ^{13}C which operate in various ways terrestrially so as to lead to a perceptible amount of differentiation of the two isotopes.

Atmospheric carbon (as CO_2) shows variations in isotopic composition which are probably temporary and local. In the ocean there are two exchange reactions:

$$(a) \quad {}^{13}CO_2 + H^{12}CO_3^- \rightleftharpoons {}^{12}CO_2 + H^{13}CO_3^-$$

for which the fractionation factor is 1·014 at 25 °C.

$$(b) \quad {}^{13}CO_2 + {}^{12}CO_3^{2-} \rightleftharpoons {}^{12}CO_2 + {}^{13}CO_3^{2-}$$

for which the factor is 1·012 at 25 °C.

It therefore follows that ^{13}C becomes concentrated in the bicarbonate and carbonate ions in solution which leads in turn to the ^{13}C enrichment in limestones.

Carbon is largely removed from the natural cycle in the form of limestone, therefore the ^{13}C content of the atmosphere should slowly decrease.

It is possible to recognize a broad division of terrestrial carbon into two groups, non-biogenic and biogenic. There is a slight but perceptible enrichment in ^{13}C in the first group and in the second it appears that plants tend to absorb $^{12}CO_2$ preferentially in photosynthesis so that some enrichment in the lighter isotope is detectable. The highest values for the $^{12}C/^{13}C$ ratio seem to occur in bituminous sediments, petroleum, and allied materials. The range of variation in the ratio is approximately 88 for some calcareous sediments to 94 for some petroleum oils.

The complete correlation of these results may, however, be difficult because calibration of the various techniques employed for the determinations is incomplete. There are several geological problems of great interest which have been studied in the light of the information on this topic already available. Interpretations made may well be rather tentative in view of some apparent irregularity in the data. Particularly in regard to biogenic carbon, it may be that the isotopic composition can be influenced to a considerable extent by local and ecological factors.

Abundance

Carbon is one of the most abundant elements cosmically, the ratio H/C atoms being of the order of 1000 : 1. The terrestrial figure of 320 g/ton for the crust places carbon fairly low down in the list—below a heavy element like barium. Carbon occurs on a wider cosmic scale than most of the other elements. Positive spectroscopic evidence is available for the existence of carbon in most types of stars and nebulae. Comets appear to be singularly rich in carbon in the form of simple molecules or radicals including CN, CH, C_2, and CO. It is known from absorption spectra that the atmosphere of Venus is rich in carbon dioxide and that the atmospheres of Jupiter and Saturn contain much methane. The thin Martian atmosphere seems to contain some carbon dioxide. Carbon atoms are also present amongst the various cosmic ray particles.

Terrestrial Occurrence

Carbon is a most ubiquitous element and is encountered in perhaps a greater variety of forms and locations than any other. As a free element it is found in the lithosphere and perhaps in the core of the Earth. Carbon dioxide is present in the biosphere, the atmosphere, the hydrosphere, and the lithosphere.

The carbonate and bicarbonate ions are present in the biosphere, the hydrosphere, and the lithosphere. Hydrocarbons are found in the biosphere, the lithosphere, and the atmosphere, and finally organic molecules in great complexity and variety are met with in the biosphere but also to some extent in the hydrosphere and the lithosphere.

Elemental forms. Carbon occurs in its two allotropic forms in Nature.

(1) *Diamond.* This mineral is both rare and valuable and also of great utility. Hence its origin has been much studied. Diamonds occur primarily in ' pipes ' of an ultrabasic rock called kimberlite in the Kimberley district of the Republic of South Africa and it is only in this locality that diamonds are won from the primary rock on a massive scale. Kimberlite pipes are found elsewhere but are not in general used as a direct source of diamonds—most other localities operate on alluvial deposits.

The total production of diamonds throughout human history is thought to be of the order of 100 tons and whilst India was in former times a great producer and in modern times there is a minor output from Venezuela, U.S.S.R., and a few Asian countries, there is no doubt that Africa is by far the greatest source of diamonds.

The rising output is largely due to the demand for industrial stones. The kimberlite ' pipes ' have diameters from 50 ft to 2000 ft and depths in excess of 3000 ft. They are not always diamantiferous, however. There are various views as to the origin of diamonds. It seems on the whole that they probably crystallized in the original magma chamber and were carried upward as crystals as the kimberlite magma welled upward in the pipe, subsequent to and associated with explosive volcanic activity. The necessary conditions of temperature and pressure for the formation of diamond are of course consistent with a deep-seated origin. In the case of alluvial deposits it is assumed that the diamonds have weathered out of a similar type of rock and have subsequently been transported so that in many cases the original source is unknown.

Diamond has also been recorded as a rare constituent of meteorites. The peculiar conditions of formation and occurrence of diamond are probably rather unusual and it is therefore likely to be a rare mineral under any circumstances. The more profitable rock at Kimberley only yields 1 part by weight of diamond per 20 000 000.

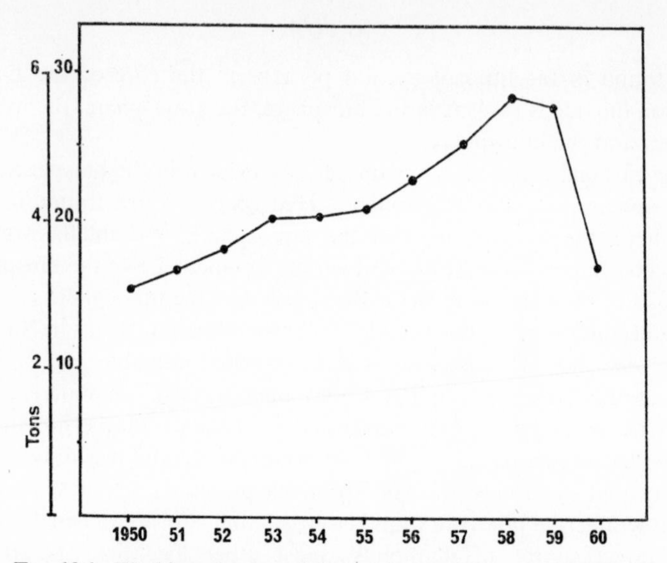

FIG. 13.1. World production of diamonds, in millions of metric carats.

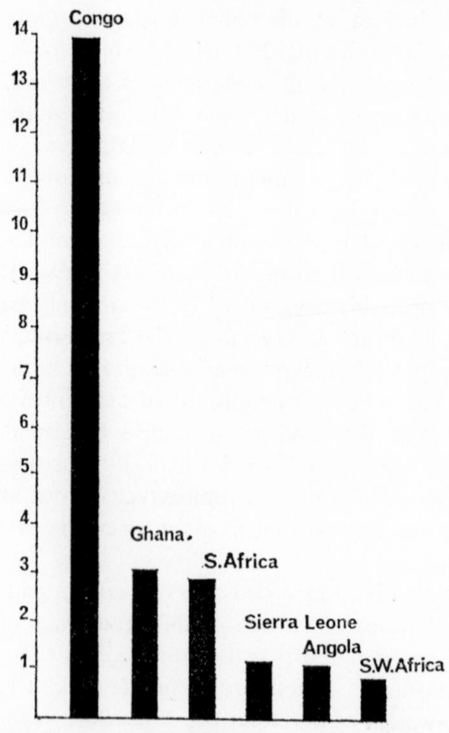

FIG. 13.2. Chief producers of diamonds, 1959, in millions of metric carats.

The production figures do not in general distinguish between gem and industrial stones, this is, however, reflected in the relative values of the outputs from different sources.

(2) *Graphite*. This form of carbon is of widespread occurrence under

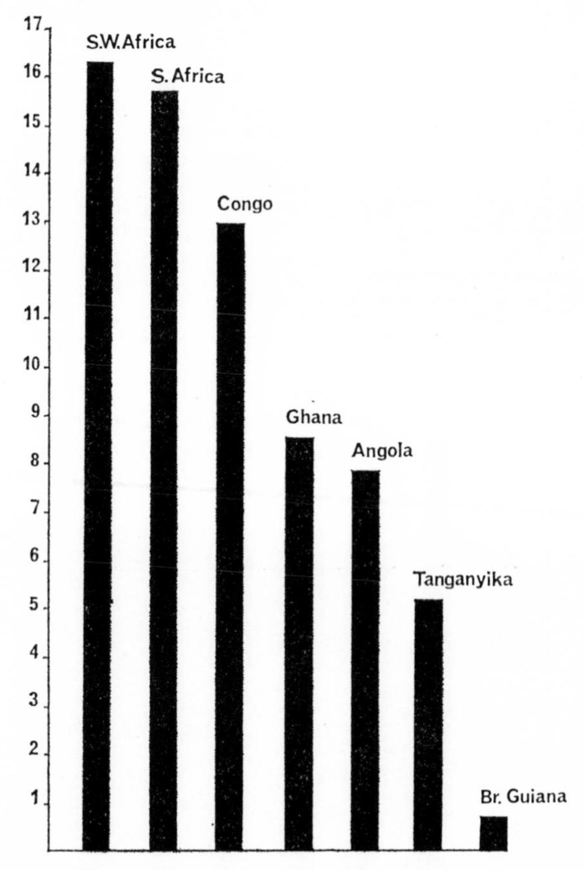

FIG. 13.3. Chief producers of diamonds, 1959. Value in £ million sterling.

circumstances which generally indicate that it has resulted from the metamorphism of carbonaceous material which is presumed to be of biogenic origin. The actual source of the original carbon, however, is often uncertain and obscure. It is possible that in some cases it may have resulted from the crystallization of original magmatic carbon. The following examples indicate the various conditions of occurrence.

Locality	*Type of Deposit*
Alibert, Irkutsk, Siberia	Associated with augite amphiboles and nepheline syenite
Ceylon	Fissure veins in gneiss and marble with intruded granite
Korea	Layers and lenses in schists near to granite. Also metamorphosed coals
Madagascar	Flakes or veins in schists and gneisses
Alabama, U.S.A.	Graphite bands in mica schist
Ovifak, Greenland	With native iron in basalt associated with coal
Borrowdale, Cumberland, England (This once famous deposit, long since abandoned, produced very pure graphite and was very profitable about 1780–1830)	Pipes or sops associated with veins in altered basic rocks

Apart from workable deposits, graphite occurs in a disseminated form in slates, schists, and other metamorphic rocks. Some coals are found to contain graphite in interstitial solid solution. Graphite is also a rather common minor constituent of meteorites, usually in amounts not exceeding 0·1 %.

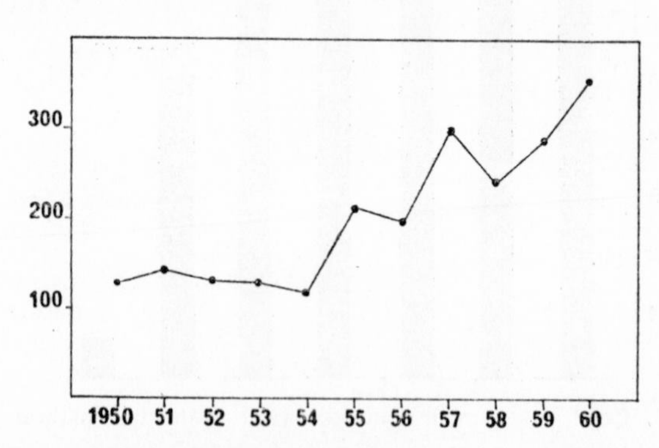

FIG. 13.4. World production of natural graphite, in thousands of tons.

Carbon Dioxide. This important compound is present in the atmosphere only to the average amount of 0·04 % by weight, corresponding to a total weight of nearly 0·022 Gg or about 0·0068 Gg of carbon. The amount of carbon dioxide in solution in the hydrosphere is of the order of 100 times that present in the atmosphere. According to estimates of

Goldschmidt, Kalle, and others this carbon dioxide is derived from three sources, *viz.*:

(*a*) 'Juvenile' carbon dioxide from volcanic activities.
(*b*) Carbon dioxide from human activities, combustion, etc.
(*c*) Carbon dioxide from respiration and decay.

At the same time carbon dioxide is removed by the following processes:

(*d*) Photosynthesis.
(*e*) Weathering of rocks.
(*f*) Formation of carbonaceous deposits.

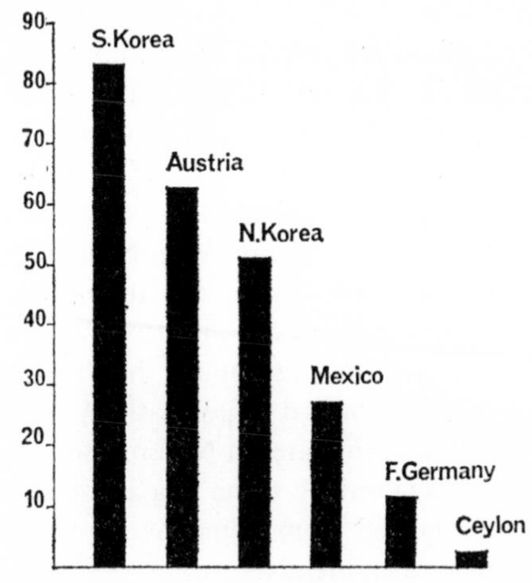

FIG. 13.5. Chief producers of natural graphite, 1959, in thousands of tons.

It would appear that there is almost a balance between the two groups of processes, with, however, a slight bias in favour of increase of carbon dioxide in the atmosphere.

This presumed rise in the carbon dioxide content of the atmosphere will, however, be gradually taken up to a large extent by solution in the ocean. In fact the hydrosphere acts as a gigantic buffer in maintaining the approximately constant carbon dioxide content of the atmosphere. There are local variations in time and place related to the various factors involved but in the long run the rate of change of carbon dioxide content in the atmosphere will be slight.

It seems unlikely that human contributions will have much permanent geochemical significance although locally and temporarily they may.

It should also be observed that the consumption of carbon dioxide by photosynthesis in the oceans is on a greater scale than was at one time appreciated.

The interchange of carbon dioxide between the atmosphere and hydrosphere is an important process and is determined by Henry's law and the law of mass action. The equilibria involve the following actions:

$$\begin{array}{l} CO_2 \\ \quad\searrow \\ \qquad CO_2 + H_2O \rightleftharpoons H_2CO_3 \rightleftharpoons HCO_3^- \rightleftharpoons CO_3^{2-} \\ \quad\nearrow \\ H_2O \end{array}$$

The partial pressure of carbon dioxide which controls the first step is governed by changes in pressure, temperature, salinity, respiration, chemical solution, and precipitation of calcium, and by photosynthesis. In this connexion see the behaviour of calcium (p. 169).

Ionic forms. From the geochemical standpoint the most important and almost the only ionic forms of carbon in Nature are the carbonate and bicarbonate ions, in which state vast quantities of carbon are immobilized in the upper part of the lithosphere. The predominant substance is of course calcium carbonate, but many other metallic carbonate minerals are known and a few of them are useful ores.

There are very few acid carbonates in Nature and the only one of any great significance is the mineral trona $Na_3H(CO_3)_2.2H_2O$ which is rather abundant in evaporites from saline lakes. The following metals form normal carbonates as minerals:

Na, Ca, Sr, Ba, Zn, Mg, Cd, Pb, Mn, Fe, and Co

The following only form basic carbonates:

Cu, Bi, U, and Ni

Aluminium and the rare earth metals are only found in double and complex carbonate minerals. The following metals do not seem to form carbonates in Nature:

Au, Ag, Sb, As, Sn, K, Rb, Cs, Zr, Th, W, and Mo

The list below represents a selection of the various types of carbonate minerals. The presence of mineral carbonates in certain types of igneous rocks has led to the idea that at least in some cases the carbon is not of indirect organic origin. It would seem, however, that in general the uprising carbon dioxide in volcanic emanations is the primary source of carbon to the upper zones of the Earth.

CARBONATE MINERALS

Name	*Formula*	*Occurrence*
Trona	$Na_3H(CO_3)_2.2H_2O$	Evaporite deposits
Calcite	$CaCO_3$	Limestones, H.T. veins, carbonatite rocks
Magnesite	$MgCO_3$	Carbonation of olivine rocks
Siderite	$FeCO_3$	Ferruginous sediments, H.T. veins
Smithsonite	$ZnCO_3$	Oxidized zinc deposits
Witherite	$BaCO_3$	H.T. veins
Cerussite	$PbCO_3$	Oxidized lead deposits
Dolomite	$CaMg(CO_3)_2$	Sedimentary rocks, H.T. veins
Natron	$Na_2CO_3.10H_2O$	Evaporite deposits
Gaylussite	$Na_2Ca(CO_3)_2.5H_2O$	Evaporite deposits
Lanthanite	$(La,Ce)_2(CO_3)_3.9H_2O$	Oxidized pegmatites
Aurichalcite	$(Zn,Cu)_5(CO_3)_2(OH)_6$	Oxidized Zn and Cu deposits
Malachite	$Cu_2(CO_3)(OH)_2$	Oxidized Cu deposits
Phosgenite	$Pb_2(CO_3)Cl_2$	Oxidized Pb deposits
Alumohydrocalcite	$CaAl_2(CO_3)_2(OH)_4.3H_2O$	Carbonation of allophane
Cancrinite	$Na_3CaAl_3Si_3O_{12}CO_3$	Igneous rocks
Carbonate–apatite	$Ca_{10}(PO_4)_6(CO_3).H_2O$	Phosphate rocks

Hydrocarbons. The naturally occurring compounds of carbon and hydrogen are numerous, abundant, and widespread. It will be convenient to classify them in accordance with their states of aggregation.

State	*Typical Formulae*	*Modes of Occurrence*
Gases	CH_4	Earth's atmosphere, trace. Atmospheres of Jupiter and Saturn. Anaerobic decay product in swamps, etc. In coal seams
	CH_4, C_2H_6, C_3H_8	Occluded in rocks and trapped under domes of impervious cap-rock. In solution in petroleum
Liquids	Paraffins C_4H_{10} upward. Cycloparaffins, aromatics	Disseminated in sedimentary rocks and accumulated in impervious strata as oil pools under anticlines—crude petroleum
Solids	$C_{20}H_{34}$, $C_{18}H_{30}$, $C_{13}H_{10}$, $C_{24}H_{18}$	Solid hydrocarbons found as natural waxes, etc.

The origin of some of these products is complex and obscure. It is presumed that the methane of the giant planets is of inorganic origin and the action of water on metallic carbides has been suggested as a possible source.

Much, if not all methane and practically all of the other hydrocarbons on Earth may be supposed to be biogenic. They may, however, have passed through various non-biological processes also. Specialized

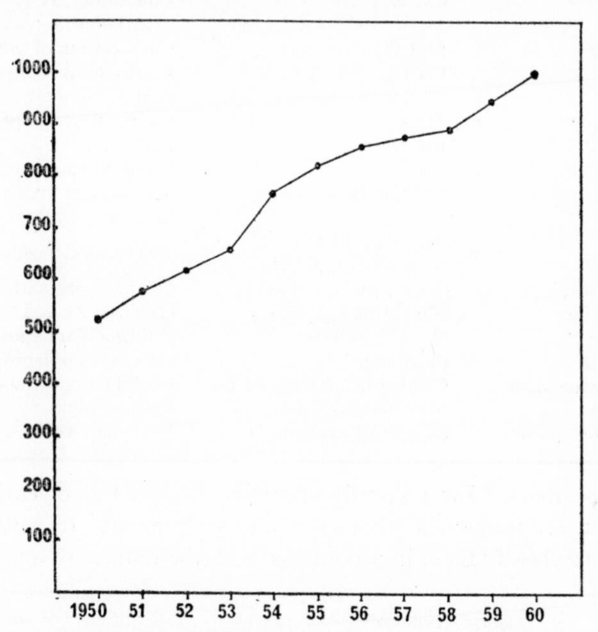

FIG. 13.6. World production of crude petroleum, in millions of tons.

works should be consulted for a full discussion of the origin of petroleum. It will be sufficient here to note the main aspects of the problem, which can now be regarded as to a large extent resolved.

(*a*) Most, if not all, petroleum is derived from aquatic and mainly marine algae, remains of which accumulate in the bottom muds or sapropels.

(*b*) The organic material in these residues has undergone changes, partly at least due to bacterial action and partly perhaps to catalytic processes leading to hydrocarbon formation.

(*c*) There is an enormous reserve of petroleum in disseminated form in sedimentary rocks.

(*d*) Petroleogenesis is still proceeding at a rate far in excess of human inroads on the material.

(*e*) Accumulation of petroleum and gaseous hydrocarbons depends on earth movements and in some cases possibly on distillation.

(*f*) Considerable variation in the nature of the raw material, and the sequence of processes to which it has been subjected, causes great variation in types of petroleum deposit, both physically and chemically.

It is to be noted that of the numerous and widespread workable deposits of hydrocarbons, some consist of liquids only, many consist of liquid and gas, and there are now many fields under exploitation which yield gas alone. Petroleum products seem to occur in all of the continents except Australasia. A summary of petroleum production is given on pp. 218 and 220.

Organic molecules. It is not within the scope of this book to discuss at length the multiplicity of molecular complexes which build up the structure and determine the functions of living organisms—this subject is adequately dealt with in the numerous works on biochemistry.

It is nevertheless desirable to consider briefly some of those carbon compounds, which, whilst not immediately biogenic, have at some stage in their formation been involved in biological processes. It has already been observed that in general the natural hydrocarbons are primarily biogenic. We now consider various other loosely classified materials which are also carbon compounds of a more complex type. They may be noted under two headings, ignoring an assortment of minor substances which are of varied and often indefinite composition.

(*a*) *Resins.* These are compounds of carbon, hydrogen, and oxygen. They are in general to be regarded as fossilized plant resins. They are natural polymers and are not generally susceptible of exact formulation. Examples are amber, copalite, and retinite. The first two of these give succinic acid when heated. The third does not.

(*b*) *Coals.* Under this heading we include those substances of a highly carbonaceous character containing also hydrogen and oxygen, together with subsidiary quantities of sulphur, nitrogen, and other elements. They have been formed by decay, dehydration, and consolidation of vegetable residues.

It is generally agreed that the series peat→lignite→bituminous coal →anthracite, with their variants, represents different stages of the same general process. Coals are to be regarded as mixtures of rather complex molecules which include carbocyclic structures with both aromatic and hydroaromatic rings, together with oxygen in the form of carbonyl, hydroxyl, and occasional methoxy-groups. The substances concerned may have molecular weights from a few hundreds to a few thousands.

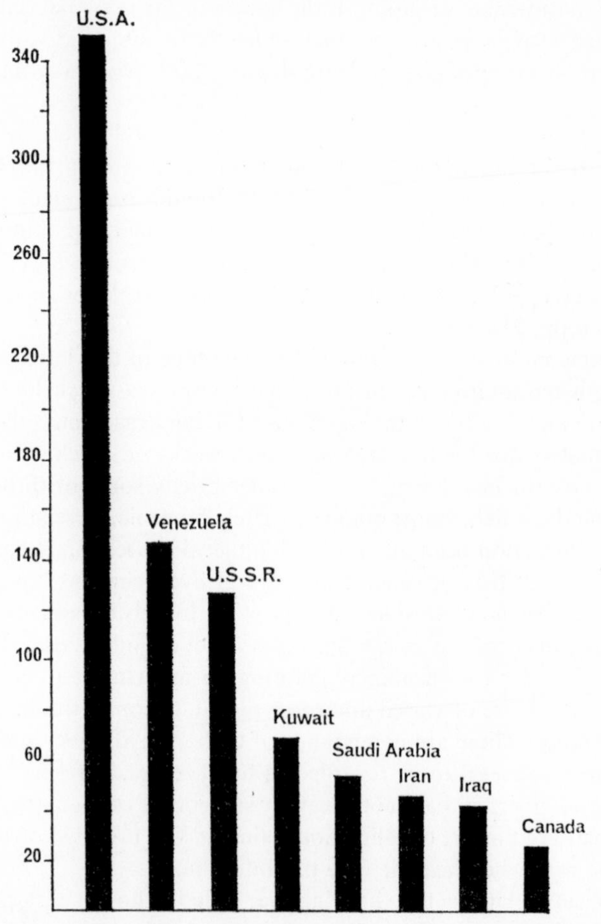

FIG. 13.7. Chief producers of crude petroleum, 1959, in millions of tons.

The full details of structure of the components of coal are still a matter of discussion as also is the origin of such materials.

It is pointed out that woody fibre contains both cellulose with its five-carbon, one-oxygen glucose rings and the complex called lignin which has aromatic rings. It is a matter of contention how far the coal substances are derived from cellulose or lignin or both. In the case of lignin it may be supposed that the aromatic rings in the coal were elaborated by the original plants from which the coal was formed. In

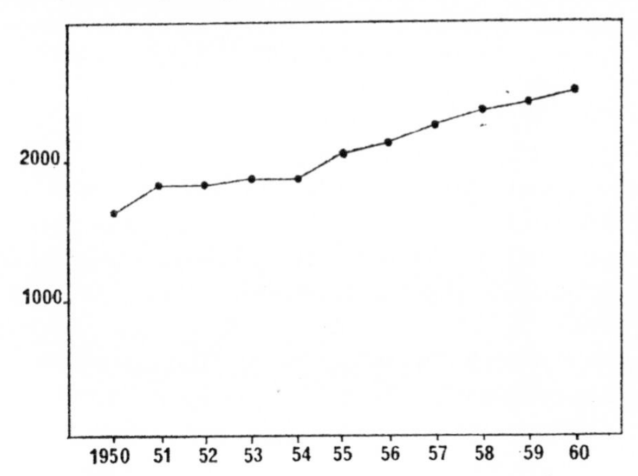

FIG. 13.8. World production of coal, in millions of tons.

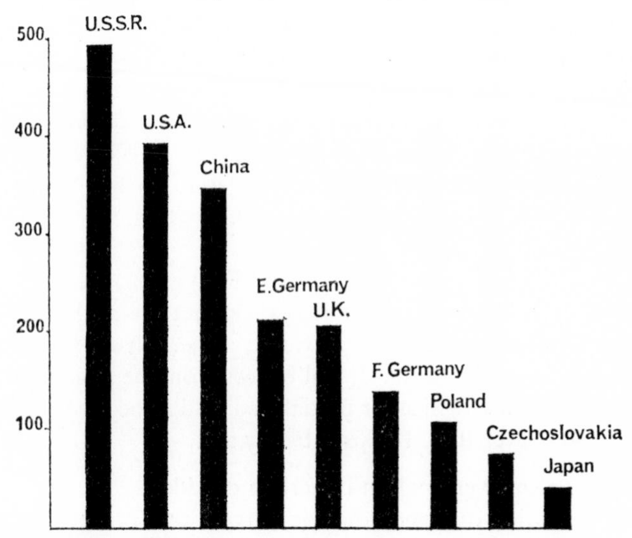

FIG. 13.9. Chief producers of coal, 1959, in millions of tons.

such a case the cellulosic part of the material would be lost by microbiological decay—as CO_2 if aerobic and as CH_4 if anaerobic. There is, however, evidence that cellulose also contributed to the structure of the coal substances.

The four macroscopically distinguishable components of bituminous coals are generally recognized under the terms fusain, clarain, vitrain, and durain. Extensive and elaborate classifications have been developed by coal petrologists.

In extreme cases the process of coalification leads to anthracite containing perhaps 94 % carbon, but the later stages in general involve some degree of thermal and regional metamorphism. The substance jet is a compact and lustrous form of lignite.

It is not apparent that coal contains any elemental carbon. Natural cokes, called carbonite, are occasionally encountered; they probably contain some graphitic carbon. The sulphur, nitrogen, and other elements found in coals are to be regarded as incidental minor components, present in various forms.

Composition of coals. In passing through the series wood→peat→lignite→coal→anthracite, the basic elements vary as follows: carbon, 50 % increasing to 95 %; hydrogen, 6 % decreasing to 2·5 %; and oxygen, 43 % decreasing to 2·5 %.

Occurrence of coals. Coals are a form of sedimentary rock. They are found stratified in seams from a fraction of an inch to 30 ft in thickness. Some coal has been formed *in situ*; in many cases, however, the material has been carried by water from its original source. The age of coals ranges from about 10^6 years for the lignites of New Zealand to 3×10^8 years for the Devonian coals of Bear Island. Nevertheless the great majority of coals were laid down in the Permo-Carboniferous period, from 200 million to 300 million years ago.

Biogeochemistry

As carbon is the basis of life it is clear that the subject of its biochemical relationships is a very extensive one. Figure 13.10 indicates the main cycles which occur. It will be seen from the figure that several of the processes are cyclical or partially so, and some are virtually non-cyclical and irreversible. In this category are:

(*a*) The volcanic emanation of carbon dioxide.
(*b*) The fixation of carbon dioxide in the form of limestone rocks.

In regard to (*a*) there has been speculation as to whether the emission of the gas has been approximately constant during geological time as might be expected in a uniformitarian scheme. It has been observed, however, that the great coal-forming periods were also in general periods of major orogeny or mountain-building during which there was enhanced volcanic activity and therefore presumably increased emission of carbon dioxide. It has in such cases been assumed that the rate of photosynthesis and building of vegetable tissue would be related to the increased amount of carbon dioxide. There are other factors involved,

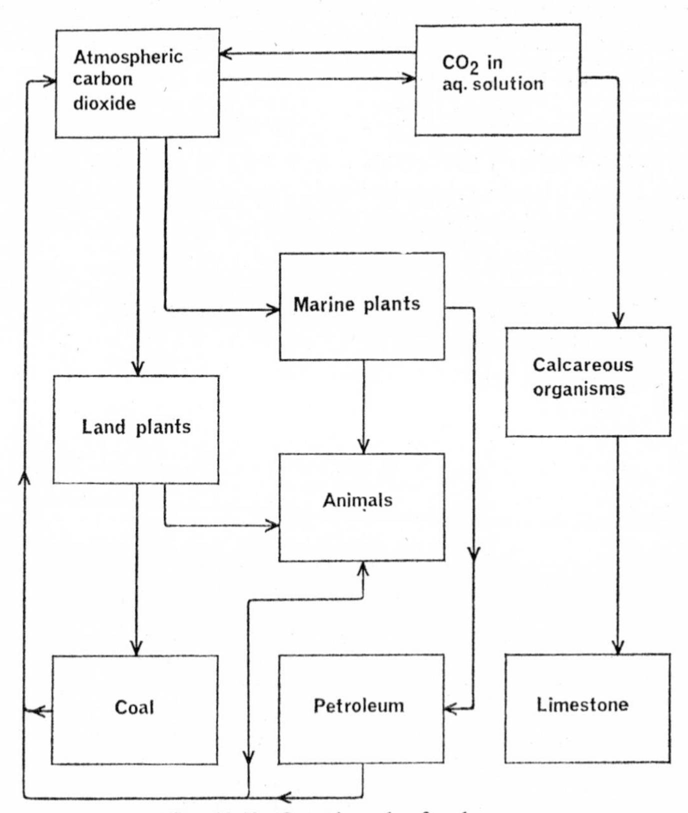

FIG. 13.10. Organic cycle of carbon.

however, which include light intensity, presence of water, and temperature. It is known that increased carbon dioxide content of the atmosphere, by giving greater absorption of the infra-red, would raise the temperature of the atmosphere. It does not appear very likely that there have been any great changes in solar radiation over the last few hundred million years so that the present level which in full sunlight approaches the light saturation value of some plants would perhaps be a limiting factor to greatly increased photosynthesis apart from a rise in temperature.

It should be observed that the optimal conditions for the plants of the coal forests are unlikely to be known with any accuracy. It is also necessary to take into account the massive buffering effect of the dissolved carbon dioxide in ocean water.

In regard to (*b*) we can consider the overall result of the action of carbon dioxide as the replacement of silicate ions by carbonate ions in the rocks of the crust. Thus in a simplified form

$$CaSiO_3 + CO_2 = CaCO_3 + SiO_2$$

If we take a relatively short-term view it may be said that the carbon dioxide is thus irretrievably locked up in the limestone rocks. However, we must also consider the reverse reaction

$$CaCO_3 + SiO_2 = CaSiO_3 + CO_2$$

as typical of the metamorphic changes which can occur when, as a result of great earth movements, masses of strata are downfolded to great depths. It is therefore evident that some at least of the carbon dioxide released from the crust is not necessarily 'juvenile', it may in fact have taken part in a major geochemical cycle.

We conclude this chapter with a summary of the estimated quantities of the various forms of carbon in the Earth ($1\,Gg = 10^{20}\,g$).

TERRESTRIAL CARBON

Form	*Location*	*Quantity* (Gg)
CO_2	Atmosphere	0·006
CO_2	Hydrosphere	0·164
Living matter	Biosphere	0·007
Coals, petroleum	Lithosphere	0·058
Limestones	Lithosphere	99·0

CHAPTER 14

Silicon

There are three stable isotopes of this element in Nature:

$$^{28}Si = 92{\cdot}27\%$$
$$^{29}Si = 4{\cdot}68\%$$
$$^{30}Si = 3{\cdot}05\%$$

There are indications that natural isotopic fractionation processes, both biological and non-biological, take place but the amount of information at present does not allow any important deductions to be made. It is possible to account for the synthesis of silicon atoms in the later stages of the α-process in ' helium burning ' stars at temperatures of the order of $1{\cdot}3 \times 10^9$ °K.

Abundance

Cosmically the H : Si ratio is about 10000 : 1, and silicon is perhaps seventh in order of abundance of the elements. In the crust of the Earth silicon is the second most abundant element after oxygen and is present to the extent of 277200 g/ton. For the Earth as a whole the value would be about half of this, making the element third after iron and oxygen. The ratio of silicon to oxygen in regard to the crust is rather significant. In terms of atoms it is Si : O = 1 : 2·92, so that the amount of oxygen is well in excess of that needed to saturate the silicon.

Forms of Occurrence

From the above observation it will be apparent that silicon is unlikely to be found in the Earth's crust in anything but its higher state of oxidation. There is no evidence for the existence of the free element in Nature, even in meteorites. The rare meteoric species, moissanite SiC, does, however, occur and represents the lowest known oxidation state of silicon in Nature. There are very few non-oxygen compounds of the element. They are indicated below.

All the other very numerous silicon compounds involve structures with oxygen. It is possible to look on them all in the light of silicates, even the oxide. The silicates were at one time regarded as a group of

Name	*Formula*	*Occurrence*
Proidonite	SiF_4	Gaseous emanation in eruption of Vesuvius, 1872
Hieratite	K_2SiF_6	In volcanic regions around Vesuvius
Malladrite	Na_2SiF_6	
Bararite	$(NH_4)SiF_6$	

minerals of very great complexity, and methods of classification seemed to be unsatisfactory and relationships obscure. It has now been established that all the structures so far met with involve silicon in four-fold co-ordination with oxygen, the bonds being of a partial covalent character and giving rise to a compact and stable tetrahedron. It is the relation of the different SiO_4 tetrahedra to each other which determines the structure of the different types of silicates. Six classes of silicates can be recognized.

Name of Class	*Way in which SiO_4 Groups are Arranged*	*Silicate Group*
Nesosilicates	Independent tetrahedra	SiO_4
Sorosilicates	Pairs of tetrahedra sharing an O atom	Si_2O_7
Cyclosilicates	Rings of tetrahedra sharing 2 O atoms	Si_3O_9,Si_4O_{12},Si_6O_{18}
Inosilicates	Single or double chains of tetrahedra sharing 2 or 3 O atoms	SiO_3 Si_4O_{11}
Phyllosilicates	Sheets of tetrahedra by sharing 3 O atoms each	Si_2O_5
Tektosilicates	Three-dimensional networks, tetrahedra sharing all 4 of their O atoms	SiO_2

EXAMPLES OF TYPICAL SILICATE MINERALS

Type	*Name*	*Formula*
Nesosilicate	Zircon	$ZrSiO_4$
Sorosilicate	Hemimorphite	$Zn_4Si_2O_7(OH)_2.H_2O$
Cyclosilicate	Beryl	$Be_3Al_2Si_6O_{18}$
Inosilicate	Enstatite	$MgSiO_3$
	Tremolite	$Ca_2(Mg,Fe)_5Si_8O_{22}(OH)_2$
Phyllosilicate	Muscovite	$KAl_2(AlSi_3O_{10})(OH)_2$
Tektosilicate	Feldspar	$KAlSi_3O_8$
	Quartz	SiO_2

The physical properties of these types are very characteristic. As we pass down the series the structure becomes increasingly open with a resulting decrease of density as the following examples show:

Name	*Formula*	*Density*
Olivine	$(Mg,Fe)_2SiO_4$	3·3
Enstatite	$MgSiO_3$	3·18
Anthophyllite	$Mg_7(Si_4O_{11})_2(OH)_2$	2·96
Talc	$Mg_3(Si_4O_{10})(OH)_2$	2·82

The number of silicate minerals which have been described is very large, being of the order of one thousand, but the number of species which are actually predominant in the crust of the Earth is quite small. The three most abundant elements in the crust are oxygen, silicon, and aluminium and it is not surprising therefore that the non-silicate mineral content of the crust is only about 5%.

The large number of silicates is related to the numerous ways in which their varied structures can be built up, but at the same time the invariable four-fold co-ordination of silicon to oxygen imposes certain limitations. Thus the number of metals that enter into silicate lattices as essential components is rather small and the number which form simple silicates is quite limited.

SIMPLE ANHYDROUS SILICATES

Name	*Formula*
Phenakite	Be_2SiO_4
Forsterite	Mg_2SiO_4
Enstatite	$MgSiO_3$
Wollastonite	$CaSiO_3$
Thalenite	$Yt_2Si_2O_7$
Zircon	$ZrSiO_4$
Thorite	$ThSiO_4$
Willemite	Zn_2SiO_4
Alamosite	$PbSiO_3$
Barysilite	$Pb_3Si_2O_7$
Eulytine	$Bi_4Si_3O_{12}$
Coffinite	$USiO_4$?
Rhodonite	$MnSiO_3$
Tephroite	Mn_2SiO_4
Fayalite	Fe_2SiO_4
Clinoferrosilite	$FeSiO_3$
Andalusite	Al_2SiO_5

It will be seen that the majority of these minerals fall in the neso-silicate class. In the more complex structures it frequently happens that a variety of cations is accommodated in the lattice. It will be appreciated that a high proportion of the silicates are of mineralogical interest but make no great contribution to the Earth's crust.

The following are the estimated proportions of the important rock-forming silicates in the crust: feldspars, 60 %, quartz, 12 %, amphiboles, 17 %, pyroxenes, 17 %, biotite, 4 %.

Free silica in Nature. There are five naturally occurring forms of silica, the trimorphic crystalline forms being tektosilicates in which the ionic charges are neutralized and no cations are required.

Name	*Formula*	*Mode of Occurrence*
Quartz	SiO_2	Primary igneous rock mineral. Pegmatic and hydrothermal mineral. Residual in sands and sandstones. Recrystallized in quartzites
Tridymite } Cristobalite }	SiO_2	Scarce minerals in volcanic rocks
Lechatelierite		Very rare natural silica glass
Opal	$SiO_2.nH_2O$	Solidified silica gel. Low-temperature deposit from siliceous waters

There are many varieties of quartz and of cryptocrystalline silica. Ornamental varieties of quartz and cryptocrystalline silica which depend mainly on colour, transparency, and hardness include rock crystal, rose quartz, amethyst, citrine, cairngorm, carnelian, agate, opal, and many others.

Natural glasses. There are cases when the material of a magma cools down rather quickly and solidifies without crystallizing. In such cases the rock is wholly or partly vitreous. Obsidian provides the best example of a natural glass and it has a chemical composition similar to that of granite, thus having an excess of silica over bases. The much less common glasses called tachylytes are of a more basic composition, similar to basalt. Glasses are not inherently stable in Nature. There are often indications of incipient crystallization and in glassy rocks of high geological age devitrification has largely taken place. Most obsidians are products of recent volcanic activity.

Chemical aspects of silica and silicates in Nature. In Chapter 3 a general indication was given of the conditions determining the separation of the various silicate minerals from a magma. The factors involved are of course quite complex but it is very noticeable that the order of crystallization is much influenced by the type of structure—the compact nesosilicates being early to separate and the open tektosilicates being relatively late. It is noteworthy that only in the ultrabasic rocks do we find a complete absence of combined water or OH groups. The function of water as a fluxing agent is very important and even with silica alone its presence greatly lowers the melting point. For the crystallization of magmatic silicates it is not in general necessary to postulate temperatures as high as some of those used in technical processes. Once the various silicates have been formed in a rock they are virtually stable under normal conditions. However, when the rock is exposed at the surface of the Earth the conditions are changed and weathering processes operate. The following represents the approximate order of reactivity:

olivine > pyroxenes > amphiboles > feldspars > micas

In general the dark-coloured 'ferromagnesian' minerals tend to weather rapidly. Thus we usually find the surfaces of basaltic rocks far more weathered than for example those of granite. The comparative ease of oxidation of ferrous ions released from these silicates is an important factor in the rate of weathering.

The feldspars which are such an important group of silicates do not all weather at the same rate. The lime-rich plagioclases are more readily attacked than pure albite or orthoclase. On the other hand the alkali-rich feldspathoid minerals like nepheline and leucite weather rather easily. The group of silicates called zeolites present special points of interest. They are tektosilicates with very open wide-meshed $(Si,Al)O_2$ lattices. The cavities in the framework contain various cations which are in many cases capable of being exchanged for such cations as K^+, Ag^+, Tl^+, and Na^+. In this process the silicate framework remains intact so that the crystal retains its shape. Similarly the water molecules may be removed by heat without much disturbance of the crystal structure.

Secondary changes in silicate minerals. The weathering processes leading ultimately to sedimentary rocks were outlined in Chapter 3. We now look into the behaviour of silicates in rather more detail.

In processes whereby the igneous silicates, which are in the main minerals of the type of olivine, pyroxenes, amphiboles, and feldspars, are converted into phyllosilicates of the clay mineral type it is fair to assume that the crystal lattice is initially completely disintegrated. The various ions then pass into solution. Their subsequent behaviour depends on their ionic potentials and the solubility products of the various possible combinations of anions and cations.

In Goldschmidt's scheme the Si^{4+} ion comes very near to the borderline between hydrolysate elements and those which form soluble complex anions. None of the other common cations comes very close to silicon. Thus, $Fe^{3+} = 4{\cdot}7$, $Al^{3+} = 3{\cdot}9$, whereas $Si^{4+} = 9{\cdot}5$. These values enable us to understand how the silica can be largely leached out in the formation of laterite soils under tropical conditions.

When actually in solution the silicon is presumably either in the form of SiO_4 ions or as colloidal silica. Both forms may be present and the content in solution is primarily governed by the hydrogen ion concentration. Thus at pH 4 the solubility of SiO_2 is about 70 p.p.m. rising to 360 p.p.m. at pH 9. Natural waters largely fall within these values and in this range the solubility of $Al(OH)_3$ and $Fe(OH)_3$ is much lower so that perceptible differentiation from silica is possible.

It is noticeable that the SiO_2/Al_2O_3 ratio in igneous silicates like feldspars is about 3·5 : 1, whereas in clay minerals like kaolinite it is as low as 1 : 1. It is evident that whilst some of the dissolved silica is rapidly combined with alumina, etc., and deposited as a clay mineral, yet immense amounts continue in solution and probably reach the ocean. There is evidence that in the presence of dissolved calcium carbonate the solubility of SiO_2 in water is reduced and cryptocrystalline silica may be precipitated. In this way Goldschmidt accounted for the abundant concretions of flint in the Chalk, so familiar in Southern and Eastern England and in Northern France. Analogous deposits are found in other geological horizons. The fact that silica is largely deposited in a marine environment is shown by its low content in ocean water compared with river waters. According to Goldschmidt the geochemical balance of silica in sea water is very low—only about 20 parts of silica remain in solution for every million parts supplied by the weathering of igneous rocks.

Biogeochemistry

Silicon is quite considerably involved in biological processes, in plants more than in animals. The silica content of some plants is quite high and one may note particularly the grasses, including bamboos, and horsetails as mentioned in Chapter 4. The group of algae called diatoms are very important in their use of silicon to build up a complex skeleton. This absorption of silica from the water in which they live produces very significant changes of a seasonal nature which are particularly apparent in freshwater lakes. The net result is the gradual accumulation of a bed of ooze rich in the siliceous remains of the dead organisms. It is from such formations that commercial deposits of diatomite are obtained. Radiolarian ooze is an analogous deposit. Among animals, sponges are noteworthy for their high silicon content. It has been suggested that silica plays a part in the rigidity of feathers.

The exact form in which the silicon exists in the various organisms is not very clear. It may well be in the form of SiO_2 itself. Complex silicon compounds analogous to organic carbon molecules are quite unknown in Nature.

Silicon Minerals of Economic Interest

Large quantities of rocks and rock materials are used for purely mechanical purposes. They are not considered here. The chief materials concerned are tabulated below.

Name	*Application*	*Approximate annual output* (tons)
Quartz, rock crystal	Optical, electrical, ornamental	
Sand, flints	Glassmaking	
Asbestos, etc.	Thermal and electrical insulation	2 million
Bentonite	Drilling mud, oil refining	1·3 million
Diatomite	Absorbent, filler	800000
Feldspar	Ceramics, glass	1·2 million
Garnets	Abrasives	11000
Mica	Electrical, thermal	135000
Pumice, Volcanic glass	Abrasives	35000
Nepheline syenite	Glass, ceramics	200000
Sillimanite	Refractories	125000
Talc	Pigment, filler, etc.	1·8 million

CHAPTER 15

The Metals of Group IV

General

The elements to be considered in this chapter fall naturally into two groups. The first sub-group comprises three transitional metals with pronounced oxyphile and lithophile characters.

The second sub-group comprises three metals in which lithophile characters are less obvious and where thiophile characters may be very marked. Titanium, zirconium, and hafnium are rather widely dispersed but their actual abundance is no guide to their availability.

The metals germanium, tin, and lead differ rather perceptibly in their modes of occurrence, tin in particular being a comparatively scarce metal of quite high availability. Germanium on the other hand is fairly abundant but widely dispersed and therefore not very readily accessible. Figure 15.1 also includes, for the sake of completeness, the two other elements of the group, carbon and silicon.

Titanium

There are five stable isotopes of this element in Nature:

$^{46}Ti = 7{\cdot}95\%$ $\quad$ $^{49}Ti = 5{\cdot}51\%$

$^{47}Ti = 7{\cdot}75\%$ $\quad$ $^{50}Ti = 5{\cdot}34\%$

$^{48}Ti = 73{\cdot}45\%$

There is no present indication of any natural fractionation of these isotopes.

According to Hoyle *et al.* titanium is one of the elements of the ‘ iron peak ’ of abundance which would be synthesized by the ‘ equilibrium ’ process in the very rapid evolution of a star beyond the point when its central temperature reaches 3×10^9 °K. In this phase there is a statistical equilibrium between nuclei, protons, and neutrons. It is thought to be a consequence of this stage of stellar evolution that nuclides of mass numbers 46–66 have so high an abundance. Titanium is also probably formed by the α-process and by neutron capture processes.

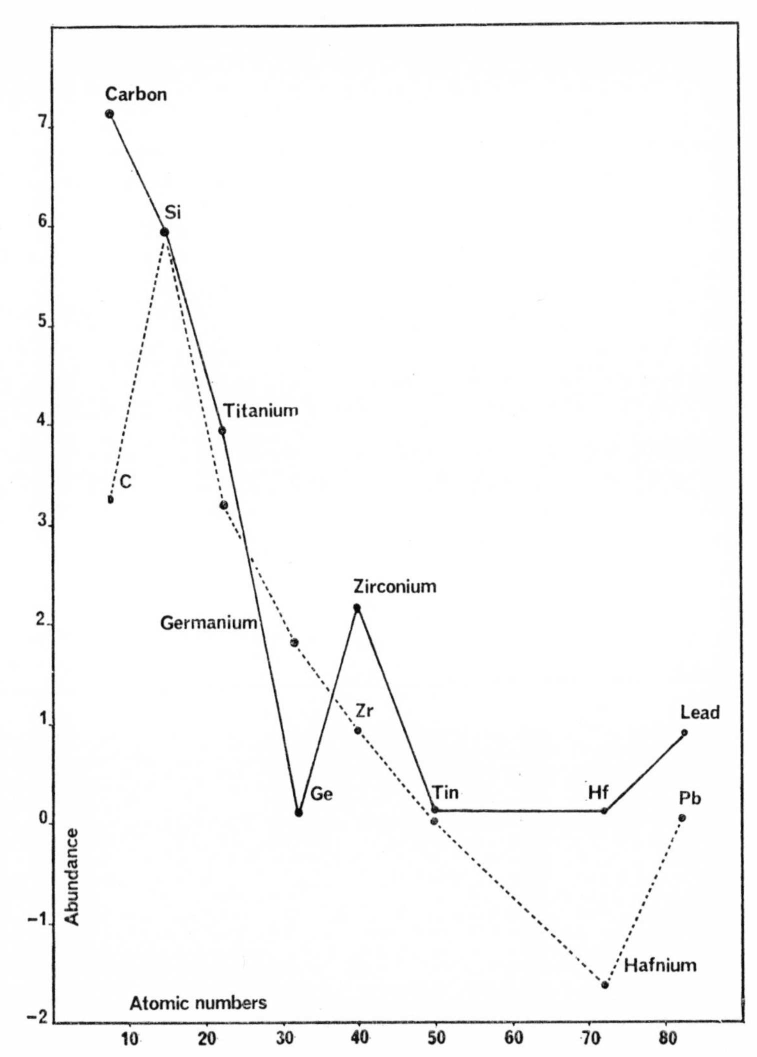

FIG. 15.1. The relative abundance of Group IV elements (logarithmic). Continuous line, cosmic. Dotted line, terrestrial.

Not only are lines of titanium itself observed in stellar spectra but in some of the cooler stars at *circa* 3800 °K the band spectrum of TiO is recognized. Another extra-terrestrial occurrence is the presence in meteorites of the very stable but rare titanium nitride TiN called osbornite.

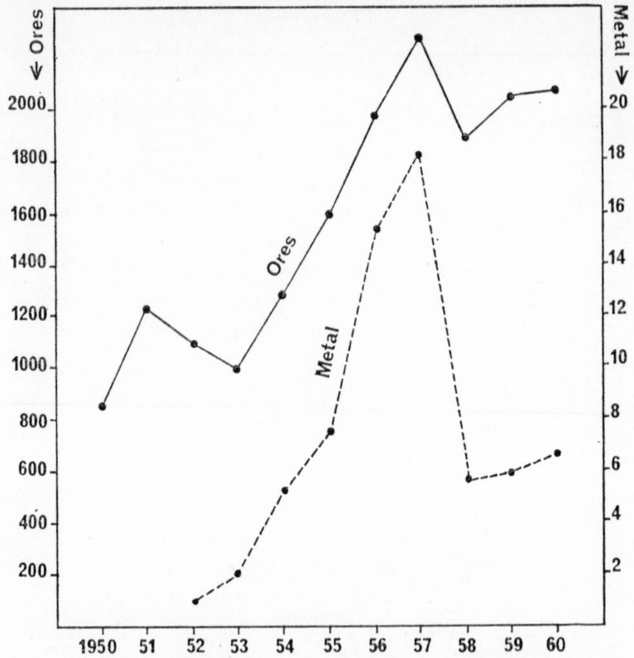

FIG. 15.2. World production of titanium metal and titanium ores, in thousands of tons.

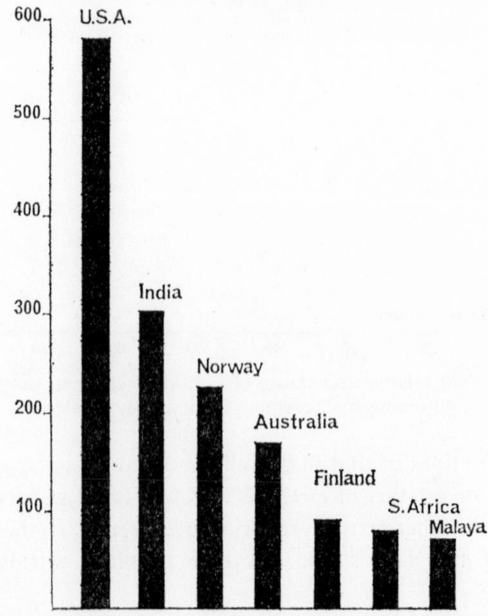

FIG. 15.3. Chief producers of titanium minerals, 1959, in thousands of tons.

Geochemistry

Titanium is very markedly lithophile in character. It occurs frequently in the form of oxide minerals, some of which tend to separate from a cooling magma at a fairly early stage. A limited number of independent titanium minerals are known and titanium is concealed in small quantities in many rock minerals. It appears that the commonest mineral, ilmenite, is a compound oxide in which the structure is as hematite Fe_2O_3, but with every other iron atom replaced by one of titanium.

Goldschmidt, however, considered ilmenite to be a titanate of ferrous iron $Fe^{2+}Ti^{4+}O_3$. The determination of the valency of titanium in its minerals presents difficulties. In the mineral perovskite the structure is also that of a complex oxide. In sphene we have a calcium titanyl orthosilicate $CaTiO(SiO_4)$. It seems unlikely that titanium anions occur in Nature.

TITANIUM MINERALS

Name	*Formula*	*Occurrence*
Rutile Anatase Brookite	TiO_2	Accessory minerals in igneous and metamorphic rocks and cavities and veins in the same
Perovskite Ilmenite Sphene	$CaTiO_3$ $FeTiO_3$ $CaTiSiO_5$	Accessory minerals in igneous rocks, also in dykes and pegmatites. May be residual in sands
Pyrophanite Benitoite Leucosphenite	$MnTiO_3$ $BaTiSi_3O_9$ $Na_3CaBaBTi_3Si_9O_{29}$	Rare rock minerals

Zirconium

There are five stable isotopes of this element:

$^{90}Zr = 51{\cdot}46\%$ $^{94}Zr = 17{\cdot}4\%$
$^{91}Zr = 11{\cdot}23\%$ $^{96}Zr = 2{\cdot}80\%$
$^{92}Zr = 17{\cdot}11\%$

They call for no special comment. There is no very certain evidence for the existence in Nature of the radioactive ^{93}Zr, which is a fission product of uranium.

Like titanium, zirconium is characteristic of certain types of stars, particularly the S-type, and bands of ZrO are observed in some stellar spectra.

Geochemistry

Zirconium is not a rare element—with a crustal abundance of 160 g/ton it is more than twice as abundant as 'common' metals like copper and zinc. It is, however, widely dispersed and only occasionally is it in concentrations of economic interest. The availability is therefore rather low. The number of independent zirconium minerals is small. It exhibits marked oxyphile and lithophile characters and all of its natural compounds appear to contain oxygen.

Zirconium compounds do not as a rule separate in the earlier stages of magmatic crystallization, they are therefore encountered as frequent accessory minerals in acid rocks like granites and in alkalic rocks of the syenite and nepheline syenite group.

ZIRCONIUM MINERALS

Name	*Formula*	*Occurrence*
Baddeleyite	ZrO_2	In gem gravels and as an accessory mineral in magnetite–pyroxene rock. May be an ore (Brazil)
Zirkelite	Complex oxide of Zr, Ti, Y, etc.	With baddeleyite and perovskite
Zircon	$ZrSiO_4$	Commonest mineral. Accessory in acidic and alkalic rocks. Detrital mineral in sands and sediments
Elpidite	$NaZrSi_6O_{15}.3H_2O$	Rare accessory minerals in igneous rocks
Dalyite	$K_2ZrSi_6O_{15}$	Rare accessory minerals in igneous rocks
Rosenbuschite	$(Ca,Na)_3(Zr,Ti)Si_2O_8F$	Rare accessory minerals in igneous rocks
Wöhlerite	$NaCa_2(Zr,Nb)Si_2O_8(O,OH,F)$	Rare accessory minerals in igneous rocks

The figure of 460 g/ton quoted for the average zirconium content of granites suggests the most obvious primary source of the element. Such material would, however, be a very intractable ore. Zircon, which is the usual mineral present in granite, is characterized by a high resistance to mechanical attrition and chemical alteration. Being also moderately dense ($d = 4{\cdot}6$) zircon has properties favourable for its concentration in resistate material derived from the weathering of

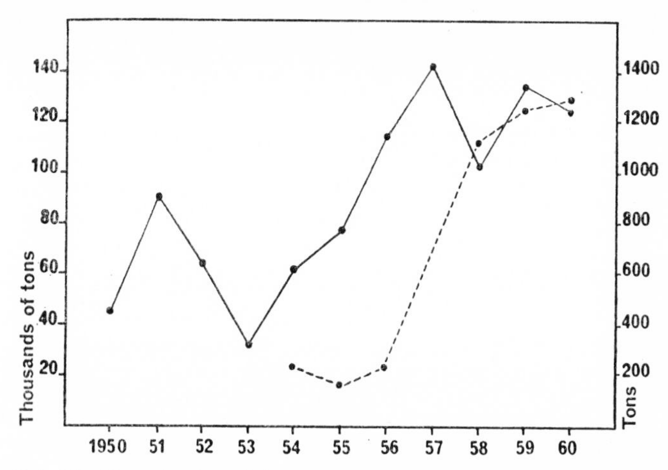

FIG. 15.4. World production of zirconium minerals in thousands of tons (continuous line) and zirconium metal in tons (broken line).

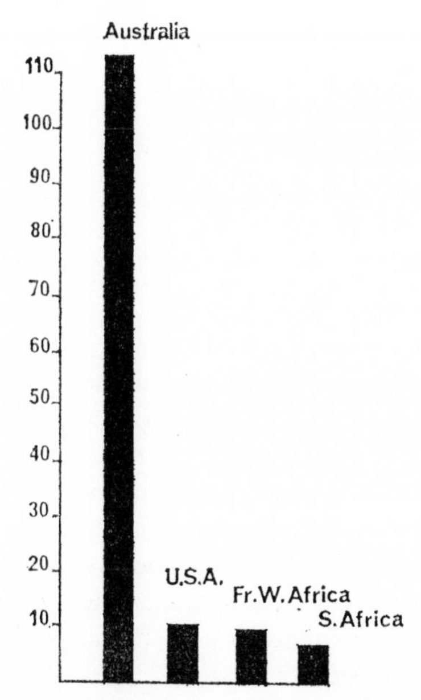

FIG. 15.5. Chief producers of zirconium mineral, 1959, in thousands of tons.

granite and syenitic rocks. Transparent crystalline zircon is also fairly common and may be of gem quality.

There is no indication that zirconium performs any significant biological role. It is not uncommonly present as an element in plant material but is probably adventitious.

Hafnium

There are six stable isotopes of this element in Nature:

$^{174}Hf = 0{\cdot}18\%$	$^{178}Hf = 27{\cdot}08\%$
$^{176}Hf = 5{\cdot}15\%$	$^{179}Hf = 13{\cdot}78\%$
$^{177}Hf = 18{\cdot}39\%$	$^{180}Hf = 35{\cdot}44\%$

There is no clear evidence for the existence of any natural radioactive isotopes of hafnium.

Geochemistry

The abundance of hafnium in the Earth's crust is estimated at 5 g/ton. It may be regarded as a moderately scarce element, but is in fact much more abundant than such 'common' metals as antimony and cadmium.

Hafnium provides perhaps the most perfect example of a dispersed and concealed element. On account of the effect of the lanthanide contraction, the atomic and ionic radii of hafnium are very slightly *smaller* than those of zirconium.

	Atomic Radius (Å)	*Ionic Radius* (Å)
Zirconium	1·60	0·79
Hafnium	1·59	0·78

Thus the geochemistry of hafnium is nearly identical with that of zirconium. The two elements always occur together, and the main source, zircon, could well be formulated $(Zr,Hf)SiO_4$. Some slight differentiation of the two elements must, however, occur in Nature because the ratio Hf : Zr varies over fairly wide limits. In nepheline syenites it is perhaps 1 : 100 and in granites about 3 : 100.

The mineral thortveitite $Sc_2Si_2O_7$ carries a few per cent of zirconium and is relatively enriched in hafnium. Certain metamict minerals derived from zircon are found to have the highest ratios of hafnium, up to 4 : 100 in the mineral cyrtolite.

As might be supposed there is no independent mineral of hafnium.

Germanium

There are five stable isotopes of this element in Nature:

$^{70}Ge = 20{\cdot}55\%$	$^{74}Ge = 36{\cdot}74\%$
$^{72}Ge = 27{\cdot}37\%$	$^{76}Ge = 7{\cdot}67\%$
$^{73}Ge = 7{\cdot}67\%$	

Slight variations in the isotopic constitution of germanium from different sources have been reported. There is some evidence that a slight degree of natural fractionation may take place.

Geochemistry

Germanium is rather a scarce element in the Earth's crust, being estimated at 2 g/ton. It exhibits some variation in its geochemical affinities. From meteorite analyses it appears that germanium may occur in the metallic phase to the extent of 500 g/ton and in the sulphide phase to the extent of 30 g/ton. It would therefore seem that it has a distinct siderophile character and a lesser thiophile character, but it is also to a certain extent lithophile. It seems probable that in the Earth as a whole germanium is a fairly common element, as its even atomic number and fairly low atomic weight would suggest.

Goldschmidt and his co-workers were able to show considerable analogy between silicon and germanium, and the fact that the element could be ' concealed ' in silicate minerals to a wide extent. The few independent germanium minerals are, however, sulphur compounds of very local occurrence.

GERMANIUM MINERALS

Name	*Formula*	*Occurrence*
Canfieldite	$Ag_8(Ge,Sn)S_6$	Vein mineral with sulphides Freiburg, Saxony and Bolivia
Germanite	$Cu_3(Ge,Ga,Fe,Zn)(As,S)_4$	With sulphide ores, Tsumeb, S.W. Africa
Argyrodite	Ag_8GeS_6	Vein mineral, Saxony and Bolivia

The ionic radius of Ge^{4+} is 0·53 against 0·42 Å for Si^{4+} so it might be expected that GeO_4 tetrahedra similar to SiO_4 tetrahedra could enter into crystal structures of similar type. It would be expected that the germanium would be accepted in the later stages. This appears in general to be the case, so that granitic rocks are richer in the element than basic ones. The exact behaviour of germanium in the weathering processes is somewhat obscure but it may be expected that GeO_2 will be more mobile than SiO_2. Germanium in soils and sediments may have been co-precipitated with $Fe(OH)_3$. It was shown by Goldschmidt that

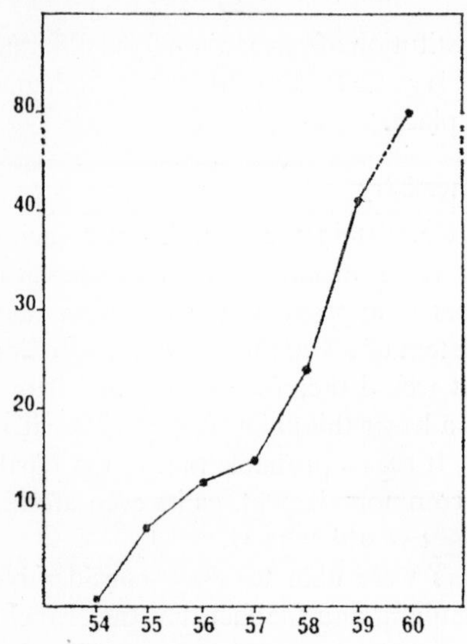

FIG. 15.6. World production of germanium metal, in tons.

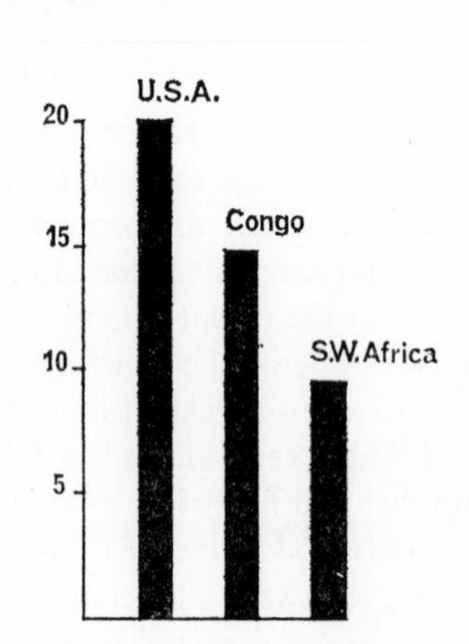

FIG. 15.7. Chief producers of germanium, 1959, in tons.

the germanium absorbed by plants tended to be concentrated in humus and ultimately in certain coals. Thus in his earlier work, Goldschmidt found in the ash of a coal from Hartley, Northumberland, England, 1·6% of GeO_2. Subsequently, many other coal ashes were found to be enriched in the element.

Thus the present source of germanium is in fact mainly from selected flue dusts from producer gas manufacture. It is also obtained from flue dusts from copper smelting—in this case the germanium is derived from the traces of sulpho-salts present in the original ore.

Tin

There are ten stable isotopes of tin in Nature, which is a greater number than for any other element:

^{112}Sn = 0·95%	^{118}Sn = 24·01%
^{114}Sn = 0·65%	^{119}Sn = 8·58%
^{115}Sn = 0·34%	^{120}Sn = 32·97%
^{116}Sn = 14·24%	^{122}Sn = 4·71%
^{117}Sn = 7·57%	^{124}Sn = 5·98%

The existence of these numerous isotopes is no doubt related to the varied processes by which they may have been formed which probably include both the slow and rapid neutron capture processes. In this connexion reference should be made to tellurium (p. 288) for what are probably analogous processes.

Geochemistry

Tin is a relatively rare element in the Earth's crust, the abundance being estimated at 3 g/ton. Its natural chemistry presents interesting features some of which require clarification.

Meteorite analyses suggest that tin is concentrated in the metallic phase and should be a markedly siderophile element. The content in the sulphide phase is also higher than in the silicate phase so that in the accessible crust it might be expected that tin would show both thiophile and lithophile characters. It is indeed met with in the form of sulphide minerals but these are local and rather rare. By far the most important natural compound is the dioxide cassiterite SnO_2.

The localization of tin deposits is remarkable. The chief regions are:

Central Europe—Bohemia, Saxony.
Western Europe—England (Cornwall), France, Spain.
S.E. Asia—Malaya, Indonesia, S.W. China.
South America—Bolivia.
Australasia—N.S. Wales, Tasmania.
Africa—Nigeria, Congo.

There are, however, vast tracts such as the North American Cordillera, the Canadian Shield, the Urals and North Western Europe which are virtually devoid of tin, although rich in other metals.

Tin in the form of its volatile fluoride or chloride SnF_4 or $SnCl_4$ may have been transported upward from the sub-crustal zone of the Earth and ultimately, near the surface, converted by hydrolysis into the very stable oxide. There does not seem to be much evidence of Sn^{2+} compounds in Nature. The formation of sulphide ores is much more local; they are, however, rather typical in Bolivia.

Cassiterite is sufficiently stable and resistant to accumulate in alluvial deposits and such placers are the principal source of Malayan tin. Cornwall, England, once the major source, has now almost ceased to produce. It is estimated that since 500 B.C. the Cornish peninsula has in fact yielded more than three million tons of tin. It can be considered to have been one of the most heavily mineralized regions in the world and competent authorities do not consider that it is yet exhausted.

In contrast with lead, tin does not form very many independent minerals. Some tin (which is present more particularly in acidic rocks), may be concealed in such silicate minerals as sphene $CaTiSiO_5$. The much larger ionic radius ($Sn^{4+} = 0{\cdot}71$ Å) is not favourable to the replacement of silicon by tin in common silicate minerals but it might be able to replace Ti^{4+}.

The occurrence of native tin appears to be rather doubtful as also is the occurrence of the sulphide SnS herzenbergite.

TIN MINERALS

Name	*Formula*	*Occurrence*
Stannite	Cu_2SnFeS_4	Hydrothermal veins, Cornwall, etc.
Teallite	$PbSnS_2$	Silver–tin veins, Bolivia
Montesite	$PbSn_4S_5$	
Cylindrite	$Pb_3Sn_4Sb_2S_{14}$?	
Franckeïte	$Pb_5Sn_3Sb_2S_{14}$?	
Cassiterite	SnO_2	Hydrothermal and pneumatolytic deposits, widespread
Nordenskiöldine	$CaSnB_2O_6$	South Norway
Stokesite	$CaSnSi_3O_9.2H_2O$	S.W. Africa
Arandisite	$Sn_5Si_3O_{16}.4H_2O$	

The processes of weathering of tin compounds are somewhat uncertain but the oxide is a very persistent mineral and is found in trace quantities in many sands and other sediments. Some beach sands have been worked for tin—*e.g.*, in Cornwall and are probably still potentially valuable.

There is no indication of any biological significance of tin.

Tin is a valuable metal, the world output is relatively small, and the price is comparatively high. For this reason placers carrying less than

1 part in 10000 of ore can be worked and vein material with about 1 % of tin is also profitable.

Typical Deposits of Tin

Malaya. Over 500 mines yielding chiefly alluvial cassiterite but veins and stockworks are also worked. The tin ores are probably associated with granites.

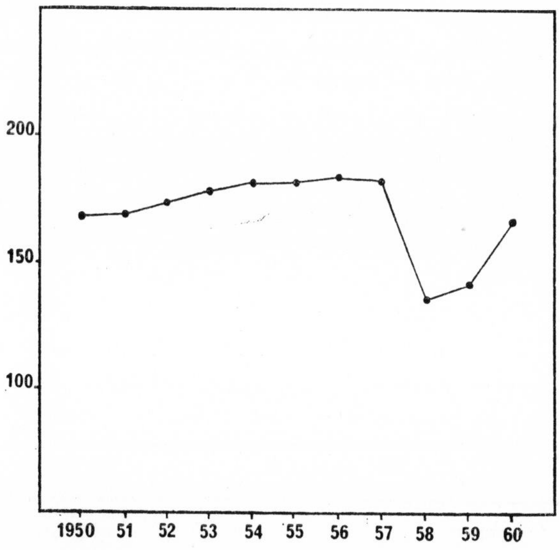

FIG. 15.8. World production of tin ore (metal content), in thousands of tons.

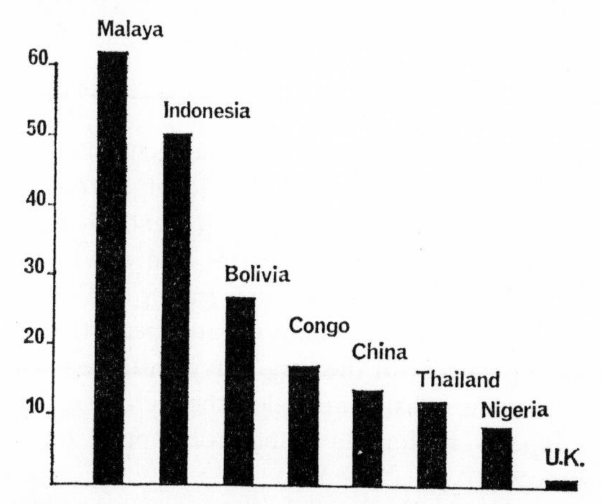

FIG. 15.9. Chief producers of tin ore (metal content), 1956, in thousands of tons.

Bolivia. Fissure or fracture veins with cassiterite, sulphide ores including marcasite, bismuthine, pyrrhotite, etc. Also Tertiary fissure veins with stannite, cassiterite, and silver ores.

Cornwall. The veins are in clay–slate (killas) with dykes of felsite (elvan) related to the granite intrusions. Near the surface were lead–silver ores, then copper, passing down into tin ore with wolfram. There is considerable vapour-phase dissemination into the wall-rock and the ore is associated with fluorine minerals.

Nigeria. Dissemination in decayed granite and in alluvials derived from it. It occurs with columbite.

Lead

There are four natural stable isotopes of this element:

$^{204}Pb = 1{\cdot}36\%$ $^{207}Pb = 21{\cdot}11\%$
$^{206}Pb = 25{\cdot}42\%$ $^{208}Pb = 52{\cdot}10\%$

There are perceptible variations in the above conventional figures and a number of causes may be cited, some of which are referred to below. Lead is intimately associated with radioactive decay processes. Not only is it the stable end-product of three radioactive families but several radioactive isotopes of the element occur as members of these families.

Parent		*Intermediate*	*End product*
Uranium	^{238}U	^{214}Pb (26·8 m)	^{206}Pb
Thorium	^{232}Th	^{212}Pb (10·6 h)	^{208}Pb
Actinouranium	^{235}U	^{211}Pb (36·1 m)	^{207}Pb

It will be seen that in lead associated with radioactive minerals there are likely to be noticeable variations in isotopic composition. It should be noted that of the various isotopes, only ^{204}Pb is entirely non-radiogenic. Nevertheless, much of ordinary terrestrial lead must be of primary non-radiogenic origin. It is presumed that lead would be formed in the more advanced stages of stellar evolution. It appears that lead isotopes could result from the decay of the very heavy unstable nuclei produced by neutron capture on a fast time-scale—the r-process. Lead isotopes could also, it appears, be formed by neutron capture on a slow time-scale—the s-process.

Isotopic variations in terrestrial lead specimens can be related in suitable cases to the recognized processes of radioactive decay of which lead is the end-product. Therefore it has been possible to devise methods of determination of geological age which depend on the isotopic constitution of the specimen.

There are several ways in which these studies can be carried out. Originally, calculations were based on the Pb/U ratio in selected minerals in which it was believed that all the lead was of radiogenic origin. Later the $^{206}Pb/U$ ratio, involving atomic weight determination, was used. The more recent mass-spectrometric methods of determining the isotopic constitution of radiogenic lead make allowance for the non-radiogenic lead present and are susceptible of a higher degree of accuracy. Geological ages determined by this method may be accurate within a few per cent. It must be remembered that these methods presuppose that a suitable mineral is available and they therefore have some limitations as compared, for example, with methods based on decay of radioactive potassium, which is much more widespread in common rock minerals.

Terrestrial Abundance

At 15 g/ton, lead is a relatively common element in the Earth's crust and this value is also comparable with its relative cosmic abundance. Lead may well be markedly siderophile but in the accessible crust we know that it is more particularly associated with sulphide bodies. Nevertheless perceptible amounts are also dispersed in magmatic minerals.

Geochemistry

The element is noticeably ubiquitous in the Earth. It is virtually absent from the atmosphere and rare in the hydrosphere but is otherwise widespread. Lead is also characterized by the large variety and number of its natural compounds. This may be ascribed to the following causes:

(*a*) Ability to exist (usually) in the divalent but also occasionally in the tetravalent state.
(*b*) Fairly basic character of divalent lead which allows of the existence of many stable oxy-salts.
(*c*) Many sparingly soluble compounds which are easily immobilized and preserved as local accumulations of possibly rare species.
(*d*) Ionic radius (Pb^{2+} = 1·20 Å) which enables it to deputize for K^+,

etc., and therefore to enter many silicate mineral lattices in trace quantities.

Lead Minerals

The majority of lead minerals have been formed from low-temperature hydrothermal solutions. They are, as previously noted in Chapter 11, very commonly associated with ores of zinc. In weathering processes, however, lead and zinc compounds behave differently and become separated to a large extent. Of the many minerals, practically only three—galena, cerussite, and anglesite—are used as ores. The others are mainly of mineralogical interest.

LEAD MINERALS

Name	*Formula*
Native Lead	Pb
Galena	PbS
Sartorite	$PbAs_2S_4$ (about 37 related species)
Bournonite	$PbCuSbS_3$ (about 25 related species)
Seligmannite	$CuPbAsS_3$
Litharge	PbO
Minium	Pb_3O_4
Plattnerite	PbO_2
Matlockite	$PbFCl$
Cotunnite	$PbCl_2$
Cerussite	$PbCO_3$
Hydrocerussite	$Pb_3(CO_3)_2(OH)_2$
Phosgenite	$Pb_2CO_3Cl_2$
Leadhillite	$Pb_4SO_4(CO_3)_2(OH)_2$
Alamosite	$PbSiO_3$
Plumbogummite	$PbAl_3(PO_4)_2(OH)_5.H_2O$
Pyromorphite	$Pb_5(PO_4)_3Cl$
Mimetite	$Pb_5(AsO_4)_3Cl$
Vanadinite	$Pb_5(VO_4)_3Cl$
Anglesite	$PbSO_4$
Wulfenite	$PbMoO_4$
Crocoite	$PbCrO_4$
Stolzite	$PbWO_4$

The list of lead minerals indicates the great diversity of natural compounds of lead. To a large extent these are secondary minerals. The primary mineral would in general be the simple sulphide galena and perhaps a few of the sulpho-salts. Sub-aerial processes would cause oxidation to sulphate, in some cases carbonation would give mainly cerussite which is perhaps the most abundant and widespread secondary lead mineral. In some districts, however, a variety of mineral solutions have interacted with the lead compounds so that many different oxy-salts are encountered. The complex phosphate pyromorphite is in some localities remarkably abundant.

Silicates are few and rather rare. Many of the complex secondary minerals are coloured, and when well crystallized provide some of the most beautiful mineralogical specimens, the more so on account of the high refractive index of most lead compounds.

In primary magmatic minerals lead is found in small amounts in potassic silicates such as feldspars and tends to be accepted in the earlier stages of crystallization as would be expected from its greater charge and slightly smaller size than the potassium ion. In soils, lead ions can be adsorbed on clay minerals in the same way as potassium ions. Under reducing conditions in humic soils and in sapropels lead is eventually re-converted to sulphide and a few workable deposits of lead ore have originated in this way.

Typical Lead Deposits

Reference should also be made to the zinc deposits described in Chapter 11.

South Eastern Missouri, U.S.A. Disseminated replacements in a flat-lying Cambrian dolomite. Probably a hydrothermal deposit from an undisclosed igneous source.

Cœur d'Alene, Idaho, U.S.A. Fillings of small faults in pre-Cambrian quartzite related to intrusions of monzonite. Lead and zinc outputs about equal, silver about 14 oz/ton.

Linares-Carolina, Spain. Rich silver–lead veins with ore-shoots in quartzite, slate, and also in granite. Zinc is absent.

Stantrg, Yugoslavia. Replacement deposits in limestone with associated Palaeozoic schist and Tertiary volcanic rocks. Variety of base metal sulphides, etc. It is a combination of mesothermal and hypothermal metallization.

Bawdwin, Burma. Replacement deposits in a shear zone between feldspathic grits and rhyolite, and tuffs of early Palaeozoic age. Lead and zinc and a little copper.

British localities. These are now worked on a very restricted scale but some at least are far from being exhausted. They are mostly fissure or vein deposits with flats in limestone of Carboniferous age as in the Mendips and the Pennine region (Derbyshire and Alston Moor region). In older Palaeozoic rocks, mainly slates and volcanic tuffs, in Wales

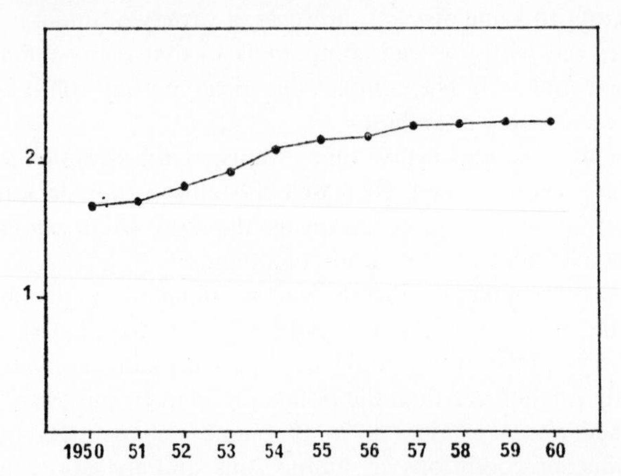

FIG. 15.10. World production of lead ore (metal content), in millions of tons.

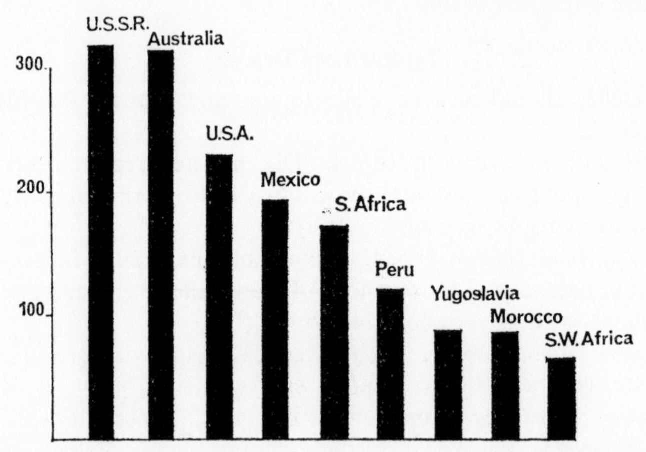

FIG. 15.11. Chief producers of lead ore (metal content), 1959, in thousands of tons.

and the Lake District. In Scotland the Leadhills region was productive with secondary minerals in great variety.

There are several localities which may still have economic interest in Ireland: for instance, the recently prospected deposit at Tynagh, Co. Galway.

CHAPTER 16

Nitrogen and Phosphorus

Nitrogen

There are two stable isotopes of this element in Nature:

$$^{14}N = 99{\cdot}635\%$$
$$^{15}N = 0{\cdot}365\%$$

so that the ratio $^{14}N/^{15}N = 273 : 1$.

Nitrogen stands sixth in order of cosmic abundance of the elements. It does not appear that nitrogen nuclei could be synthesized in a pure 'hydrogen-burning' star, but assuming a certain amount of carbon present, then the well-known carbon cycle could become established and this involves (as indicated in Chapter 7, p. 109) the formation of both of the above isotopes but the $^{14}N/^{15}N$ ratio is estimated at 100000 : 1. The mode of formation is proton capture by ^{12}C nuclei and the reaction chain is affected by temperature so that the proportions of carbon and nitrogen formed vary with the special conditions found in different stars. Thus the central star of 'Campbell's hydrogen envelope star' seems to contain no nitrogen but the surrounding nebula is rich in it. There is a group known as 'nitrogen stars' in which the abundance of the element is ten to fifteen times that of carbon. Nitrogen is also found to be abundant in novae and in the Crab Nebula (an exploded supernova). The only nuclear reactions of nitrogen occurring on the Earth involve the capture of cosmic ray neutrons:

$$^{14}N + n = ^{14}C + ^{1}H \qquad \text{and} \qquad ^{14}N + n = ^{12}C + ^{3}H$$

and nitrogen atoms are by far the strongest absorbers of neutrons in the upper atmosphere.

It has been claimed that fractionation of the two nitrogen isotopes by gravitation takes place in the atmosphere, slightly but perceptibly, but the results are doubtful. A number of extra-terrestrial occurrences of nitrogen compounds can be reported:

(*a*) In super-giant stars—NO, CN, NH, and N_2.

(*b*) In comets—NH, CN, N_2^+, NH_2.

(*c*) In planetary atmospheres—Mars, N_2 (?), Jupiter, Saturn, Titan, NH_3.

Geochemistry

This subject may be treated conveniently under the following headings:

(1) Atmospheric chemistry of nitrogen.
(2) Nitrogen in the lithosphere and hydrosphere.
(3) Nitrogen in relation to the biosphere.
(4) Natural cycles of nitrogen.

(1) The nitrogen content of the atmosphere is quoted as 75·15% by weight, which represents a total mass of 36·648 Gg of the element as N_2. Reactions of nitrogen in the atmosphere are due to electrical or photochemical causes. Electrical discharges (lightning, etc.) can cause the combination

$$N_2 + O_2 = 2NO$$

which will ultimately be followed by

$$2NO + O_2 = 2NO_2$$

and the dioxide formed can be washed down by rain as HNO_2 and HNO_3, and so removed from the atmosphere. The importance of this method of fixation as compared with photochemical methods is by no means certain, and may be much less important than was at one time thought.

Nitrogen differs from oxygen in its atmospheric photochemistry in having a very small cross-section for dissociation of its molecules by ultra-violet absorption. The only important process leading to dissociation involves the band 1200–1250 Å. Atomic nitrogen can also be formed by such reactions as

$$N_2^+ + O = NO^+ + N \quad \text{and} \quad O_2^+ + N_2 = NO^+ + N$$

The principal reaction by which atomic nitrogen is removed is probably

$$N + O_2 = NO + O$$

It is estimated that at 100 km height the atomic nitrogen concentration might be 5×10^6 atoms/cm³.

This last reaction is the main photochemical source of nitric oxide. It can be destroyed by the reaction

$$N + NO = N_2 + O$$

At night, the reactions

$$NO + O = NO_2 + h\nu \qquad \text{and} \qquad NO_2 + O = NO + O_2$$

occur. The rates of these two reactions are approximately equal so that an equilibrium between NO and NO_2 in about equal amounts is established. In daytime the ionization of NO becomes important due to solar X-rays of 1–10 Å, which are able to reach the upper atmosphere. The minute nitrous oxide content of the atmosphere may be due to

$$N + NO_2 = N_2O + O$$

(2) Rainwater always contains nitrogen compounds, mainly ammonia and nitrates, and the ammoniacal content is generally in excess. The total precipitation of fixed nitrogen to the soil per annum may be of the order of 1 ton per square mile in cool temperate regions and up to more than three times this figure in regions of tropical rainfall. It is not clear how far the ammonia is re-cycled from the soil. Nitrogen compounds will pass ultimately to the sea and this process is aided by the readiness with which nitrates are leached out from soils.

There are three inorganic nitrogen species in the ocean, *viz.*, NH_3, NO_2^-, and NO_3^- ions. Their amounts are variable in latitude, in depth, and also seasonally. They are intimately connected with the organic cycle of nitrogen and are further considered below.

The following figures have been quoted:

N as NO_3^-, 0·001—0·7 g/ton.
N as NO_2^-, 0·0001–0·05 g/ton.
N as NH_3, 0·005–0·05 g/ton.

In the evaporite deposits derived from sea water the content of nitrogen compounds is negligible. There are, however, some very specialized deposits undoubtedly formed by evaporation which are rich in nitrates. The most important of these are the very remarkable nitrate deposits of the desert region of Chile, South America, not far inland from the Pacific seaboard. The deposits, known as *caliche*, occur in a narrow strip 450 miles long and 10–50 miles wide, at elevations of 1000–3000 ft on gentle valley slopes in layers from a few inches to several feet in thickness. The beds are not generally more than 40ft below the surface and may be worked open-cast. The sodium nitrate content is about 25% and is accompanied by 2–3% of potassium nitrate together with sodium chloride, and an extraordinary assembly of salts including iodates, borates, phosphates, and even chromates.

As previously stated, this seems to be the locality of highest oxidation potential in the Earth. It is now generally agreed that these salts have been transported by ground water and deposited by evaporation. The origin of the deposit is very controversial. The presence of borates and other non-nitrogenous matter lends support to the theory that volcanic material containing ammonium salts provided the substance for oxidation to nitrate under rather specialized climatic conditions.

Lewis and Randall (1923) drew attention to the reaction

$$\tfrac{1}{2}N_2 + \tfrac{1}{2}H_2O + \tfrac{5}{4}O_2 = HNO_3 \text{ aq. } (\Delta F = 753 \text{ cal})$$

and pointed out the small energy intake required to favour the reaction. They suggested that with a suitable catalyst the whole of the atmospheric oxygen and much of the nitrogen could contribute to making the ocean a dilute nitric acid solution.

Goldschmidt thought that in the special conditions of the Chilean desert such a catalyst might be operative. The nitrate formation is then ascribed to the peculiar combination of ecology, meteorology, and topography which exists in this region.

Much less important accumulations of nitrates are found in saline lakes in Western U.S.A., such as Searle's Lake and in Death Valley in Nevada. In some of these cases a volcanic origin seems probable.

The possible occurrence of nitrogen in magmatic material is very obscure. There is no certain evidence of nitrates in igneous rocks and the ammonium compounds associated with some volcanic emanations have been attributed to organic material sources. It is possible, however, that binary nitrides may occur in the sub-crustal regions of the Earth and by reaction with steam would form ammonia. Titanium nitride TiN is known in some meteorites. It may also occur deep in the Earth. Goldschmidt also suggested boron nitride as a possible source of ammonia. If the Earth's nitrogen was originally in the form of ammonia (as seems to be the case with the larger planets), then the gas may have been partly lost into space and partly converted into elemental nitrogen such as now exists. On the present evidence it would not appear that nitrogen can be regarded as a lithophile element.

Biological Aspects

Some reference to this topic has already been made in Chapter 4 (p. 69). Nitrogen is one of the most important elements in biological processes. The nitrogen content (dry weight) of living organisms is considerable, being about 12% for a mammal and 3% for a plant. Although these amounts show great variation, yet in animals the nitrogen content is generally much higher than in plants because of the very much larger amount of protein. In the case of plants, the protein, although essential, does not contribute to the structure of the organism as in animals, its place being taken by the non-nitrogenous polysaccharides.

Nitrogen in the biosphere is very largely involved in cyclical processes as indicated in Figs. 16.1–16.3. As compared with carbon, there are not many substances in which nitrogen can be fixed in a permanent or semi-permanent form. There are, for example, no predominantly nitrogenous sediments comparable to the carbonaceous bioliths. Furthermore, the ocean acts less as reservoir and buffer for nitrogen compounds than it does for carbon as carbon dioxide. In general, protein material is very susceptible to bacterial attack, therefore it soon returns its nitrogen to the cycle. Mention should, however, be made of the nitrogenous carbohydrate chitin which is largely produced by arthropod animals and which is comparatively resistant to decay. There is also a small nitrogen content in coal and petroleum. It is of the order of 1 % or less. Such nitrogen is present in various forms, including a number of heterocyclic substances.

Mention has been made in Chapter 4 (p. 81) of speculations on the possibility of a life system in which liquid ammonia replaces water as a basic fluid medium. In connexion with the utilization of nitrogen compounds by organisms we must note the massive contribution of man-made fertilizers. The total content of nitrogen artificially fixed in this form is somewhat difficult to estimate but is probably now 4–5 million tons annually and it seems to be increasing steadily, being in fact a reflection of the intensive development of agriculture.

NITRATE MINERALS

Name	*Formula*
Nitratine	$NaNO_3$
Nitre	KNO_3
Gerhardtite	$Cu_2NO_3(OH)_3$
Nitromagnesite	$Mg(NO_3)_2.6H_2O$
Nitrocalcite	$Ca(NO_3)_2.nH_2O$
Nitrobarite	$Ba(NO_3)_2$
Buttgenbachite	$Cu_{19}Cl_4(NO_3)_2(OH)_{32}.3H_2O$
Darapskite	$Na_3NO_3SO_4.H_2O$

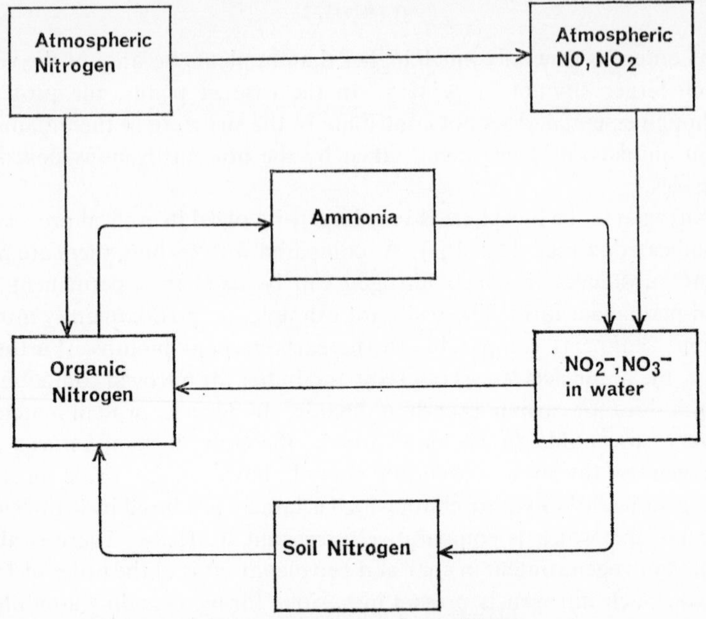

FIG. 16.1. General cycle of nitrogen.

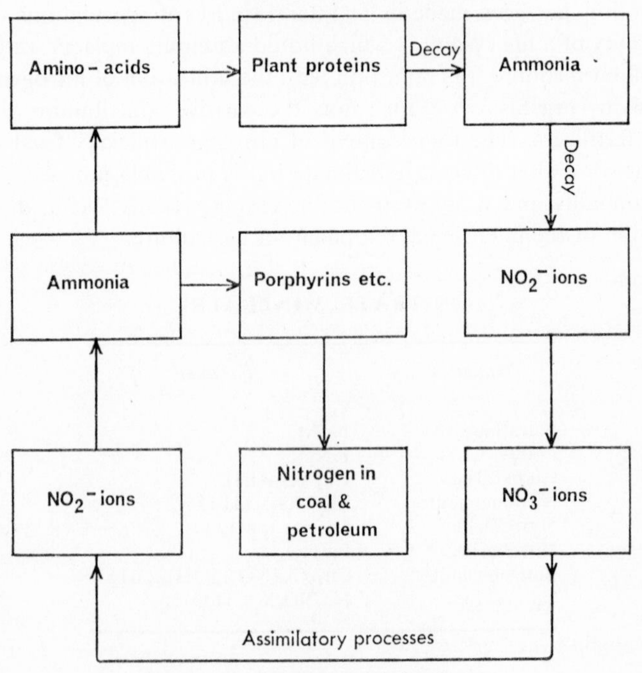

FIG. 16.2. Nitrogen cycle in plants.

Nitrate Minerals

These are mostly of rare occurrence and chiefly in evaporites and in cave earths.

Nitrogen Cycles in Nature

For the sake of clarity these are presented as a number of simple cycles rather than a single comprehensive diagram (Figs. 16.1–16.3).

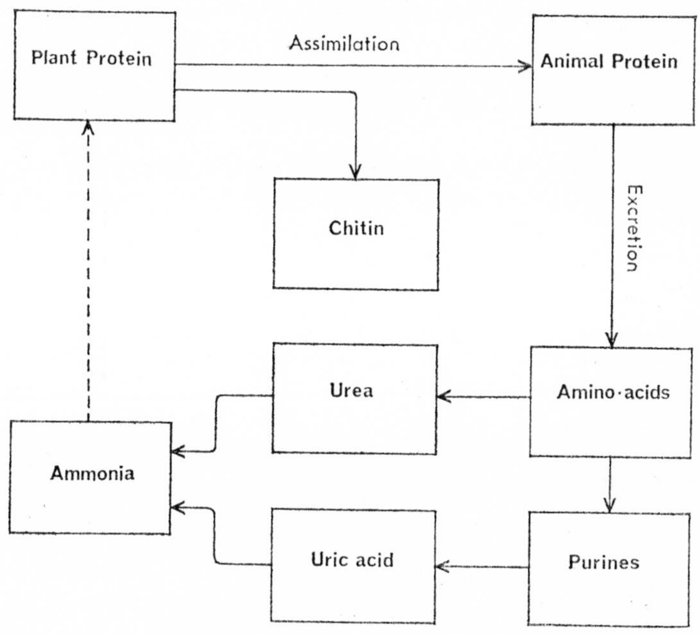

FIG. 16.3. Nitrogen cycle in animals.

Phosphorus

Only the single nuclide ^{31}P occurs in Nature. The relative cosmic abundance of phosphorus is rather low compared with other elements near to it in order of atomic number. There is no very great variation in its relative abundance in other locations, including the Earth's crust, solar atmosphere, and in meteorites.

It is thought that phosphorus atoms could be built up at the end of the 'helium burning' phase of a star when the temperature has reached

$1 \cdot 3 \times 10^9$ °K. Under these conditions a variety of reactions can take place including:

$$^{16}O + ^{16}O = ^{31}P + ^{1}H \quad \text{and} \quad ^{31}S + n = ^{31}P + \beta^+ + \nu$$

Geochemistry

The analyses of different phases of meteorites give no very conclusive indication of the geochemical affinities of the element. The most interesting aspect is its occurrence in the metallic and sulphide phases of meteorites as the phosphide schreibersite $(Fe,Ni, Co)_3P$. This seems to be the only case in Nature of the definite occurrence of phosphorus in such a low oxidation state. The oxygen-rich conditions of the Earth's

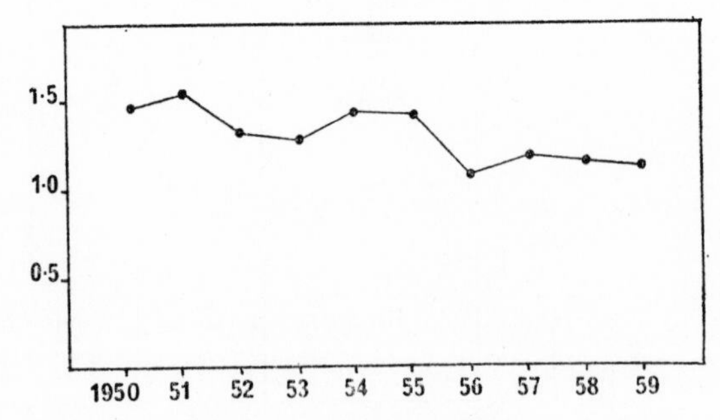

FIG. 16.4. World production of natural nitrates, in millions of tons (almost wholly from Chile).

crust lead, as far as is known, to the exclusive existence of phosphorus as the PO_4 ion. On account of the stability of this tetrahedral structure it presents analogies with the corresponding SiO_4 ion. It is in fact possible in a few cases for replacement of silicate by phosphate to take place. This only seems to happen among ortho-silicates like zircon $ZrSiO_4$, and the electrical neutrality of the lattice is then maintained by replacement of part of the tetravalent zirconium by trivalent lanthanide metals.

It is perhaps for this reason that such examples are not very common. An important difference from silicates is that the PO_4 tetrahedra in minerals apparently cannot link together by sharing oxygen atoms. The mineral chemistry of phosphorus is therefore less complex than that of silicon. There is also some analogy between PO_4 and SO_4 as is evident from the existence of a number of complex minerals containing both of these ions, which along with the silicate ion are all of about the same size, the X—O separation in these structures being generally 1·5–1·6 Å. In practice phosphorus is to be regarded as a lithophile and oxyphile element. A large number of phosphates are known but it is probable that as much as 95% of the phosphorus of the Earth's crust is contained in minerals of the apatite group. These are very common accessory minerals in igneous rocks. The general formula of apatites is $Ca_5(PO_4)_3X$, where X is an anion such as F, Cl, or OH. The commonest form is probably $Ca_5(PO_4)_3F$ or fluorapatite. The crystal lattice of this mineral is rather complex, involving a hexagonal network and arrangement of Ca and F ions which gives a very stable structure and allows of a remarkable amount of substitution of both cations and anions.

PHOSPHATE MINERALS

Name	*Formula*	*Occurrence*
Monetite	$CaHPO_4$	Underlying guano deposits
Alluaudite	$(Na,Fe,Mn)PO_4$	Altered pegmatites
Whitlockite	$Ca_3(PO_4)_2$	Pegmatites and phosphate deposits
Xenotime	YPO_4	Minor accessory rock minerals
Monazite	$(Ce, etc.)PO_4$	(Minor accessory rock minerals)
Belinite	$AlPO_4$	Associated with certain iron ores
Brushite	$CaHPO_4.2H_2O$	Organic phosphates and cave earths
Vivianite	$Fe_3(PO_4)_2.8H_2O$	Altered ores, fossils
Cornetite	$Cu_3PO_4(OH)_3$	Oxidized copper ores
Pseudomalachite	$Cu_5(PO_4)_2(OH)_4.H_2O$	(Oxidized copper ores)
Fluorapatite	$Ca_5(PO_4)_3F$	Accessory rock minerals
Chlorapatite	$Ca_5(PO_4)_3Cl$	(Accessory rock minerals)
Hydroxyapatite	$Ca_5(PO_4)_3OH$	Organic phosphate residues
Pyromorphite	$Pb_5(PO_4)_3Cl$	Oxidized lead ores
Lazulite	$(Mg,Fe)Al_2(PO_4)_2(OH)_2$	Metamorphic aluminous, rocks
Turquoise	$CuAl_6(PO_4)_4(OH)_8.5H_2O$	Secondary minerals
Wavellite	$Al_6(PO_4)_4(OH)_6.9H_2O$	(Secondary minerals)

The phosphorus content of igneous rocks is generally well known and ranges from about 2000 g/ton for basic rocks like gabbro and basalt to 600 g/ton for acid rocks like granite and rhyolite. It is evident that a considerable proportion of the phosphorus crystallizes from the magma as an independent mineral and at a fairly early stage. The relation of this to some of the other components of the rocks is shown in Fig. 16.5.

It was at one time thought that much of the apatite associated with basic, gabbroic rocks was chloroapatite, whilst the apatite of acidic, granitic rocks was mainly fluoroapatite, but this view is not now maintained. The OH content of apatites may be considerable. Apatite is also met with in hydrothermal veins to a minor extent, and it may occur as a detrital mineral in sedimentary rocks.

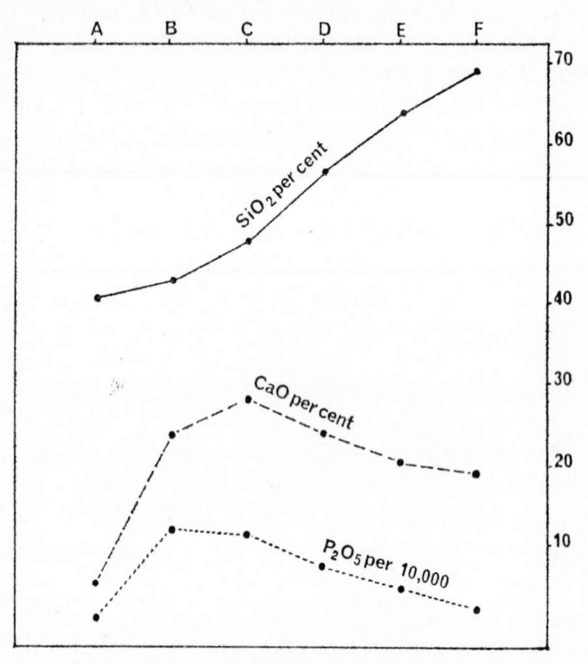

FIG. 16.5. A, dunites. B, hornblendites. C, gabbros. D, diorites. E, granodiorites. F, granites.

Biogeochemistry

The persistence of the PO_4 radical is remarkable. It appears that all naturally occurring organic derivatives of the element are esters and amides of phosphoric acid, and the esters of polyphosphoric acids containing up to three PO_4 units. These compounds are found in organisms as actual structural units, as coenzymes, and as enzyme substrates. Phosphorus is found in every cell of the vertebrate body.

Phosphate residues in metabolites facilitate breakdown, but are also prospective carriers of energy. The ‘ energy-rich ’ phosphate bonds may liberate 11000cal on fission, as compared with only about 3000cal for a normal phosphoric ester. Types of compounds possessing such bonds include linkages with carbonyl, enol groups, nitrogen, and oxygen in polyphosphate radicals.

Bone is an important animal material rich in phosphorus. The formation of bone depends on the supply of PO_4 through hexose phosphate. The bone salts may belong to the apatite group and approximate to a carbonate–apatite $\{Ca_3(PO_4)_2\}_n.CaCO_3$; both OH and F ions may be present. The mineral material is contained in the cartilaginous matrix and does not seem to form part of the molecular structure of the bone.

Biological processes are very largely concerned in the phosphorus cycle in Nature and they involve both concentration and dispersal as will be seen from the diagram Fig. 16.6. The processes include:

(*a*) Solution of PO_4 ions in the weathering of primary rock phosphates.
(*b*) Absorption of PO_4 ions by organisms from soils and from sea water.
(*c*) Absorption of phosphorus by animals both as organic food and as mineral salts.
(*d*) Excretion of phosphates and their dispersal in soils, etc.
(*e*) Localized concentration of phosphates in the excreta of animals.

The chief biological concentrations of phosphorus are in fact due mainly to excreta of sea birds on oceanic islands and of mammals such as cave-dwelling bats. The chief human concentration is the result of burial as practised in civilized communities. On the other hand sewage disposal which accounts for large amounts of soluble phosphates leads in general to complete dispersal of the element, mainly into the sea.

There is a considerable co-precipitation of PO_4^{3-} ions along with the positive colloid $Fe(OH)_3$ in marine sediments, which produces some of the types of phosphatic ores. The oceanic red clay is also enriched in phosphate which is thereby largely removed from the cycle in the abyssal regions of the ocean.

The one possible example of the terrestrial occurrence of trivalent phosphorus is the reputed formation of phosphines (PH_3, etc.) by anaerobic bacterial action on phosphatic organic residues. Thus the alleged ' Will o' the Wisp ' of swamps and the ' corpse light ' of damp graveyards have been attributed to this cause. The exact nature of the bacteria responsible for such action seems to be obscure, but Rudakov

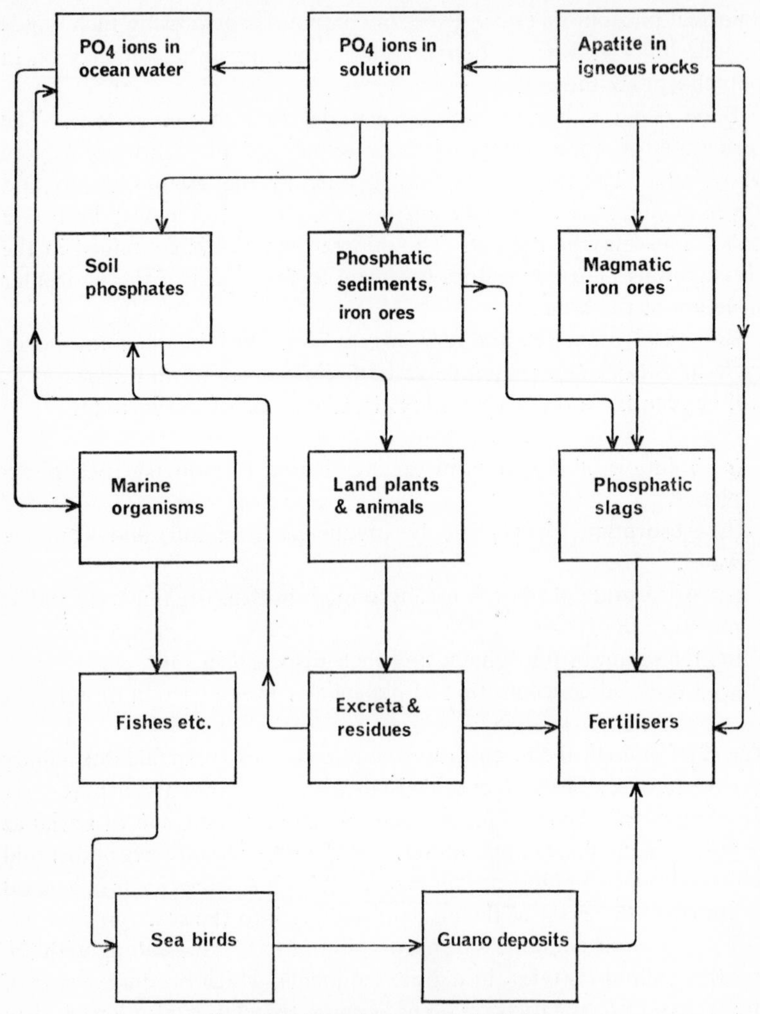

FIG. 16.6. Phosphorus cycle in Nature.

in 1927 reported the isolation of an organism able to reduce phosphate to phosphite, hypophosphite, and finally to phosphine.

Production of Phosphates

There are three main sources of phosphates at present:

(1) Organic deposits, mainly on oceanic islands and chiefly from the excreta of sea birds. Initially ammonium and sodium salts, these by

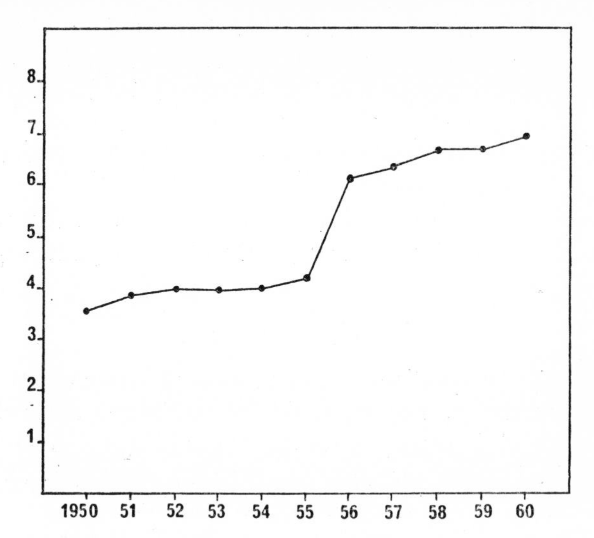

FIG. 16.7. World production of phosphates, $Ca_3(PO_4)_3$ content, in millions of tons.

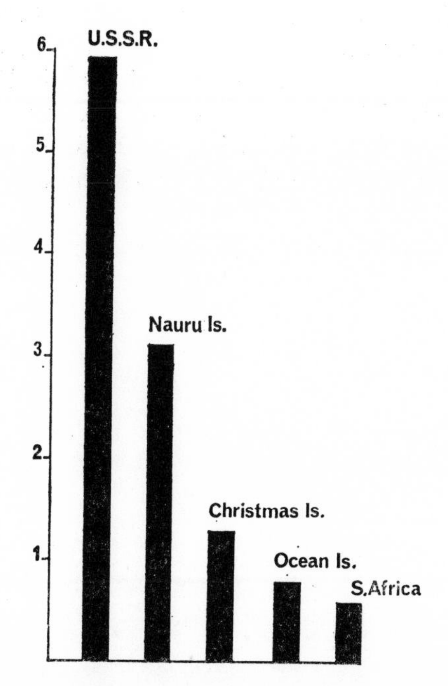

FIG. 16.8. Chief producers of phosphates, 1959, in millions of tons.

leaching processes tend to be converted largely into calcium salts which constitute much of the earthy phosphates of this type.

(2) Sedimentary phosphate deposits, found in many geological horizons, and generally shales derived from the weathering of phosphatic limestones or as nodular concretions of phosphates of marine origin in calcareous rocks.

(3) Rock phosphates, apatites in igneous rocks, particularly pegmatites, some in gabbros (Norway) or in nepheline syenite pegmatites (U.S.S.R.).

Ultimate Forms

A very high proportion of the phosphorus compounds used is employed as fertilizer. The application of fertilizers to the land affords a high degree of dispersal. The proportion of phosphorus taken up by plants in food crops, and thus indirectly by animals, is difficult to estimate. It is, however, interesting to examine the figures for the United Kingdom. The annual consumption of phosphates is of the order of 1 million tons. In respect of the human population probably not more than 100000 tons can be accounted for in the form of soluble phosphatic matter in sewage, which probably mainly reaches the sea and is lost. About one-tenth of this amount is deposited in the ground in cemeteries. It would thus seem that a very large proportion of the phosphates remain in the soil, having been rendered less readily available to vegetation in agricultural land by frequent dressings of lime.

CHAPTER 17

The Metals of Group V

General

The six metals of Group V fall into two sub-groups of the fifth periodic group and this constitutes also a natural division geochemically between the three oxyphile metals, vanadium, niobium, and tantalum, of Group Va and the three thiophile metals (or semi-metals), arsenic, antimony, and bismuth of Group Vb. The relative cosmic and terrestrial abundances of these elements show that in the case of niobium and tantalum they are relatively more abundant terrestrially than cosmically, which suggests that they have undergone a degree of concentration in the crust of the Earth.

Vanadium

There are two stable isotopes of this element in Nature:

$$^{50}V = 0{\cdot}24\%$$
$$^{51}V = 99{\cdot}76\%$$

The observed and predicted abundances of vanadium are consistent with the theory that, as one of the elements of the short mass range 46–66, its origin may be attributed to the ‘ equilibrium process ’, occurring in the phase of stellar evolution when the α-process has ended and the central temperature has risen to 3×10^9 °K.

Geochemistry

The element, as observed above, is lithophile. It has a crustal abundance of 110g/ton and cannot be regarded as a rare element. Vanadium is, however, rather widely dispersed and is mainly found as a fairly abundant trace element in rocks. It also forms a variety of independent minerals. These characters may be related to the existence in Nature of at least three oxidation states, 5, 4, and 3, and the tendency to the formation of complexes with oxygen and sometimes with sulphur. The ionic radius of trivalent vanadium (0·65 Å) is near to that of ferric iron (0·67 Å) so that there is some degree of replacement in iron minerals, particularly in the magnetite Fe_3O_4 associated with gabbroic magmas, which may contain 2000 p.p.m. The vanadium content of the more siliceous diorites and granites is appreciably lower. To a limited extent the vanadium in igneous rocks may occur as the VO_4 ion, as a replacement in silicate minerals. Vanadium is not commonly present in hydrothermal deposits except occasionally as in the form of secondary minerals of oxidized zones—*e.g.*, vanadinite.

VANADIUM MINERALS

Name	*Formula*	*Occurrence*
Patronite	VS_4	In bituminous veins
Navajoite	$(V_2O_5).3H_2O$	
Montroseite	$(V,Fe)O.OH$	
Vanoxite	$(VO)_4V_2O_9.8H_2O$?	
Ardennite	$Mn_5Al_5(As,V)O_4Si_5O_{20}(OH)_2.2H_2O$	Hydrothermal veins
Turanite	$Cu_5(VO_4)_2(OH)_4$	Sec. mineral in limestones
Tangeïte	$CuCaVO_4(OH)$	Sec. mineral in sandstones
Pintadoite	$Ca_2V_2O_7.9H_2O$	Sec. mineral in sandstones
Steigerite	$AlVO_4.3\frac{1}{4}H_2O$	Sec. mineral in sandstones
Carnotite	$K_2(UO_2)_2(VO_4)_2.3H_2O$	Sec. mineral in sandstones
Vanadinite	$Pb_5(VO_4)_3Cl$	Oxidized lead deposits
Roscoelite	Muscovite with V^{3+} replacing Al^{3+}	Rock mineral
Descloizite	$(Zn,Cu)Pb(VO_4)(OH)$	Oxidized zone of ore deposits
Mottramite	$(Cu,Zn)Pb(VO_4)(OH)$	Oxidized zone of ore deposits
Pucherite	$BiVO_4$	Oxidized zone of ore deposits

Biogeochemistry

It appears that in the process of weathering of rocks, vanadium passes largely into the form of the vanadate ion VO_4, which can remain in solution over a wide range of pH values. Hence vanadium can be carried over considerable distances. It is present in most soils to the extent of a few parts to a few hundred parts per million. It tends to be immobilized in ferruginous layers and zones. There do not appear to be any notable cases of concentration of the element by terrestrial plants.

The vanadium content of sea water is not more than 0·0003 g/ton, nevertheless some marine organisms contain remarkable amounts of the element. Thus some ascidians and holothurians contain it as an apparently essential element in their blood, *Ascidia mentula* contains 0·186% vanadium on dry weight. The bottom muds of the ocean are enriched in vanadium and the bottom-dwelling holothurians may obtain it from this source, the ascidians possibly from algal plankton enriched in the element. From such sources it would seem that vanadium passes, perhaps still in the form of organic complexes, into bituminous and asphaltic residues and ultimately into petroleum. It is notable that some petroleum produces on combustion an ash or soot

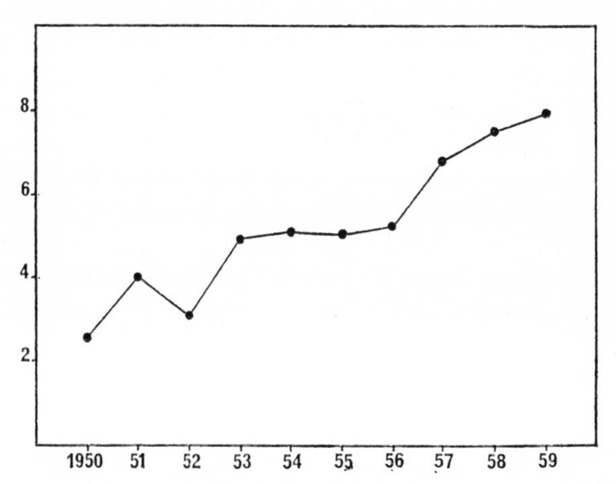

FIG. 17.1. World production of vanadium (metal content), in thousands of tons.

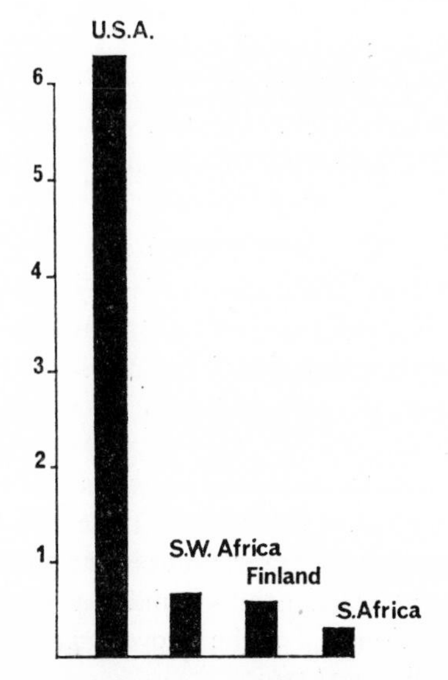

FIG. 17.2. Chief producers of vanadium ore (metal content), 1959, in thousands of tons.

very rich (over 50 % in some cases) in V_2O_3. It is therefore a useful secondary source of vanadium.

Production of Vanadium Ores

One of the largest deposits is at Minasagra in Peru where calcium vanadate and patronite occur, filling cracks and fissures in bituminous shale of Cretaceous age. The deposits in the U.S.A. are either vanadium-bearing carnotite deposits in Colorado and Utah, roscoelite ores in the same region or oxidized silver–lead–molybdenum ores in Arizona.

The ores of Broken Hill, Northern Rhodesia consist of descloizite and vanadinite, and at Tsumeb, South West Africa, vanadium ores of a similar type are carried in oxidized replacement deposits of copper–lead–zinc ores. Very little vanadium seems to occur in Great Britain. Vanadinite specimens were found in the Leadhills district, Scotland and a little vanadium is sometimes associated with pyromorphite in the Lake District.

Niobium

There is only one stable nuclide, ^{93}Nb, in Nature. The abundance of the element in the Earth's crust is relatively greater than cosmically—in the ratio of about 25 : 1—which suggests a degree of differentiation in the Earth. However, in stars of the S-type, such as R-Andromedae, niobium is ' over-abundant '. Atoms of niobium would probably be synthesized by the process of neutron capture on a slow time-scale.

Geochemistry

The crustal abundance of the element is 24 g/ton but it is largely in a state of dispersal. Both niobium and tantalum seem to occur naturally only in the form of their pentavalent ions and tend to replace tetravalent titanium, particularly in such minerals as sphene $CaTiSiO_5$ and perovskite $CaTiO_3$.

Niobium seems not to form any silicates, probably due to the difficulty of fitting a pentavalent ion into these structures. Along with its sister element tantalum, however, it does form a series of independent minerals which are characteristically found in pegmatites. On account of the very close relationship of the two elements a joint list of their minerals is given on the next page.

Tantalum

Like niobium, only a single nuclide, ^{181}Ta, occurs in Nature. The element is of lower abundance than niobium and like it is relatively more abundant in the Earth's crust than it is cosmically. It is probably also a product of nucleo-synthesis by the s-process.

Geochemistry

Niobium and tantalum show very markedly the effect of the lanthanide contraction so that their ions are of very nearly the same size (Nb = 0·69 Å, Ta = 0·68 Å). They are therefore almost identical in their modes of occurrence. The reasons for the slight differentiation which does occur as between acidic and alkalic magmatic rocks are somewhat obscure, but on the whole it seems that tantalum tends to separate at a slightly later stage of magmatic crystallization. Niobates and tantalates characteristically often contain lanthanide cations.

In the weathering of rocks both of these metals tend to accumulate in hydrolysate sediments and are therefore present in clays and shales rather than limestones and sandstones. There is no indication that they have any particular biological significance.

Niobium and Tantalum Minerals

In general these minerals occur in pegmatites or alluvial deposits derived from them. Many are of very rare occurrence.

NIOBIUM AND TANTALUM MINERALS

Name	*Formula*
Microlite	$(Ca,Na)_2(Ta,Nb)_2(O,OH,F)_7$
Dysanalite	$(Ca,Na,Ce)(Ti,Nb,Ta)O_3$
Simpsonite	$Al_5Ta_3O_{15}$
Stibiotantalite	$Sb(Ta,Nb)O_4$
Bismutotantalite	$BiTaO_4$
Columbite	$(Fe,Mn)(Nb,Ta)_2O_6$
Fergusonite	$(Yt,Er)(Ta,Nb)O_4$
Pyrochlore	$(Ca,Na,Ce)(Nb,Ti,Ta)_2(O,OH,F)_7$
Mossite	$(Fe)(Nb,Ta)_2O_6$
Fersmite	$(Ca,Ce)(Nb,Ti)_2(O,F)_6$
Hatchettolite	$(Ca,Fe,U)(Nb,Ta,Ti)_2(O,OH,F)_7$
Euxenite	$(Yt,Er,Ce,La,U)(Nb,Ta,Ti)_2(O,OH)_6$
Samarskite	$(Yt,U,Fe,Th)(Nb,Ta)_2O_6$

Production of Niobium and Tantalum

In general, the production is from pegmatites or the alluvial or placer deposits derived from them. There has also been some recovery

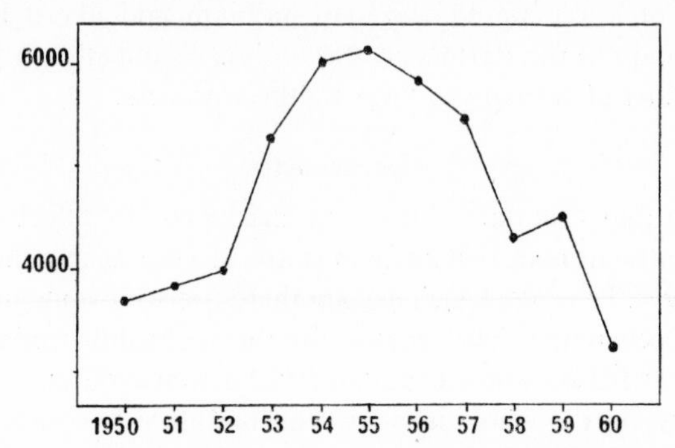

FIG. 17.3. World production of niobium and tantalum minerals, in tons.

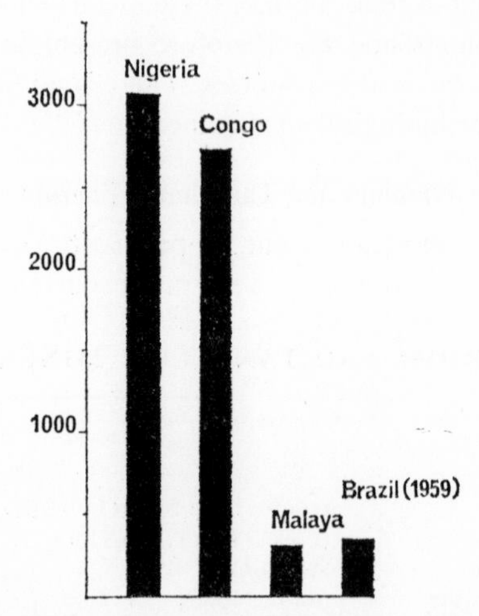

FIG. 17.4. Chief producers of niobium and tantalum minerals, 1955, in tons.

from the slags of tin smelting. Small amounts have been extracted in many countries, including: Congo, Nigeria, Brazil, India, U.S.A., and U.S.S.R. The ratio Nb/Ta varies in different localities, being about 1 : 1 in the Congo but in Nigeria it is about 7 : 1.

Arsenic

There is only the single stable nuclide ^{75}As in Nature. The relative cosmic and terrestrial abundances of the element are about the same. Arsenic is probably synthesized in the s-process of nuclear reactions in stars.

Geochemistry

The analysis of meteorites suggests that arsenic is essentially a thiophile element and this is strictly in accordance with its known modes of occurrence in the Earth's crust, where it is found predominantly in sulphide ore bodies. It is a good example of an element, comparatively rare in the Earth (2 g/ton), which is in fact quite readily accessible on account of its concentration in such deposits. The small amount of arsenic in magmatic rocks is probably in the form of arsenides or sulpharsenides, ' hidden ' in some of the sulphide minerals which are present to a small extent in such rocks. Arsenic occurs in the elemental state to a slight extent. There are many examples of binary compounds with metals, some of which have the structure of alloys.

There are sulphides and thio-salts of arsenic which exhibit the element in both its trivalent and pentavalent states. Natural arsenites are rare but pentavalent arsenic in the orthoarsenates is very common.

ARSENIC MINERALS

Name	*Formula*	*Occurrence*
Arsenic	As	Hydrothermal veins with Ag, Cu, and Ni ores
Domeykite	Cu_3As	As above with Cu ores
Realgar	As_4S_4	Hydrothermal veins and volcanic sublimates
Orpiment	As_2S_3	L.T. hydrothermal veins and hot springs
Dimorphite	As_4S_3	Solfataric deposits
Lollingite	$FeAs_2$	Mesothermal veins with Cu and Fe sulphides
Arsenopyrite	FeAsS	Commonest arsenical mineral. Hydrothermal veins
Smaltite	$CoAs_2$	In Cu and Ni veins
Cobaltite	CoAsS	In Cu and Ni veins
Niccolite	NiAs	In Cu and Ni veins
Chloantite	$NiAs_2$	In Cu and Ni veins
Sperrylite	$PtAs_2$	Associated with igneous rocks—*e.g.*, norite
Tennantite	Cu_3AsS_3	Metalliferous vein deposits
Sartorite	$PbAs_2S_4$	Metalliferous vein deposits
Seligmannite	$CuPbAsS_3$	Metalliferous vein deposits
Enargite	Cu_3AsS_4	Metalliferous vein deposits
Arsenolite	As_2O_3	Oxidized zones of hydrothermal veins

Arsenate minerals are very numerous, about seventy species having been described. They are to a large extent secondary products derived from the oxidation of primary arsenical ores in the upper zones of the deposits.

ARSENATE MINERALS

Name	*Formula*	*Name*	*Formula*
Trichalcite	$Cu_3(AsO_4)_2.5H_2O$	Trögerite	$HUO_2.AsO_4.4H_2O$
Olivenite	Cu_2AsO_4OH	Scorodite	$FeAsO_4.2H_2O$
Hörnesite	$Mg_3(AsO_4)_2.8H_2O$	Pharmacosiderite	$H.Fe_4(AsO_4)_3(OH)_4.6H_2O$
Köttigite	$Zn_3(AsO_4)_2.8H_2O$	Erythrite	$Co_3(AsO_4)_2.8H_2O$
Mansfieldite	$AlAsO_4.2H_2O$	Mimetite	$Pb_5(AsO_4)_3Cl$
Rooseveltite	$BiAsO_4$		

Arsenic in Secondary Processes

As observed above, arsenic in its primary minerals tends to pass by oxidation to the AsO_4^{3-} ion. Under acid conditions the ion is fairly mobile and most of the mineral arsenates, like the phosphates, would tend to go into solution. Therefore some arsenate ions must ultimately reach the ocean. Goldschmidt showed that the amount actually present in sea water was in fact immensely less than that supplied to the sea by known weathering processes during geological time. In fact a large proportion of the arsenic is precipitated by colloids such as ferric hydroxide and therefore collects in hydrolysate sediments, and the arsenic content of the ocean remains well below the limit of toxicity for the organisms which live in it.

In spite of its poisonous character arsenic is found in minute amounts in many animals and some plants. In a few cases of plants growing in arsenical soils it appears that the element is collected adventitiously. There is no indication that the element performs any specific function. It has been claimed that in certain burial grounds the soil is arsenical and that corpses interred therein can absorb some of the arsenic to the extent of containing apparently toxic quantities.

Production

The major proportion of the commercial output of arsenic compounds is derived from the recovery of arsenious oxide as a by-product in the smelting of the ores of copper, lead, tin, etc. The amount so produced is in fact in excess of the normal demand.

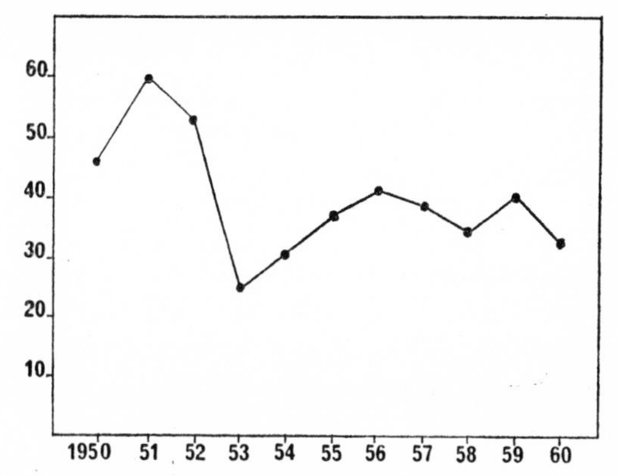

FIG. 17.5. World production of arsenic (As_2O_3), in thousands of tons.

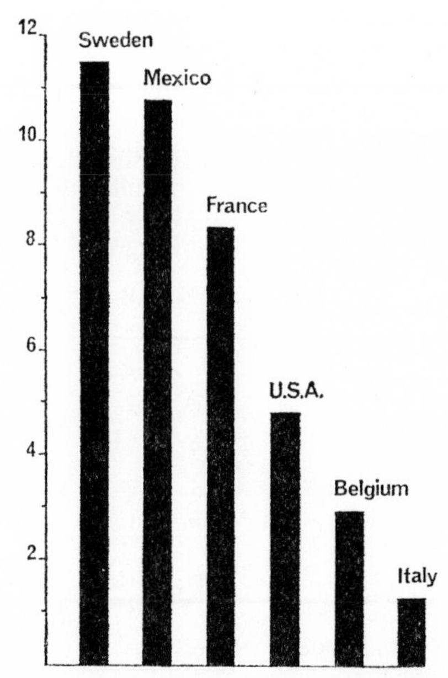

FIG. 17.6. Chief producers of arsenic (As_2O_3), 1959, in thousands of tons.

Antimony

There are two stable isotopes of antimony in Nature:

$$^{121}Sb = 57{\cdot}25\%$$
$$^{123}Sb = 42{\cdot}75\%$$

There appears to be no perceptible variation in the isotopic ratio. The relative abundances of antimony under cosmic and terrestrial conditions do not differ significantly.

Antimony nuclei could presumably be synthesized by the s-process and also by the β-decay of the neutron-rich isobar 123. Some of these nuclides could perhaps be fission-products of very heavy nuclei up to mass number 287, thought to be formed by the r-process of neutron capture on a fast time-scale under supernova conditions.

Geochemistry

Antimony must be reckoned among the rarer elements in the Earth's crust with an abundance of about 0·2 g/ton. It is very definitely thiophile in its distribution and as usual with such elements it is often found in concentrations connected with the later phases of igneous activity. Antimony minerals are generally found in low-temperature hydrothermal deposits which may often, as is also the case with mercury, be at a considerable distance from the original magmatic source. The number of antimony minerals, which includes the free element, is not large. Unlike arsenic, it does not readily form stable oxy-salts in the

ANTIMONY MINERALS

Name	*Formula*
Antimony	Sb
Allemontite	$AsSb$
Allargentum	Ag,Sb
Aurostibite	$AuSb_2$
Famatinite	Cu_3SbS_4
Stibnite	Sb_2S_3
Cylindrite	$Pb_3Sn_4Sb_2S_{14}$
Valentinite	Sb_2O_3
Swedenborgite	$NaBe_4SbO_7$
Nadorite	$PbSbO_2Cl$
Klebelsbergite	Hydrated basic sulphates
Jamesonite	$Pb_4FeSb_6S_{14}$

pentavalent state and, unlike bismuth, it is not sufficiently electropositive to form many salts with the common anions.

The amount of antimony involved in the formation of sedimentary rocks is very small and the changes it undergoes are obscure. There is some evidence that it may, like arsenic, be concentrated in hydrolysate material of ferruginous type. The antimony content of sea water is apparently beyond the limits of detection, but very small amounts (about 0·2 p.p.m.) have been reported in some marine animals.

About thirty different complex sulpho-salts containing antimony are recorded as minerals in addition to those shown on p. 272.

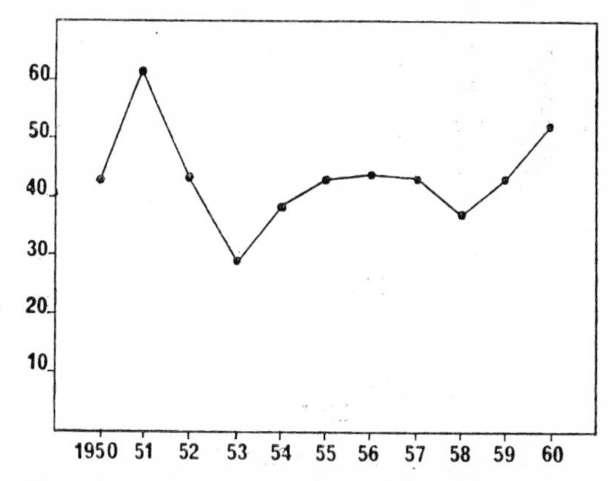

FIG. 17.7. World production of antimony ore (metal content), in thousands of tons.

Production

There are some rich deposits of antimony, mainly as sulphide.

China. The most important deposit is in the Yunnan province, South Western China, where the ore occurs in veins, seams, and pockets in Silurian sandstone.

Japan. Very fine crystalline stibnite is obtained from mines in the Island of Shikoku, Japan.

Mexico. At Potosi, Oaxaca, and Queretaro, veins in limestone yield the sulphides and some oxidation products.

Bolivia. Shallow quartz veins in Palaeozoic shale.

South Africa. At Murchison, Transvaal. There is a large output from the processing of gold–antimony ore.

Great Britain. No antimony is now produced but small deposits exist, for example in Cornwall, the Lake District, and Southern Scotland where stibnite was obtained at localities both to the east and the west of the Leadhills mineralized area—at Glendinning, Dumfriesshire and Knipes, Ayrshire.

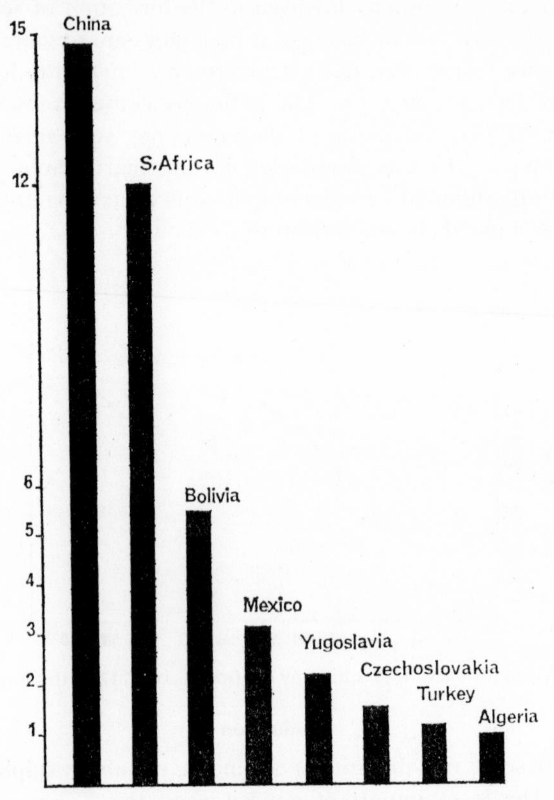

FIG. 17.8. Chief producers of antimony ore (metal content), 1959, in thousands of tons.

Bismuth

Natural bismuth consists of the single nuclide ^{209}Bi. There are in addition a number of short-lived isotopes of bismuth in the various families of natural radioactive elements.

It is now confirmed that ^{209}Bi is in fact radioactive with the very long half-life of 2×10^{17} years and decaying to ^{205}Tl by α-activity. It is probable that the bulk of terrestrial bismuth is non-radiogenic. Bismuth is interesting as being the heaviest nuclide which is virtually stable and it is therefore the element of highest mass number which could be

Family	*Nuclide*	*Radiation*	*Half-life*
Uranium	^{214}Bi	β^-,α	19·7 m
	^{210}Bi	β^-,α	5·02 d
Thorium	^{212}Bi	β^-,α	1·01 h
Actinium	^{211}Bi	β^-,α	2·16 m
Neptunium	^{213}Bi	β^-,α	47 m

formed by the capture of neutrons on a slow time-scale. It seems probable that some bismuth would also be formed by neutron capture on a fast time-scale—the r-process.

Geochemistry

Bismuth is a rare element with a crustal abundance of about 0·2 g/ton and with pronounced thiophile affinities. Unlike antimony it tends to separate in the earlier high-temperature stages of hydrothermal activity associated with granitic intrusions. The free element is not uncommon and may be associated with gold. Apart from the sulphides and sulpho-salts, compounds of bismuth with selenium and tellurium are rather characteristic. Being more electropositive than antimony,

BISMUTH MINERALS

Name	*Formula*	
Bismuth	Bi	
Maldonite	Au_2Bi	
Bismuthinite	Bi_2S_3	
Guanajuatite	$Bi_2(Se,S)_3$	
Tellurbismuth	Bi_2Te_3	
Tetradymite	Bi_2Te_2S	
Josëite	Bi_4TeS_2	
Galenobismutite	$PbBi_2S_4$	In addition up to
Beegerite	$Pb_6Bi_2S_9$	20 species of
Gladite	$CuPbBi_5S_9$	sulpho-salts have
Cosalite	$CuPb_7Bi_8S_{22}$	been described
Bismite	Bi_2O_3	
Bismutite	$(BiO)_2CO_3$	
Eulytine	$Bi_4Si_3O_{12}$	
Bismutotantalite	$BiTaO_4$	
Rooseveltite	$BiAsO_4$	
Arsenobismite	$Bi_2(AsO_4)(OH)_3$	
Pucherite	$BiVO_4$	

bismuth can form a number of independent oxy-salts, including even a silicate. Most of the minerals, which are generally found in veins associated with granitic rocks, are comparatively rare and local.

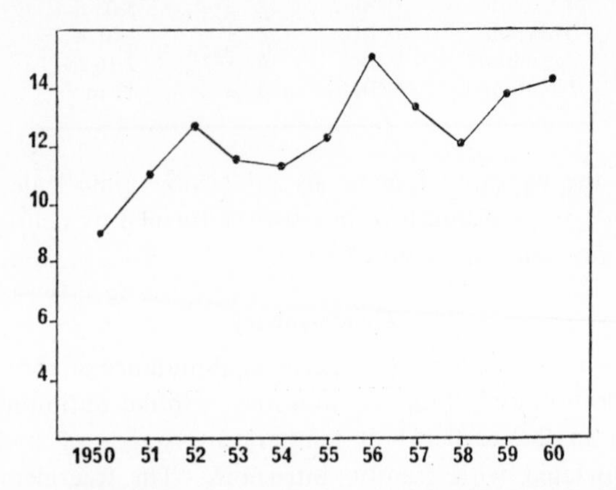

FIG. 17.9. World production of bismuth ore (metal content), in hundreds of tons.

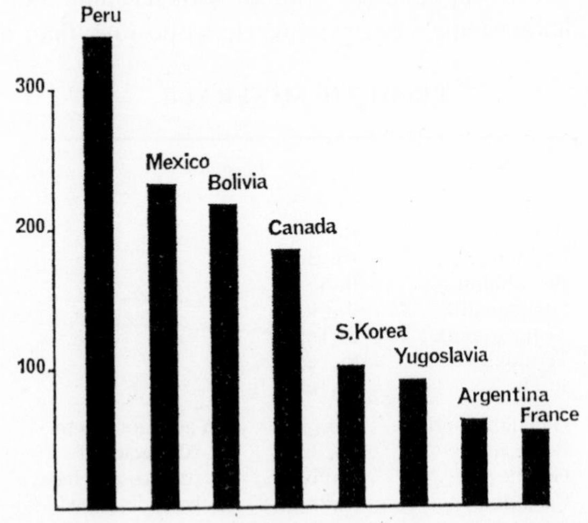

FIG. 17.10. Chief producers of bismuth ore (metal content), 1959, in tons.

Production

The most important bismuth deposits are probably in Bolivia where it is associated with tin ores (see Chapter 15, p. 244). The large output from Peru is a by-product from the lead smelter flue dust at Cerro de

Pasco. Similarly most of the U.S.A. production is from lead refinery slimes. There are small but rich deposits of bismuth ores at Cordoba, Spain.

British occurrences. Small amounts of bismuth minerals, chiefly native bismuth and the sulphide, have been found in Cornwall. In Cumberland, bismuth with sulphide and telluride minerals is associated with tungsten minerals in the small metalliferous area around the Skiddaw granite. Bismuthinite similarly occurs in the granite of Shap in Westmorland. These occurrences are of purely mineralogical interest.

CHAPTER 18

Sulphur, Selenium, and Tellurium

These three elements show a close relationship in their natural chemistry. They are of course 'thiophile' or, more properly, 'chalcophile', and on account of their association with ore bodies, the extent of which is difficult to estimate, the accurate assessment of their terrestrial abundance is necessarily difficult. The study of meteorites, however, suggests that in the Earth as a whole they may well be much more abundant than in the crust only. Sulphur, unlike the other two elements, is rather ubiquitous, being met with in a very wide range of terrestrial conditions.

Sulphur

There are four stable isotopes of sulphur in Nature:

$$^{32}S = 95{\cdot}1\% \qquad ^{34}S = 4{\cdot}2\%$$
$$^{33}S = 0{\cdot}74\% \qquad ^{36}S = 0{\cdot}016\%$$

There are variations in the composition of natural sulphur which are most significant in regard to the $^{32}S/^{34}S$ ratio. This ratio seems to vary from as low as 21·6 to as high as 22·7. Meteoritic sulphur is of practically constant composition with an isotopic ratio of 22·2.

There would appear to be three exchange reactions involving non-biogenic sulphur:

$$(a)\quad H_2{}^{34}S + {}^{32}SO_4{}^{2-} \rightleftharpoons H_2{}^{32}S + {}^{34}SO_4{}^{2-} \quad (K = 1{\cdot}074 \text{ at } 25\,^\circ C)$$
$$(b)\quad H_2{}^{32}S + H^{34}S^- \rightleftharpoons H_2{}^{34}S + H^{32}S^- \quad (K = 1{\cdot}006 \text{ at } 25\,^\circ C)$$
$$(c)\quad H_2{}^{32}S + {}^{34}S^{2-} \rightleftharpoons {}^{32}S^{2-} + H_2{}^{34}S \quad (K = 1{\cdot}013 \text{ at } 25\,^\circ C)$$

Isotopic fractionation may also occur by diffusion of hydrogen sulphide through rock strata and fissures, with the opposite effect to (*b*) and (*c*). In general, the tendency is for enrichment in ^{34}S in inorganic processes and enrichment in ^{32}S in biological processes, which is more or less analogous to the behaviour of the carbon isotopes ^{12}C and ^{13}C (Chapter 13). It is thought that sulphur would be synthesized in the α-process, taking place in 'helium burning' stars at $1{\cdot}3 \times 10^9\,^\circ K$, due to a succession of α-particle captures by ^{20}Ne. Some sulphur (including perhaps the less abundant isotopes) could also be formed by neutron capture by silicon, in the s-process.

Geochemistry

Sulphur is quite an abundant element in the Earth's crust at 520 g/ton but is probably more abundant in the Earth as a whole, the amount, according to Mason, being 2·7 %.

As previously stated (Chapter 2), Goldschmidt believed that a discrete sulphide–oxide layer or shell existed below the mantle of the Earth, but later somewhat modified this view in the direction of a supposed sulphide phase in the deeper regions of the mantle. It will be convenient to consider the occurrence of sulphur under the following headings:

(1) Elemental sulphur.
(2) Hydrogen sulphide.
(3) Sulphur oxides.
(4) Sulphur-containing ions in aqueous solution.
(5) Solid mineral sulphides.
(6) Solid mineral sulphates.
(7) Organic sulphur compounds.

(1) Elemental sulphur. Free sulphur is commonly the result of volcanic activity, being at times deposited as a direct sublimate. In some cases it has been formed by partial oxidation of hydrogen sulphide by air or possibly by sulphur dioxide. Such sulphur deposits, which are not usually of great extent, are found in volcanic regions either in active vents or as a result of fumarolic activity. Examples are at Solfatara, Pozzuoli near Naples, Italy and at Etna in Sicily. There are also examples in Iceland, Japan, Indonesia, Hawaii, Mexico, and the Andes. In 1936, in an eruption of Siretoko-Iôsan in Japan, a discharge of molten sulphur of a total volume of over 150000 m³ took place at an initial temperature of about 120 °C.

Many fumaroles and thermal waters as in Yellowstone Park, U.S.A., New Zealand, and numerous other places produce some free sulphur. Some ' sulphur springs ' in regions only remotely connected with magmatic activity yield a little free sulphur as a result of the activities of various species of sulphur bacteria on waters containing sulphides.

Most of the economically important deposits of sulphur are, however, not directly connected with volcanic or other magmatic activity. The famous Sicilian deposits are associated with gypsum, calcite, and celestine. They are of sedimentary type. Sedimentary sulphur deposits at Kuibyshev, U.S.S.R. are of Permian age in thin gypsum beds with layers of pure sulphur, laminations of sulphur and calcite, or sulphur nodules in bituminous limestone, and celestine is also present. They are thought to have been formed in lagoons. The great sulphur production of the Gulf Coast States, Texas and Louisiana, U.S.A. is from sedimentary rocks of Tertiary age with gypsum and limestone. There are also similar deposits in Mexico.

The most valuable deposits are in cavernous cap-rocks and in the overlying anhydrite covering the main body of ' salt domes '. Only a limited number of the domes contain sulphur but they may be very rich—thus Boling Dome, Texas with an area of 1200 acres is estimated to contain 40 million tons of sulphur.

The mode of formation of these sedimentary sulphur deposits is to some extent obscure. They are probably a reduction product of sulphates—*e.g.*, $CaSO_4$, and it may be that the process was mainly biogenic due to the activities of various kinds of bacteria, particularly anaerobic species (see Chapter 4, p. 79).

Small amounts of sulphur sometimes occur in sulphide ore deposits—*e.g.*, pyrite, where it may have been formed by partial oxidation of the sulphide under rather unusual conditions.

(2) Hydrogen sulphide. This gas occurs under several different environments in the Earth. It is a common component of fumarolic gases along with other acid substances such as hydrogen fluoride and hydrogen chloride. There is no indication as to whether the gas is of deep-seated origin—it may, at least in part, have been formed by the action of water or steam or acid gases on metallic sulphides. Many deep mineral springs yield sulphides, mainly hydrogen sulphide, in solution. These, long known as hepatic waters, are reputed to have therapeutic properties and are probably to be regarded as a very long sustained manifestation of magmatic activity. They are found in many parts of the world, including Britain.

It is probable that very much natural hydrogen sulphide is biogenic. It is well known that whilst surface waters are well oxygenated, the deeper stagnant layers of lakes, inlets, and fjords are active zones for anaerobic bacteria. Such species as *Desulphovibrio* obtain their oxygen from sulphate ions, which are reduced in the process to hydrogen sulphide. These deeper waters hold considerable hydrogen sulphide in solution. Thus at 1500 m depth in the Black Sea there is reported to be $6{\cdot}17\,cm^3/l$

of the gas in solution. Even higher values have been reported for the deep waters of some Norwegian fjords. Some of this hydrogen sulphide precipitates sulphides of iron and other metals in the black bottom muds. The presence of hydrogen sulphide as well as organic sulphides has long been recognized in petroleum and in petroligenic gases. It is possible that hydrogen sulphide is a by-product of the formation of petroleum. The enormous gas field of South Western France, now being exploited in the region of Lacq, is remarkably rich in hydrogen sulphide which averages over 18 % by volume. This is removed by absorption under pressure in diethanolamine, afterwards recovered and catalytically decomposed to yield sulphur at a present rate of 4000 tons per day. Thus France has rapidly become a major producer of sulphur.

(3) Sulphur oxides. Both of the oxides SO_2 and SO_3 are reported in volcanic or fumarolic gases, but apparently rather rarely. Thus the gases from the Halemaumau lava lake in Hawaii are stated to contain 7·03 % SO_2 and 1·86 % SO_3 by volume. Natural oxidation of sulphides usually occurs under hydrous conditions not favourable to the formation of the oxides but we must not ignore the considerable formation of sulphur dioxide (and presumably some of the trioxide) by human agency, that is in the combustion of fuel. Most coals contain sulphur and a figure of 2 % S is probably conservative. On this basis, the combustion of the world's annual total of, say, 2×10^9 tons of coal ($= 4 \times 10^7$ tons S) should produce 8×10^7 tons of sulphur dioxide. This tends to remain in notable amounts in the air of industrial regions, but it will ultimately be converted by oxidation to sulphuric acid which is washed down into the soil and in many cases forms calcium sulphate. The British contribution to this figure is of the order of 8 million tons of sulphur dioxide per annum or 12 million tons of sulphuric acid.

Sulphur dioxide has antiseptic properties and it is unlikely to favour bacterial activity. There do not seem to be any species which can either produce it or utilize it in Nature.

(4) Sulphur-containing ions in aqueous solution. There are probably three of these: SH^-, S^{2-}, and SO_4^{2-}, the last being by far the most important. The sulphate ion is the second most abundant anion in sea water, with a content of 2·65 parts per 1000. The mass of ocean water being 14000 Gg, it follows that the total sulphate content is of the order of 37 Gg or $3·7 \times 10^{15}$ tons. Fresh waters contain sulphate in somewhat variable amounts, but the quantity in some of the great rivers of the world is quite considerable.

Many mineral springs contain sulphates and in some special cases (*e.g.*, The Devil's Inkpot, Yellowstone Park, U.S.A.) contain free sulphuric acid. Mine drainage waters in oxidation zones of sulphide ores also occasionally contain the free acid.

(5) Solid mineral sulphides. We have previously noted (Chapter 2, p. 37) that most common ore bodies are in general masses of sulphides and it is evident that they are predominantly of magmatic origin. From the chemical viewpoint we may note that the following metals form sulphide minerals: Cu, Ag, Ca, Zn, Cd, Hg, Tl, Sn, Pb, V, As, Sb, Bi, Cr, Mo, W, Mn, Co, Ni, Ru, Pt, and Pd. Many, but not all, of these are recognized as thiophile elements. In the case of chromium the mineral daubréelite $FeCr_2S_4$ is known only to exist in meteorites and the same applies to the calcium sulphide mineral oldhamite (Ca,Mn)S. The majority of the sulphides will be found alluded to under their respective elements.

The meteoritic mineral troilite appears to be stoichiometric FeS, whereas the terrestrial mineral pyrrhotite is non-stoichiometric, due to vacant holes in the crystal lattice which should be occupied by iron atoms—it is approximately $Fe_{0\cdot8}S$.

The sulpho-salts form a numerous and assorted class of minerals and they are particularly encountered in the case of the metals Cu, Ag, Fe, Pb, Sb, As, Sn, Tl, Ge, Ga, and In.

(6) Solid mineral sulphates. Sulphate minerals are very numerous, over 130 species having been described. They can be regarded as falling into several classes.

(*a*) Minerals which are found as sedimentary bedded deposits. These are usually associated with the evaporite type of saline deposit and they include anhydrite $CaSO_4$, gypsum $CaSO_4.2H_2O$, and—less commonly—celestine $SrSO_4$ and barytes $BaSO_4$ in residual clay deposits. Numerous complex sulphates of alkali metals, etc., are also found in oceanic salt deposits of the Stassfurt type. These minerals include the following:

Kieserite	$MgSO_4.H_2O$
Vanthoffite	$Na_6Mg(SO_4)_4$
Glauberite	$Na_2Ca(SO_4)_2$
Langbeinite	$K_2Mg_2(SO_4)_3$
Mirabilite	$Na_2SO_4.10H_2O$
Bloedite	$Na_2Mg(SO_4)_2.4H_2O$
Polyhalite	$K_2Ca_2Mg(SO_4)_4.2H_2O$

(*b*) Minerals of volcanic or fumarolic origin are commonly found around volcanic vents, fumaroles, and solfatara. The region of Naples and Vesuvius provides many well-known localities. The following are typical, but they sometimes also occur in other environments:

Mercallite	$KHSO_4$
Misenite	$6KHSO_4.K_2SO_4$
Mascagnite	$(NH_4)_2SO_4$
Picromerite	$K_2Mg(SO_4)_2.6H_2O$
Ferrinatrite	$Na_3Fe(SO_4)_3.3H_2O$
Alunogen	$Al_2(SO_4)_3.18H_2O$

(*c*) A very large number of sulphate minerals occur in the upper oxidized zones of sulphide ore deposits. Some sulphates also occur on or near the surface of bedded rocks which may have contained sulphides such as pyrite FeS_2. Only a selection of typical minerals is listed below. Others will be found referred to in connexion with various metals. (The mineral barytes $BaSO_4$, mentioned above, is commonly found as a vein mineral and may be regarded as a primary mineral of barium.)

Chalcanthite	$CuSO_4.5H_2O$
Melanterite	$FeSO_4.7H_2O$
Anglesite	$PbSO_4$
Brochantite	$Cu_4(SO_4)(OH)_6$
Alunite	$KAl_3(SO_4)_2(OH)_6$
Goslarite	$ZnSO_4.7H_2O$

(7) Organic sulphur compounds. These must be considered along with the general biogeochemistry of the element. It does not seem very likely that there are any naturally occurring organic sulphur compounds which are not the result of the action of living organisms.

Ordinary plants assimilate the sulphate ion, and as noted in Chapter 4, p. 72, the so-called sulphur bacteria are able to utilize several sulphur compounds and also the element itself. Animals obtain complex sulphur compounds in vegetable food and they also take in mineral sulphates.

It may be said that sulphur, unlike phosphorus, exists in organic molecules in a variety of combinations and oxidation states. In some plants (marine algae) carbohydrate sulphates are known to be present. Proteins, both plant and animal, contain sulphur which may be in the thiol group SH and also as linked sulphur atoms S–S. Heterocyclic compounds with sulphur in the ring are also met with. The ' mustard oils ' which occur in various plants of several different families, contain the isothiocyanate group—NCS.

Sulphur is present in all animal cells primarily as protein, in particular the various keratin structures—hair, hoofs, horns, etc., which contain cystine units. Several enzymes and hormones contain it and the acetyl-Coenzyme A contains a high-energy sulphur bond equivalent to the high-energy bonds of phosphorus.

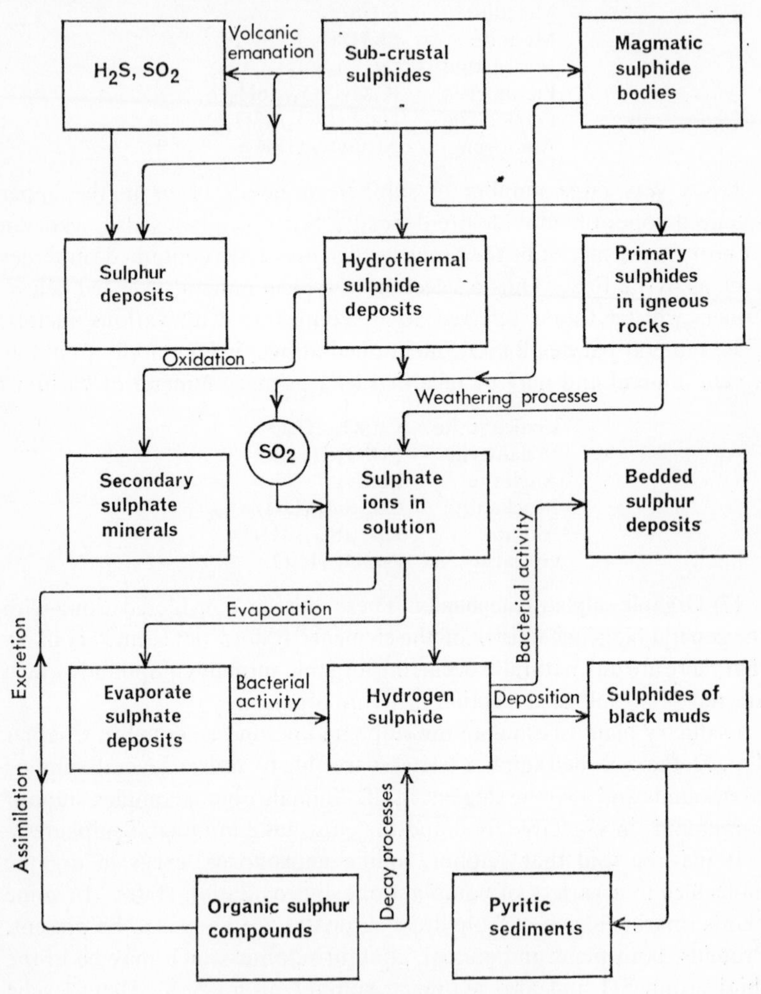

FIG. 18.1. Geochemical cycle of sulphur.

Production

Large amounts of sulphur dioxide are produced in the smelting of various sulphide ores—no further reference will be made to it here. Elemental sulphur is produced mainly as follows:

American Gulf Coast 'Salt Dome' deposits. These deposits are extracted by the Frasch hot-water process but not all of the domes are in fact suitable for this treatment.

Sicily. The sulphur is usually volatilized by heating the crude ore, and some of the sulphur itself may be used as fuel for the purpose.

Norway, Portugal, and Spain. Large amounts of pyrite FeS_2 are available and by a special smelting process a mixture of gases containing

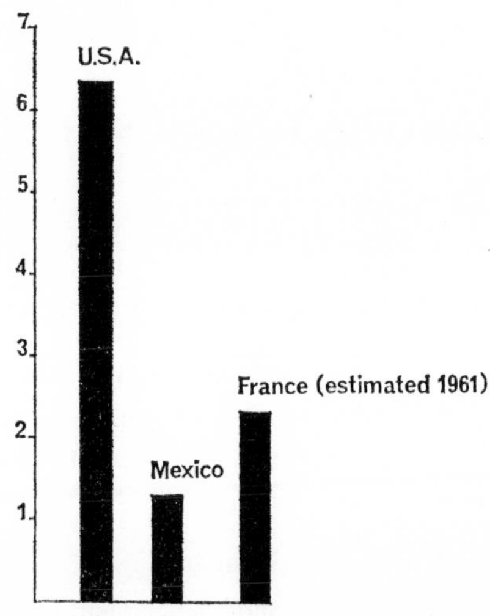

FIG. 18.2. Chief producers of sulphur, 1959, in millions of tons.

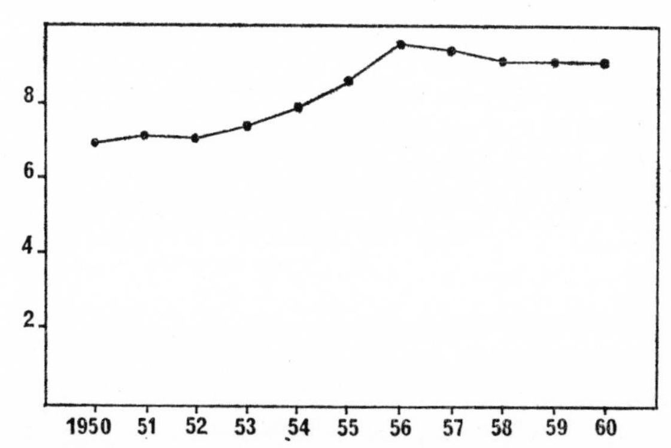

FIG. 18.3. World production of sulphur (S content), in millions of tons.

S, SO_2, CS_2, COS, H_2S, and CO_2 is produced. Suitable thermal treatment over a catalyst yields free sulphur, which may require further treatment to remove arsenic.

Japan, Indonesia. Elemental sulphur from volcanic sources is purified by flotation and other methods.

U.S.A., Canada, and other oil-producing countries. Sulphur as H_2S in petroligenic gases or refinery gases is being recovered by methods similar to that referred to above in connexion with the gas of Lacq in France.

Selenium

There are six isotopes of this element in Nature:

^{74}Se = 0·87%	^{78}Se = 23·52%
^{76}Se = 9·02%	^{80}Se = 49·82%
^{77}Se = 7·58%	^{82}Se = 9·19%

There is no indication of isotopic variation or of any natural radioactivity among these nuclides.

The synthesis of selenium atoms would probably proceed by neutron capture processes and the large number of isotopes suggests that several mechanisms would be involved in the lines exemplified by the case of tellurium (see below).

Geochemistry

The estimated crustal abundance of the element of less than 0·1 g/ton indicates that selenium is rather scarce. It is almost exclusively chalcophile, occurring in minerals where it replaces sulphur. The occurrence of the oxide SeO_2 as the mineral selenolite is somewhat doubtful. The selenites seem to be represented by a very few minerals of which only one need be mentioned, *viz.*, chalcomenite $CuSeO_3.2H_2O$. Selenium requires a higher oxidation potential than sulphur to raise it to the hexavalent state and this is not often realized in Nature—the mineral selenojarosite $KFe_3^{3+}\{(S,Se)O_4\}(OH)_6$ seems to be a rare example of genuine SeO_4 ions in a natural substance.

Other selenium minerals are indicated below:

Selenium	Se	Clausthalite	PbSe
Klockmannite	CuSe	Guanajuatite	$Bi_2(S,Se)_3$
Berzelianite	Cu_2Se	Achavalite	FeSe
Naumannite	Ag_2Se	Crookesite	$(Cu,Tl,Ag)_2Se$
Tiemannite	HgSe	Penroseite	$(Ni,Cu)Se_2$

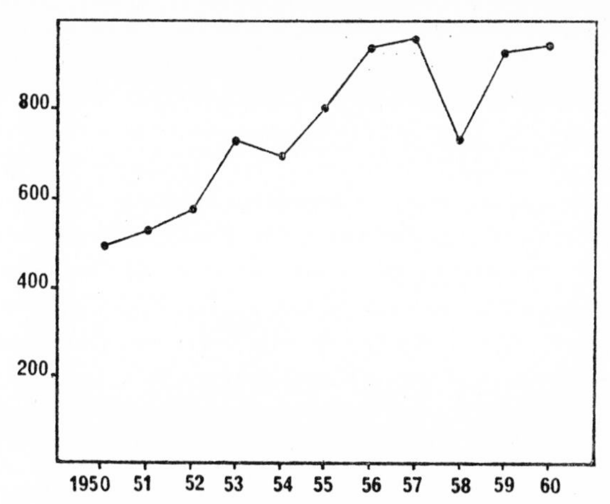

FIG. 18.4. World production of selenium, in tons.

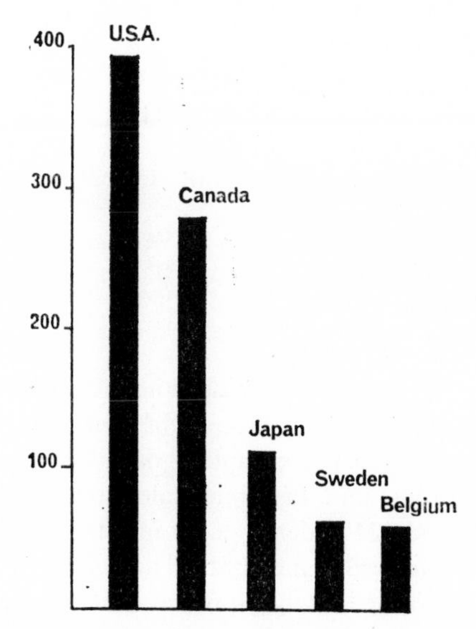

FIG. 18.5. Chief producers of selenium, 1959, in tons (not necessarily raised in same country).

Trace quantities of selenium are present in many sulphide ores and the divalent ionic radii of the two elements (S=1·83 Å, Se = 1·94 Å) make this possible. On the other hand, in weathering processes sulphur will readily pass into solution as sulphate but selenium will tend to remain in heavy metal selenides, and there is little tendency for it to be concentrated in these processes. Some small amounts of selenium which pass into solution are largely precipitated in bottom sediments of the ocean in ferruginous hydrolysates and the selenium content of sea water is therefore small. Selenium is present in soils in a form which can be absorbed by plants and in some regions there are certain plants which absorb the element and seem to require it for their metabolism. They include species of *Astragalus*, *Xylorrhiza*, *Oonopsis*, and *Stanleya*. Such plants may be toxic to animals and possess an offensive odour. Goldschmidt reported that the caliche of Chile contained up to 5 p.p.m. of selenium in a water-soluble form which he considered to be selenate owing to the very high oxidation potential of this particular environment (see Chapter 16, p. 251). It appears from some recent work that a minute amount of selenium is a necessary trace element for some animals.

Tellurium

There are eight stable isotopes of this element in Nature:

^{120}Te = 0·089 %	^{125}Te = 6·99 %
^{122}Te = 2·46 %	^{126}Te = 18·71 %
^{123}Te = 0·87 %	^{128}Te = 31·79 %
^{124}Te = 4·61 %	^{130}Te = 34·49 %

Mechanisms for the synthesis of these isotopes have been proposed by Hoyle, Burbidge, and Fowler. The light isotopes ^{122}Te, ^{123}Te, and ^{124}Te are produced by neutron capture on a slow time-scale. When neutron capture occurs on a fast time-scale the neutron-rich isobars of masses 122, 123, and 124 undergo decay into the stable isotopes of tin and antimony—^{122}Sn, ^{124}Sn, and ^{123}Sb. The heaviest tellurium isotope, ^{130}Te, can only be formed by the r-process as it is the decay product of neutron-rich isobars of mass 130. In the s-process ^{130}Xe is built up instead of ^{130}Te. The isotopes ^{125}Te, ^{126}Te, and ^{128}Te can be built by either process, but ^{128}Te is mainly formed by the r-process. The rare light isotope ^{120}Te has an excess of protons and cannot be built by either process, instead it may have resulted from the introduction of a

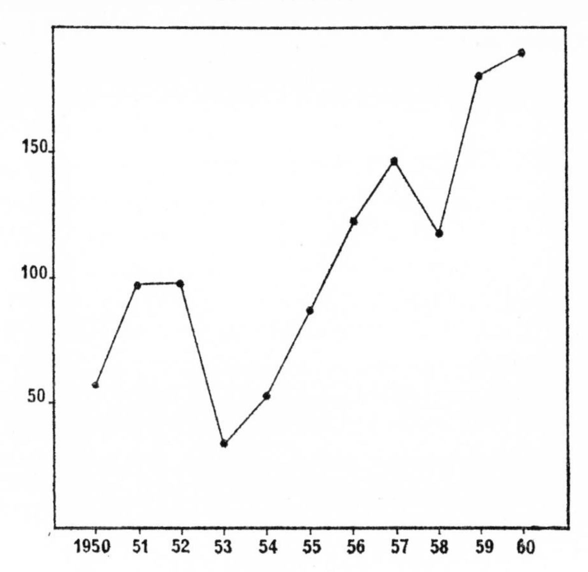

FIG. 18.6. World production of tellurium in tons.

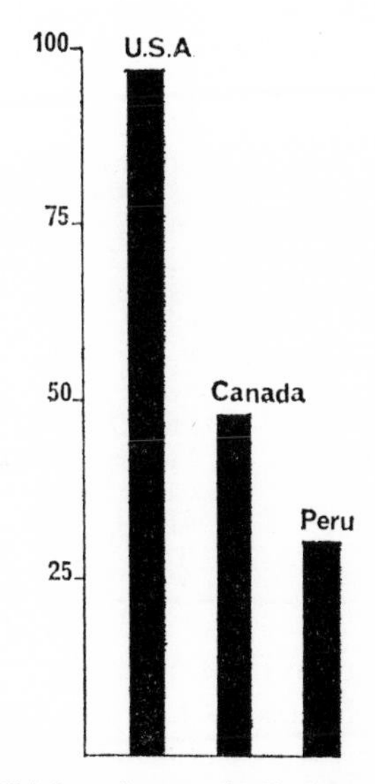

FIG. 18.7. Chief producers of tellurium, 1959, in tons.

small proportion of heavy-element material into a hydrogen-rich region or by the photo-dissociation of heavy nuclides by γ-ray absorption and emission of neutrons.

Geochemistry

The abundance figure of 0·002 g/ton for tellurium in the Earth's crust is not likely to be of high accuracy, but it indicates that the element is probably less abundant than gold or platinum, and amongst the rarest of the elements. The relative cosmic abundance appears to be much higher. The geochemistry of the element is somewhat incompletely known. It is chalcophile and, although sometimes associated with sulphur and selenium, it forms a number of binary compounds with metals and is characteristically associated with gold and with bismuth. There are several oxy-salts with a tellurium anion in which both tellurous and telluric ions are found.

TELLURIUM MINERALS

Name	*Formula*
Tellurobismuth	Bi_2Te_3
Tetradymite	Bi_2Te_2S
Joseite	Bi_4TeS_2
Wehrlite	$BiTe$
Calaverite	$AuTe_2$
Sylvanite	$AgAuTe_4$
Hessite	Ag_2Te
Weissite	Cu_2Te
Tellurium	Te
Frohbergite	$FeTe_2$
Dunhamite	$PbTeO_3$
Emmonsite	$Fe_2(TeO_3)_3.2H_2O$
Teïneite	$Cu(Te,S)O_4.2H_2O$
Montanite	$BiTeO_6.2H_2O$
Tellurite	TeO_2

Practically all the tellurium minerals are of rare occurrence and the metal, like selenium, is obtained as a by-product from the smelting of other metals or in refining processes of metals.

CHAPTER 19

The Metals of Group VI

We shall here consider the three elements chromium, molybdenum, and tungsten. In their natural chemistry they show no very strong resemblances. The terrestrial abundance of chromium is relatively higher than its cosmic abundance in the ratio of about 100 : 1, and in the case of molybdenum the relative abundances are about the same, whereas for tungsten the cosmic abundance is relatively greater than the terrestrial in the ratio of about 10 : 1. These apparent variations are probably related to some extent to the difficulties involved in assessing the terrestrial abundances.

Chromium

There are four stable isotopes of this element in Nature:

$^{50}Cr = 4{\cdot}31\%$ $\quad$ $^{53}Cr = 9{\cdot}55\%$

$^{52}Cr = 83{\cdot}76\%$ $\quad$ $^{54}Cr = 2{\cdot}38\%$

Chromium is one of the elements of the ' iron peak ' for which synthesis by the ' equilibrium process ' is postulated. The observed abundance of the element in fact agrees well with the value deduced by Burbidge, Burbidge, Fowler, and Hoyle on the basis of this process.

Geochemistry

The estimated crustal abundance of chromium is 200 g/ton and the analyses of meteorites indicate that it should be markedly thiophile. However, no sulphur compounds of chromium are known in the Earth and they could only be formed under much lower redox potentials than obtain in the crust. Goldschmidt thought that the meteoritic mineral daubréelite might contain divalent chromium and be the analogue of pentlandite $(Ni,Fe)_9S_8$; thus the formula $FeCr_2S_4$ would be replaced by $(Cr,Fe)_9S_8$. The absence of chromous sulphide from terrestrial locations was ascribed to the reaction

$$CrS \text{ (in troilites)} + FeO \text{ (in silicates)} \longrightarrow FeCr_2O_4 + FeS$$

thereby preventing the retention of chromium in sulphide ores. The silicate phase of meteorites also shows some concentration of chromium which is consistent with its general mode of terrestrial occurrence—*i.e.*, that of a lithophile element.

Chromium is rather widely distributed in the Earth and is found in both its trivalent and hexalent but not in its divalent state. The most important concentrations depend on the refractory character and rather high density of certain oxide minerals of chromium which separate at an early stage of magmatic crystallization and may form segregations in basic and ultrabasic rocks. These oxide minerals are of great stability and were considered by Goldschmidt to be largely covalent in structure. The majority of the terrestrial minerals carry the element in the trivalent state and they are not very numerous. However, owing to similarity in ionic size, some replacement of Fe^{3+} and Al^{3+} ions by Cr^{3+} can take place and chromium is therefore found in small amounts in a number of common silicate minerals.

It appears that most of the chromium in the Earth's crust is in the form of spinellid minerals of which the most important is the common ore mineral chromite $FeCr_2O_4$.

Chromium is the usual colouring matter of the green beryl called emerald. The chromium content of emeralds is of the order of 1% of Cr_2O_3. It is presumed that part of the aluminium of the mineral is replaced by chromium in its trivalent state.

The garnet, uvarovite $Ca_3Cr_2(SiO_4)_3$, is a rare mineral found mainly in the Urals. Its beautiful green crystals are unfortunately not generally suitable for use as a gem stone.

CHROMIUM MINERALS

Name	*Formula*	*Occurrence*
Magnochromite	$MgCr_2O_4$	Accessory minerals in ultrabasic and basic igneous rocks
Cromite	$FeCr_2O_4$	
Picotite	$(Fe,Mg)(Al,Cr)_2O_4$	
Stichtite	$Mg_6Cr_2CO_3(OH)_{16}.4H_2O$	In altered ultrabasic rocks
Babertonite	$Mg_6Cr_2CO_3(OH)_{16}.4H_2O$	
Uvarovite	$Ca_3Cr_2(SiO_4)_3$	
Tarapacaite	K_2CrO_4	In caliche deposits, Chile
Lopezite	$K_2Cr_2O_7$	
Crocoite	$PbCrO_4$	In oxidized zones of lead veins
Vauquelinite	$(Pb,Cu)_3\{(Cr,P)O_4\}_2$	
Beresovite	$Pb_6(CrO_4)_3O_2CO_3$	
Dietzeïte	$Ca_2(CrO_4)(IO_3)_2$	Caliche deposits, Chile

Chromium in the formal valency state of Cr^{6+} is met with in the chromates which only occur rather rarely as mineral substances. It is evident that the high oxidation potential needed to raise the element to this state is but rarely attained in Nature. It is in fact not at all easy to account for the formation of the natural chromates. They are found in small quantities associated with the remarkable suite of highly oxidized compounds of the caliche of Chile or else as secondary minerals in the oxidized zone of lead veins.

Chromium in Secondary Processes

The process of weathering of chromiferous rocks and the formation of soils and sediments is not known with certainty. It is probable that the element passes into solution and that in the form of Cr^{3+} ions is probably not very mobile. On the other hand the actual state of oxidation of chromium in such deposits is not perfectly known. It seems likely that under rather exceptional circumstances it may be oxidized to Cr^{6+} and the chromates so formed are generally more readily soluble except when immobilized as the very sparingly soluble lead chromate. Much of the chromium content in soils from serpentine rocks seems to be in the form of insoluble residual chromite.

There is a certain amount of concentration of trivalent chromium in hydrolysate sediments, particularly those containing alumina, but not to any extent in oxidate sediments—*e.g.*, certain types of iron ores. There is no evidence that chromium is a necessary element for living organisms and it is indeed toxic to plants. There are a few cases known where ‘ exchangeable ’ chromium in soils has been shown to cause lack of fertility.

Production

Chromium ores do not lend themselves to concentration methods so that economic deposits have to be relatively rich. They are exclusively of the segregated type associated with fresh (peridotitic) or altered (serpentinized) ultrabasic rocks. The only mineral of interest is chromite $FeCr_2O_4$.

Southern Rhodesia. The deposits are connected with the Great Dyke, which is 330 miles long and 4 miles wide and consists of ultrabasic rocks largely altered to serpentine. The dyke has a layered structure and chromite occurs in persistent bands up to 8 in thick. Lenses of chromite are also found adjacent to the dyke in talc schists. It is probably a late magmatic crystallization in this group of rocks.

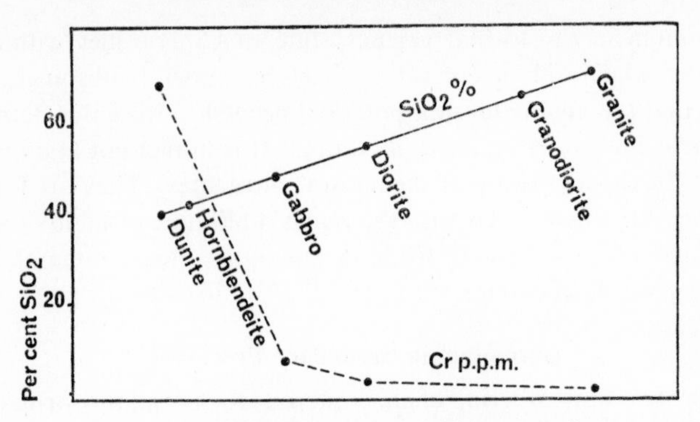

FIG. 19.1. Chromium content of igneous rocks.

South Africa. The Bush Veld igneous complex contains basic rocks such as norite carrying bands of chromite from a few inches to 3 ft thick. The chromium horizons can be traced for many miles and there are large reserves. The ore is of rather low grade, averaging only about 43 % Cr_2O_3.

U.S.S.R. In the Sverdlovsk region, east of the Urals, where Palaeozoic schists have been intruded by serpentinized ultrabasic rocks, there are bands of chromite up to 30 ft wide. The ore is low in iron and approximates to $(Fe,Mg)Cr_2O_4$. It probably crystallized at a late stage from an olivine-rich magma. There are also placer deposits in the vicinity.

Turkey. Deposits are probably magmatic segregations in ultrabasic intrusions. They are of very high grade.

Great Britain. Basic rocks of Tertiary age in the Isle of Skye contain picotite as an accessory mineral. A little chromite has been extracted from ultrabasic serpentinized rocks of pre-Cambrian age in the Island of Unst, Shetland. Vauquelinite has been described from the Leadhills region of Southern Scotland.

Molybdenum

There are seven stable isotopes of this element in Nature:

$^{92}Mo = 15{\cdot}86\%$ $^{97}Mo = 9{\cdot}45\%$
$^{94}Mo = 9{\cdot}12\%$ $^{98}Mo = 23{\cdot}75\%$
$^{95}Mo = 15{\cdot}70\%$ $^{100}Mo = 9{\cdot}62\%$
$^{96}Mo = 16{\cdot}50\%$

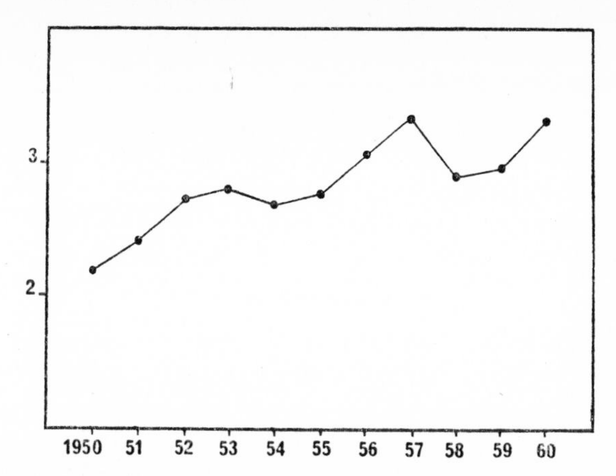

FIG. 19.2. World production of chrome ore (Cr_2O_3 content), in millions of tons.

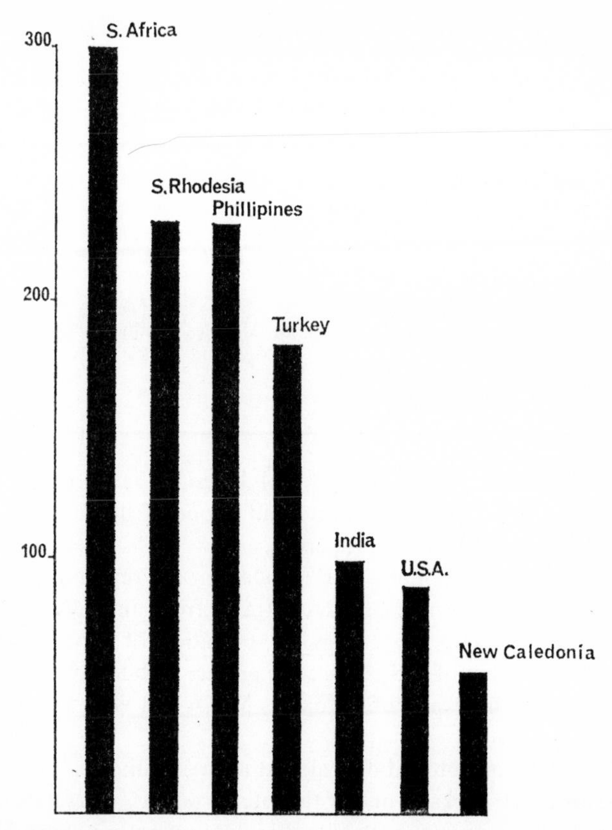

FIG. 19.3. Chief producers of chrome ore (Cr_2O_3 content), 1959, in thousands of tons.

It has not been possible to substantiate the suggestion that ^{100}Mo is radioactive and a β-emitter. If this were so the decay product would be ^{100}Tc, which should therefore exist terrestrially.

There are certain aspects of the nuclear chemistry of molybdenum which are referred to in connexion with technetium (Chapter 21).

Geochemistry

Molybdenum is not a very abundant element terrestrially at 1 g/ton in the crust. Meteorite analyses show appreciable concentration of the element in the metallic and somewhat less in the sulphide phase, which would imply a siderophile and thiophile character for the element. In the Earth's crust, however, molybdenum is to be regarded as lithophile, being mainly associated with the later stages of magmatic crystallization, *viz.*, in granitic rocks. It is not, however, oxyphile as its principal

MOLYBDENUM MINERALS

Name	*Formula*	*Occurrence*
Molybdenite	MoS_2	High-temperature veins, etc., associated with granitic rocks
Molybdite	MoO_3 ?	Oxidized zones of mineral veins associated with granitic rocks
Lindgrenite	$Cu_3(MoO_4)_2(OH)_2$	
Powellite	$CaMoO_4$	
Wulfenite	$PbMoO_4$	
Koechlinite	$(BiO)_2MoO_4$	
Ferrimolybdite	$Fe(MoO_4)_3.7\frac{1}{2}H_2O$	

natural compound is the disulphide molybdenite. This mineral has a characteristic layered lattice structure and is one of the few solid substances possessing lubricating properties.

The few oxygen compounds of molybdenum are rare and local. Its geochemistry shows some marked contrasts from that of its related element, tungsten. The free energy of formation of MoS_2 is almost 20 % greater than that of WS_2, hence the greater stability of the former. On the other hand in respect of the oxides MoO_3 and WO_3 the positions are reversed.

Molybdenite is, as noted above, almost always found associated with the acidic (granitic) rocks as one of the later products of crystallization, occurring chiefly in pneumatolytic and high-temperature hydrothermal veins, etc. In consequence of this some economically valuable concentrations of the element are available. Oxidation to the oxide MoO_3 presumably occurs in zones of weathering but the composition of the mineral molybdite is somewhat doubtful.

It is evident that the MoO_4^{2-} ion is produced under these conditions and is initially in solution but is very readily immobilized in the form of sparingly soluble molybdates of various metals. Poly-molybdates do not seem to occur in Nature. There are some rare cases of silicates of the feldspathoid group containing small percentages of molybdenum, but whether the MoO_4^{2-} ion replaces the SiO_4^{4-} ion, or whether, as in the case of sodalite, some of the chloride ions are replaced, is not yet certain. The behaviour of molybdenum in weathering processes is rather obscure; it appears, however, to be dissolved and may frequently be accumulated in oxidate sediments, particularly those containing

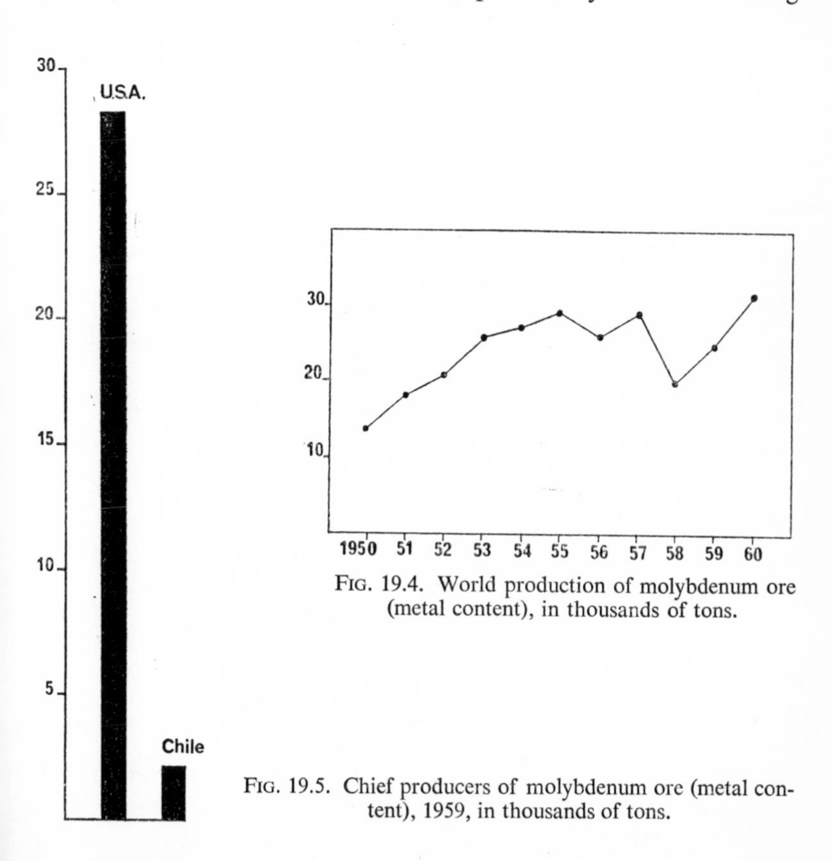

FIG. 19.4. World production of molybdenum ore (metal content), in thousands of tons.

FIG. 19.5. Chief producers of molybdenum ore (metal content), 1959, in thousands of tons.

manganese. The black bottom muds also show some concentration of molybdenum as MoS_2 and shales derived from them may contain some hundreds of parts per million of Mo. It has been established that molybdenum is an essential trace element for plants and probably enters into some of the organic complexes which function as coenzymes.

Production

U.S.A. The largest deposit in the world being worked at present is at Climax, Colorado where Tertiary quartz–monzonite dykes penetrate a pre-Cambrian granite. The altered granite is intersected by numerous ramifying veins and veinlets carrying quartz, orthoclase, fluorite, and molybdenite. The deposit is hydrothermal, presumably from the magma that supplied the Tertiary dykes. The reserves are estimated at over 200 million tons of ore which averages about 0·5 % of molybdenum.

Increasing amounts of molybdenum are being recovered also as a by-product from the working-up of copper and tungsten ores in Utah, Arizona, Nevada, California, etc. Molybdenum is similarly obtained in connexion with copper sulphide ores in Chile.

Norway. Production is mainly from the Knaben mine in Fjotland where the molybdenite occurs in quartz veins and disseminations in granite.

Ireland. There has been no production but a considerable deposit of ore is claimed to be associated with a granite mass near Roundstone in Connemara.

Great Britain. No workable deposits of molybdenum ores are known but small amounts of molybdenite are found associated with granite intrusions in many localities in Cornwall, the Lake District, and S.W. Scotland, where also some of the oxidation products are found as rare minerals.

Tungsten

There are five stable isotopes of this element in Nature:

$^{180}W = 0{\cdot}135\,\%$ $\quad$ $^{184}W = 30{\cdot}6\,\%$
$^{182}W = 26{\cdot}4\,\%$ $\quad$ $^{186}W = 28{\cdot}4\,\%$
$^{183}W = 14{\cdot}4\,\%$

It is probable that these isotopes could be formed by neutron capture processes, perhaps both by the s- and r-processes.

Geochemistry

The terrestrial abundance of tungsten is about the same as that of molybdenum: 1 g/ton for the crust. Tungsten is pronouncedly lithophile in character as would be expected from its concentration in the silicate phase of meteorites.

Unlike molybdenum, the sulphide of tungsten, WS_2, is not a common mineral. There are rather more natural compounds of tungsten than of molybdenum and they are practically all tungstates.

TUNGSTEN MINERALS

Name	*Formula*
Tungstenite	WS_2
Tungstite	H_2WO_4
Cuprotungstite	$Cu_2WO_4(OH)_2$
Scheelite	$CaWO_4$
Samartinite	$ZnWO_4$
Anthoinite	$AlWO_4(OH).H_2O$
Stolzite	$PbWO_4$
Chillagite	$Pb(Mo,W)O_4$
Hubnerite	$MnWO_4$
Ferberite	$FeWO_4$
Wolframite	$(Mn,Fe)WO_4$

There is similarity in the general mode of occurrence of molybdenum and tungsten but whilst tungsten is also associated with the residual crystallization of granitic magmas it is rather commonly found along with tin (cassiterite) and sometimes columbite under pegmatitic and pneumatolytic conditions.

There is no indication of any biological significance of tungsten and the changes it undergoes in the process of weathering and sedimentation are not yet clearly understood.

Production

There is some variety in the types of tungsten deposit but they are all related to granitic intrusions.

China. Actually and potentially, China is the largest producer. The most important areas are Kiagsi, Hunan, Kwantung, Kwangsi, and Yunan. The deposits are steeply dipping fissure veins in granite or

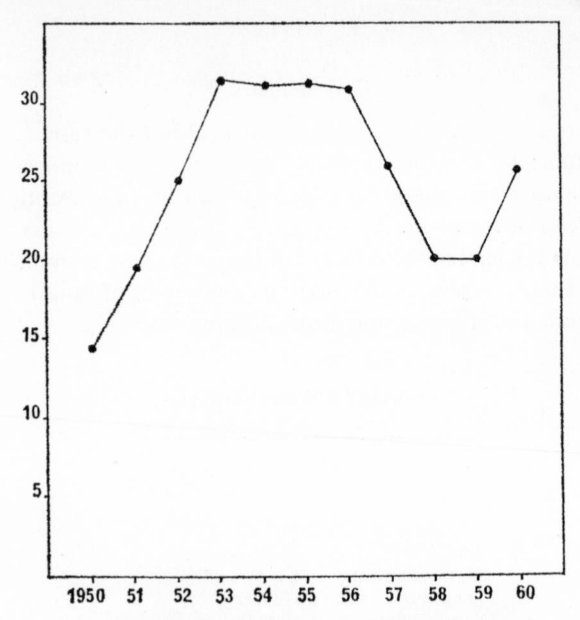

FIG. 19.6. World production of tungsten ore (metal content), in thousands of tons.

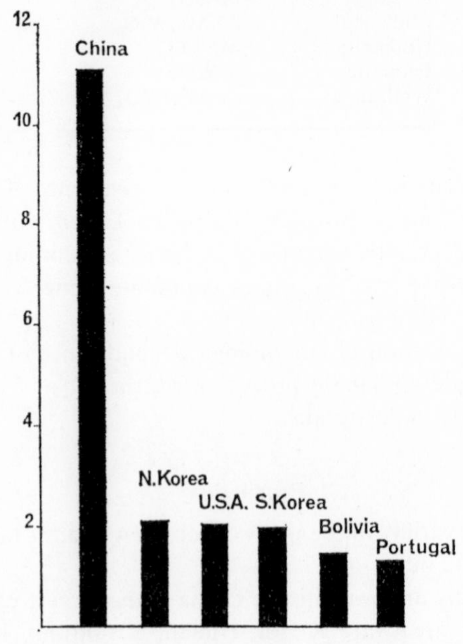

FIG. 19.7. Chief producers of tungsten ore (metal content), 1959, in thousands of tons.

adjacent sediments. There are also pegmatites. The ore is wolfram with subsidiary scheelite and accessory bismuthinite, molybdenite, topaz, fluorite, and sulphides. In some cases there is also workable cassiterite.

Burma. The chief localities are Mawchi, Merguir, and Tavoy. There are veins in tourmaline granite with quartz, wolframite, cassiterite, and some scheelite, together with base-metal sulphides, bismuthinite, and arsenopyrite.

U.S.A. Workable deposits are found in Nevada, Colorado, Idaho, California and North Carolina. The Nevada deposits are vein-like and contact-metasomatic in limestone intruded by a post-Jurassic granodiorite. Scheelite is the chief mineral, associated with garnet and epidote.

Korea. At Sangdong lenticular beds have been metasomatically altered into scheelite-bearing rocks containing also appreciable amounts of bismuth.

Brazil. The deposits are mainly contact-metasomatic yielding chiefly scheelite.

Portugal. Veins carrying chiefly wolfram with minor bismuth and in ores and some columbite.

Great Britain. Small amounts of wolfram have been raised. In Cornwall it is found in close association with cassiterite in the typical high-temperature veins of the region. In Cumberland a very small amount of wolfram and some scheelite have been produced from quartz veins in a greisenized granite together with a mineral assembly not unlike that found in the Asiatic tungsten deposits.

CHAPTER 20

The Halogens

General

The most interesting aspect of the natural occurrence of these elements is that chlorine apparently has a higher cosmic abundance than fluorine. In respect of terrestrial occurrence the position is reversed but there are various difficulties in assessing with accuracy the terrestrial abundance of the halogens so that no very sure conclusions can be drawn from the above observations.

In general the halogens form a fairly coherent group geochemically. This is to be expected of the only full family of elements showing persistent anionic characters. Recent studies indicate that there are in fact many obscure points in their geochemistry and some of the earlier views probably represent an over-simplification.

Fluorine

Only the single nuclide ^{19}F exists in Nature.

According to modern theories of nucleogenesis, fluorine atoms would be built in some of the less frequent reactions of the α-process, taking place probably towards the end of this stage of stellar evolution, when detailed mechanisms are difficult to predict.

Some of the ' white dwarf ' stars may provide examples of locations where this type of element building is taking place.

The progenitor of ^{19}F may be ^{15}N, which is one of the nuclides involved in the so-called carbon–nitrogen cycle (see p. 109).

Geochemistry

The actual terrestrial abundance of fluorine, as of the other halogens, is a matter of some doubt. The fluorine content of different rocks varies quite widely even for the same rock type. The figure of 700 g/ton should therefore be accepted with some reserve. It is, however, clear that fluorine is a comparatively abundant element in the Earth's crust.

The most electronegative of the elements, fluorine is normally found as its anion ($F^- = 1{\cdot}33$ Å) which is very nearly the same size as the ions OH^- and O^{2-} and it can therefore enter into a considerable variety of minerals.

Elemental Fluorine

It appears remarkable that the free element, which long eluded the laboratory chemist, should actually exist in Nature, but this appears to be the case. The ' stinking fluorspar '—*stinkfluss*—of Wölsendorf in Bavaria, also called antozonite, is a variety of fluorite CaF_2 which contains inclusions of uranium minerals. The fluorine is occluded in the spar and is liberated on fracture. By reaction with the moisture in the air it at once gives the characteristic odour of ozone:

$$F_2 + H_2O = 2HF + O$$
$$O + O_2 = O_3$$

It is probable that the fluorine is the result of radiation reactions on the calcium fluoride. Free fluorine to the extent of 15–16 p.p.m. has been reported also from a Canadian fluorite which likewise contains radioactive minerals.

Volatile Fluorine Compounds

At least two of these exist in Nature.

Hydrogen fluoride. This is met with as a volcanic emanation and may well be fairly abundant but has only occasionally been reported in analyses. The most famous occurrence is in the Katmai district of Alaska, where in the Valley of Ten Thousand Smokes the gas is emitted at an estimated annual rate of 200000 tons along with large quantities of hydrogen chloride. It is recorded similarly in the gases of the Devil's Kitchen, Hawaii.

Silicon fluoride. This compound was reported in the volcanic gases of Vesuvius in the 1872 eruption. The existence of such minerals as ferruccite $NaBF_4$ as well as hieratite K_2SiF_6 in fumarolic deposits suggests that, in addition to silicon fluroide, boron fluoride may also occur as a natural substance.

Fluorine Minerals

These are less numerous than those of chlorine.

Only in a few cases is fluorine associated with oxygen compounds but some of these minerals are actually very abundant and widespread.

Fluorine minerals occur mainly in four ways:

(*a*) Fumarolic deposits.
(*b*) Pneumatolytic and hydrothermal deposits.
(*c*) Accessory rock minerals.
(*d*) Pegmatitic minerals.

FLUORINE MINERALS

Name	*Formula*	*Occurrence*
Villiaumite	NaF	In cavities in alkalic rocks
Fluellite	$3AlF_3 \cdot 4H_2O$	H.T. hydrothermal veins
Cryolithionite	$Li_3Na_3Al_2F_{12}$	In granite pegmatite with cryolite
Cryolite	Na_3AlF_6	As above—very local
Sellaite	MgF_2	In dolomite rocks and fumarolic
Fluorite	CaF_2	In hydrothermal veins and accessory in granite rocks
Cerfluorite	$(Ca,Ce)F_{2-3}$	Pegmatitic mineral
Matlockite	$PbFCl$	Hydrothermal lead veins
Ferruccite	$NaBF_4$	Fumarolic
Hieratite	K_2SiF_6	Fumarolic
Fluorapatite	$Ca_5(PO_4)_3F$	Accessory rock mineral, widespread
Fluoborite	$Mg_3BO_3(F,OH)_3$	Hydrothermal mineral
Topaz	$Al_2SiO_4(OH,F)_2$	Accessory in greisenized granites

Fluorite. This mineral, commonly called fluorspar, is the only source of fluorine for industrial purposes. It is often very abundant in hydrothermal veins and may be massive or well crystallized. Very large cubic crystals are sometimes found and are well known from Weardale, England. Fluorite shows great variety of colour and is frequently fluorescent. There are probably varied causes for these effects. In some cases occluded organic matter may be responsible. The topic has been the subject of much discussion.

Fluorine in Rocks

Much of the fluorine in igneous rocks is present in the form of the mineral fluorapatite $Ca_5(PO_4)_3F$, which in the pure state carries 3·77 % F. But most apatites have some replacement of F^- by OH^- and the fluorine content is usually less. Chlorine is also commonly present and there are continuous isomorphous series between the three minerals fluorapatite, chlorapatite, and hydroxyapatite.

It was at one time considered that in acid rocks fluorapatite was the characteristic mineral whereas in basic rocks chlorapatite was said to predominate. In considering rock analyses it is quite frequently found that the fluorine (and chlorine) content is appreciably in excess of the amount corresponding to the known apatite content of the rock. This is because fluorine in particular is present also in various silicates where it replaces OH^- ions in such minerals as the micas and amphiboles.

There is replacement by chlorine as well as fluorine, but in much smaller amounts because, no doubt, the larger Cl^- ion (radius 1·81 Å) can less easily enter the lattices. The mineral sphene $Ca(TiO)SiO_4$ also sometimes has some replacement of oxygen by fluorine. The following are examples of silicates containing variable small amounts of fluorine.

Leucophane	$(Ca,Na)_2BeSi_2(O,OH,F)_7$
Meliphane	$(Ca,Na)_2Be(Si,Al)_2(O,F)_7$
Norbergite	$Mg_3SiO_4(F,OH)_2$
Fluorbiotite	$K(Mg,Fe)_3AlSi_3O_{10}F_2$
Clinohumite	$Mg_9Si_4O_{16}(F,OH)_2$

These fluorine-bearing minerals are numerous, over fifty species having been described. They often contain metals such as lithium, beryllium, and the rare earths. They are generally accessory rock minerals or occur in the residuals of magmatic crystallization—*e.g.*, in pegmatites. The fluorine in the original magma has probably considerable fluxing properties and favours the highly fluid melts or solutions which are characteristic of the formation of pegmatites.

In many cases it is possible to account completely for the distribution of the fluorine in a rock.

Thus, to quote an analysis given by Koritnig for the Brocken granite, Harz Mountains:

Total fluorine by analysis (%)	0·086 %
7·0 % biotite with 0·415 % F (%)	0·029
0·11 % fluorite (by difference) (%)	0·053
Apatite, F content (%)	0·004
Total (%)	0·086

Pegmatitic Minerals

The most interesting of these deposits is the remarkable pegmatite body in a granite stock intruded into gneiss at Ivigtut in the district of Frederikshaab, West Greenland. This deposit contains abundant cryolite with numerous other minerals including fluorite and various rare fluorides mainly derived from the alteration of the cryolite. This occurrence of fluorine compounds is almost unique, the only other important locality for cryolite being Miask in the Ilmen Mountains, U.S.S.R.

Fluorine in Weathering Processes

The weathering of rocks containing fluorine compounds is susceptible to different climatic conditions and to the nature of the original fluorine-bearing minerals. Thus the weathered products of a Cornish granite (quoted by Butler) show that the fluorine content has been halved so that much of the fluorine has been leached out whereas, in the case of a syenite from Norway, the derived soil seems to have retained practically all of the fluorine of the original rock.

The fluorine content of soils is connected with the apatite content, suggesting that the actual component is fluorapatite to a large extent. The actual fluorine values for soils are very variable but the average is of the order of 200 p.p.m. Clays and argillaceous rocks generally seem to be enriched in fluorine, often in excess of 1000 p.p.m. which is in part due to apatite and more particularly to micaceous minerals. The true clay minerals—*e.g.*, kaolinite and bentonite—often contain considerable amounts of fluorine (2000–4000 p.p.m.).

Fluoride Ions in Solution

Some at least of the fluorine in rocks is dissolved in the process of weathering and passes ultimately into the ocean, the fluorine content of which is 1·4 p.p.m., which represents over five times the average content of river water. The presence of fluorine in drinking water is of significance to human health and has been widely investigated. A desirable content is stated to be 0·3–0·5 p.p.m. which is adequate to safeguard against dental caries but not high enough to have harmful effects. There appear to be considerable areas of the Earth where the water supplies are deficient in fluoride. Some fluorine compounds must pass into the air from volcanic emanations but the amount is not perceptible in the atmosphere as a whole. On the other hand, local accumulations of volatile fluorine compounds from industrial processes are claimed to have caused toxic effects to vegetation and animals in the vicinity. Such pollution seems to be derived particularly from high-temperature processes involving coal and ceramics.

Biogeochemistry

It is known that plants absorb fluorine from the soil, a noteworthy case being the tea plant, which may contain 400 p.p.m. One of the most remarkable occurrences of fluorine is the presence of the potassium salt of fluoroacetic acid in the plant *Dichapetalum cymosum*, as a toxic principle. This plant may carry up to 150 p.p.m. fluorine, evidently in the covalent state.

In general, fluoride ions are inimical to biological processes and the enzyme adenosine triphosphatase is known to be inhibited by them. This enzyme is important in metabolic processes.

In the case of animals most of the fluorine is found in bones and teeth and it is evidently related to the apatite which constitutes a major part of the structure of these materials.

It has been shown that bones buried in the soil take up more fluorine, perhaps by replacement of OH groups in apatite by fluorine, and determination of the fluorine content has been used as a means of estimating the age of the bone. It is a remarkable fact that the bones of marine animals contain greatly increased quantities of fluorine, the factors being almost 15 : 1 for fishes, 11 : 1 for mammals and 3 : 1 for birds. This is evidently due to the direct or indirect absorption of fluoride ions from sea water. The highest fluorine content is met with in the shells of some brachiopods, where it may reach 1·52 % in the ash, corresponding to a high proportion of fluorapatite in the original shell. On the other hand, many crustaceans and brachiopods appear to have no very large fluorine content in their shells. The nudibranch *Archidoris* and the brittle star *Ophiocomina* contain notable amounts of fluorine in their tissues.

Production of Fluorspar

Fluorite or fluorspar is the only fluorine mineral of economic importance and the demand has considerably increased during the last decades as a result of the development of the nuclear energy industry and the increasing production of organic fluorine derivatives.

Formerly the compound was regarded as a gangue mineral in some areas, particularly in lead mining, and even used as road metal in some cases. Most of the commercial deposits consist of replacement bodies or fissure veins, often in limestone rocks. Where vugs or cavities occur in the veins they may yield fine crystallized specimens of the mineral. The more massive varieties, often having a banded structure, have been used as ornamental stone—*e.g.*, the ‘Blue John’ of Derbyshire, England.

Fluorspar for chemical purposes should be low in silica and in sulphur and a purity of 98 % is desirable and is attainable in some localities.

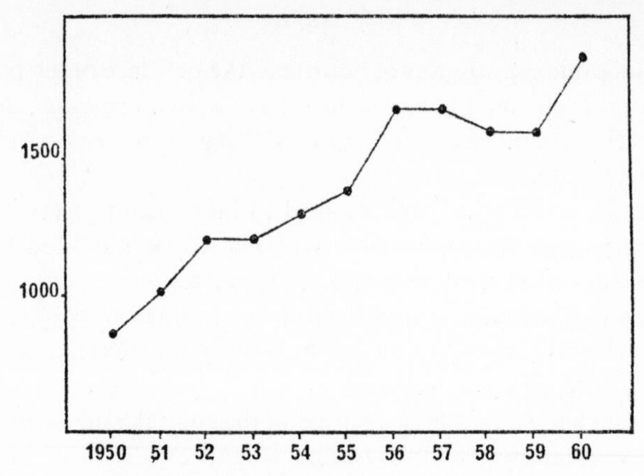

FIG. 20.1. World production of fluorspar, in thousands of tons.

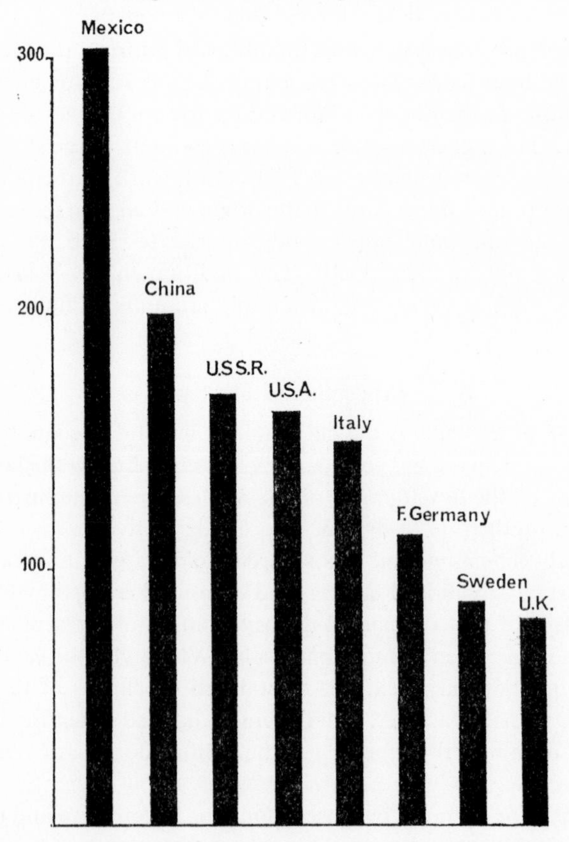

FIG. 20.2. Chief producers of fluorspar, 1959, in thousands of tons.

Important producing countries include Mexico, U.S.A., Germany, U.S.S.R., Italy, and many other localities where metalliferous veins are worked.

Great Britain. Workable deposits occur in Derbyshire, as at Eyam, and in County Durham in Weardale. Mineralogically interesting fluorite is found in connexion with the tin lodes of Cornwall, in some of the veins of the Alston Moor area, Cumberland, and associated with hematite deposits in West Cumberland.

Chlorine

There are two stable isotopes of this element in Nature:

$$^{35}Cl = 75{\cdot}4\%$$
$$^{37}Cl = 24{\cdot}6\%$$

There is no convincing evidence that any natural fractionation of these isotopes takes place.

The cosmic abundance of chlorine seems to be somewhat greater than that of fluorine by a factor of about 1·8, but the figures quoted for these two elements are subject to some uncertainty.

Chlorine atoms may be synthesized near the end of the α-process and also perhaps in the s-process of neutron capture on a slow time-scale taking place during the course of stellar evolution.

Geochemistry

Although the estimated abundance of chlorine (200 g/ton) in the Earth's crust turns out to be appreciably lower than that of fluorine, it is, on account of the vast quantities present in the ocean, an element of much greater accessibility, fluorine being available in bulk only from the mineral fluorite which is of comparatively local occurrence.

It will be convenient to consider the occurrence of chlorine under the following headings:

(1) Hydrogen chloride and free chlorine.
(2) Chloride ions in solution and as evaporites.
(3) Other minerals containing chlorine.

(1) Hydrogen chloride and free chlorine. Hydrogen chloride is a relatively common gas in connexion with volcanic activity. Estimates of the output from a number of localities have been made. The most noteworthy example is the case of the Valley of Ten Thousand Smokes, mentioned above in connexion with hydrogen fluoride, where the hydrogen chloride produced annually is estimated at 1¼ million tons. No other known occurrences seem to be on such a large scale but there are numerous minor sources and in the aggregate they must be considerable. Examples are Kilauea, Hawaii, Yellowstone Park and Steamboat Springs, U.S.A., and Arima, Japan. The total contribution made to the chlorinity of the ocean by these emanations is quite significant (*vide infra*).

Free chlorine seems to occur less commonly in volcanic gases. It is, however, reported in an analysis of Kilauea gases at 0·41 %, with hydrogen chloride apparently absent. It appears that the composition of gases from active vents is very variable whereas that from fumaroles is more constant. The origin of this free chlorine and of hydrogen chloride is of some interest. In view of the character of some of the salts in fumarolic deposits and the presence of great excess of water vapour in volcanic emanations generally, it may well be that most of the hydrogen chloride is a product of hydrolysis of various metallic chlorides liberated in vapour form by de-gassing of deep-seated zones in the crust. Other reactions leading to the liberation of HCl include possibly:

$$2FeCl_3 + 3H_2O = Fe_2O_3 + 6HCl$$
$$H_2S + 3Cl_2 + 2H_2O = SO_2 + 6HCl$$
$$S + 2Cl_2 + 2H_2O = SO_2 + 4HCl$$

The free chlorine is unlikely to have been directly liberated by de-gassing —it is more probably the result of some secondary reactions.

It is appropriate to mention here the occurrence of hydrochloric acid in the gastric juice of animals.

(2) Chloride ions in solution. The principal anion in the sea is of course chloride and the total amount at present is 280 Gg for the whole of the oceans. If this quantity of chloride has been supplied during geological time, say, 3×10^9 years, then the average annual contribution should be 10^{13} g or 10 million tons. This may well be the case, for the amount supplied by the few examples of volcanic gases which have been estimated is more than one-eighth of this requirement. Water from rivers is relatively low in chloride—the numerous cations are mainly partnered by the HCO_3^- ion and the content of Cl is about 8·3 p.p.m. which is less than that in sea water by a factor of over 2000.

It was pointed out by Goldschmidt that the chlorine content of the hydrosphere was far greater than could be accounted for by the weathering of rocks and leaching out of the products. Therefore the contribution made by volcanic gases (as referred to above) is probably of considerable importance. There is also the possibility of a primordial atmosphere, rich in chlorine compounds, which was washed down into the ocean to provide much of its chloride content. However, the evidence is probably more favourable to the view that progressive emission of gases has taken place (and is still taking place). Some mineral springs are rich in chloride. It is not, however, always certain whether all of this is 'juvenile' chloride or whether indirectly derived from the ocean—*e.g.*, by leaching out of evaporite salt deposits. There are considerable concentrations of chlorides in some inland seas. Thus the Dead Sea has a total salinity of 30 % and the salts consist of about two-thirds of their weight of chloride. The Great Salt Lake of Utah has about 20 % salinity and the salts contain 55 % chloride whereas the salts of Owen's Lake, California contain only 25 % chloride. These variations are probably due to differing character of the rocks of the surrounding country from which the lakes receive drainage.

It is evident that there is an inexhaustible supply of chloride in the ocean and it is in general also oceanic salt which provides the material of the extensive evaporite deposits from which sodium chloride, etc., are largely obtained. The production of chlorine is essentially the production of common salt and therefore reference should also be made to Chapter 9, dealing with sodium compounds, where the evaporite deposits are dealt with. It may be noted that in the case of lakes where evaporation has proceeded to a high concentration, as in the Dead Sea, the chloride content can be related predominantly to magnesium chloride rather than sodium chloride in a ratio of almost 2 : 1, whereas for normal oceanic salts the sodium–magnesium ratio is about 9 : 1.

Apart from evaporation there is no geochemical process having any appreciable effect on the chloride content of the ocean. The effect of biological processes is negligible. There is a certain amount of cycling of chloride ions from sea to land and back by way of salt spray in the atmosphere, through rain and rivers back to the sea again, but it is relatively local and confined to maritime regions. The chloride content of the atmosphere due to this cause is reflected in the corrosion of copper roofs, etc., in seaside places where the green deposit is basic chloride similar to atacamite, whereas in inland places the corrosion product is mainly a basic carbonate similar to malachite.

In the evaporite deposits of the caliche in Chile it has been established that the perchlorate ion ClO_4^- is present. The mechanism of its formation is obscure; it is one more indication of the remarkably high oxidation potential attained in this region.

(3) Other minerals containing chlorine. Chlorine minerals, which probably all contain the chloride ion, are more numerous than those of fluorine, but not more abundant, many of them being only mineralogically interesting. In the following list the evaporite minerals are also included for the sake of completeness.

The most significant chlorine-bearing minerals are chlorapatite and sodalite. They are essentially rock minerals and probably account in the aggregate for a very considerable proportion of the chlorine in the Earth's crust. Relatively little chlorine is carried by such minerals as

CHLORINE MINERALS

Name	*Formula*
In evaporites	
Halite	$NaCl$
Hydrohalite	$NaCl.2H_2O$
Sylvine	KCl
Bischofite	$MgCl_2.6H_2O$
Carnallite	$KMgCl_3.6H_2O$
Chlorocalcite	$KCaCl_3$
Douglasite	$K_2FeCl_4.H_2O$
Rinneite	$K_3NaFeCl_6$
Sulphohalite	$Na_6(SO_4)_2ClF$
Kainite	$KMgSO_4Cl.3H_2O$
In volcanic deposits	
Salammoniac	NH_4Cl
Scacchite	$MnCl_2$
Lawrencite	$FeCl_2$
Erythrosiderite	$K_2FeCl_5.H_2O$
Mitscherlichite	$K_2CuCl_4.2H_2O$
In oxidized zones of metalliferous deposits	
Nantokite	$CuCl$
Eriochalcite	$CuCl_2.2H_2O$
Atacamite	$Cu_2Cl(OH)_3$
Chloroargyrite	$AgCl$
Calomel	Hg_2Cl_2
Cotunnite	$PbCl_2$
Pyromorphite	$Pb_5(PO_4)_3Cl$
In rocks as main or accessory minerals	
Sodalite	$Na_4Al_3Si_3O_{12}Cl$
Chlorapatite	$Ca_5(PO_4)_3Cl$

micas and amphiboles which often have on the other hand a notable fluorine content.

Biogeochemistry

The presence of chlorine in plants suggests that it may be an essential element but its functions seem to be of minor importance. A few biologically important substances contain chlorine, probably mainly or wholly as anion as in the anthocyanins, but the chlorine content of plant material is only of the order of a few parts per million. Coals contain chlorine which may be mainly derived from plants.

In animals the chloride ion is particularly characteristic of the body fluids. The general similarity, quantitatively and qualitatively, between the salinity of such fluids and that of the ocean has led to the theory that this reflects the physiological development of terrestrial animals from aquatic marine forms. It also implies some fair degree of constancy of salinity of the ocean during biological time which runs somewhat counter to the view that there has been a progressive emission of chloride by volcanic emanation over geological time. It is not clear how far this view is of great significance.

Production

As far as saline substances are concerned reference should be made to Chapter 9. The world production of elemental chlorine for industrial purposes has risen from 150000 tons in 1919 to 3800000 tons in 1951 and is probably still rising.

Ultimate Forms

As far as human activities are concerned there seems little doubt that most of the chlorine used for all kinds of purposes, including food, must ultimately find its way into the ocean. It therefore forms part of a much larger geochemical cycle.

Bromine

There are two stable isotopes of this element in Nature:

$$^{79}Br = 50{\cdot}52\,\%$$
$$^{81}Br = 49{\cdot}48\,\%$$

Both of these isotopes could be derived from long-lived fission products of uranium, ^{79}Se by β-decay to ^{79}Br and ^{81}Kr by K-capture to ^{81}Br.

These fission products are not known with any certainty in uranium minerals so that the existence of radiogenic bromine in the Earth is a matter of speculation. It would appear that bromine atoms could be built up both by the s-process and the r-process in the nuclear reactions in stellar interiors.

Geochemistry

Bromine is a rather scarce element in the Earth's crust at 3 g/ton and it is also to a large extent dispersed. There are few natural concentrations of the element. The bromine content of igneous rocks is rarely more than double the crustal average. The form in which bromine is mainly present in rocks is therefore difficult to ascertain, but as its ion (radius 1·96 Å) is not greatly larger than that of chlorine (1·81 Å) it can probably replace the latter to some extent. Thus amounts of up to 5 or 6 p.p.m. of bromine have been reported in some chlorapatites and in micas. It may be, however, that bromine remains behind in residual fluids in magmatic crystallization and enters into fluid inclusions with the mother liquors.

In the weathering of igneous rocks it appears that bromine is enriched and it has been suggested that some kind of combination with organic matter occurs. In the case of a soil containing 8·3 p.p.m. of bromine it was found by Behne that 69 % was retained after leaching with water. Vinogradov showed that the Cl/Br ratio was in general lowest in humus-rich soils and highest in desert earths.

There is no indication that inorganic bromine ever occurs in any form but the bromide ion. There are very few independent bromine minerals, the most important being silver compounds of very low solubility, *viz.*, bromargyrite AgBr and embolite Ag(Cl,Br). Both of these are secondary minerals of silver deposits.

The bromine content of volcanic emanations is of very minor importance but some thermal springs have been found to be quite rich in bromine. The following may be noted:

Geyser at Reykjanes, Iceland	77 p.p.m.
Hot springs, Lac Abbé, French Somaliland	2·7 p.p.m.

The water of the River Jordan before entering the Dead Sea contains 2 p.p.m. whereas the river before entering the Sea of Galilee is virtually free of the element. It appears that thermal springs in or near the lake supply the bromine. The salts of the Dead Sea contain approximately 2·7 % bromides and this is one of the most important natural concentrations of the element. The bromine content of sea water is 65 p.p.m., which compared with the average river water represents a concentration by a factor of over 10000—much higher than with any of the other halogens.

Sea water is now the most important source of bromine. The extraction process, which depends on the liberation of the free element by action of chlorine after acidulation, is carried out in several places. Suitable sites require that an adequate intake of fresh sea water is

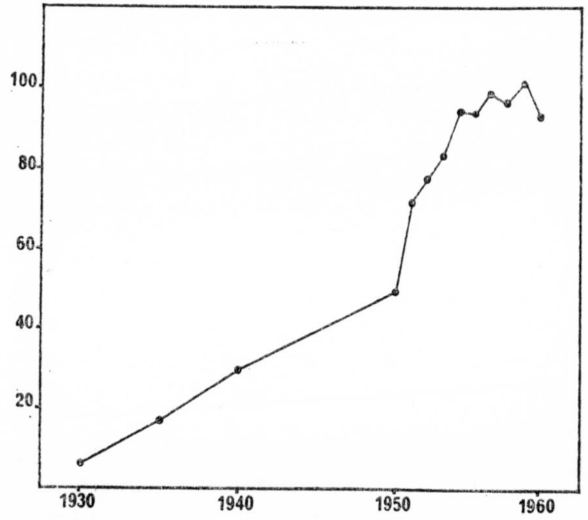

FIG. 20.3. World production of bromine, in thousands of tons.

assured and at the same time the effluent must be capable of being discharged without contaminating the intake. Localities where bromine is or has been extracted include Wilmington, North Carolina, U.S.A., Hayle, Cornwall, England, and Amlwch, Wales.

The other traditional source of bromine is from the mother liquors or bitterns from the manufacture of solar salt and from the extraction of salts from evaporite deposits such as those of the Stassfurt region in Germany.

Biogeochemistry

Little is known of the role of bromine which is present as a microconstituent of most plants; its presence may be adventitious. A few plants, however, show a marked concentration of bromine—*e.g.*,

Raphanus sativus	8 p.p.m.
Boletus scraber	36 p.p.m.
Curcuma citrullus	262 p.p.m.

In marine organisms, both plant and animal, bromine is present in more perceptible amounts and in some the bromine content may exceed that of the chlorine. Two organic bromine compounds may be mentioned—in certain sponges, corals, and annelids dibromotyrosine is found, and in the gastropods *Murex brandaris* and *M. purpureus* is a substance which on appropriate treatment yields dibromoindigo, the famous Purple of the Ancients.

Iodine

There is only the single stable nuclide, ^{127}I, existing in Nature. The isotope ^{129}I is a fission product of uranium formed by β-decay from ^{129}Te, and itself decays to ^{129}Xe. It has been thought possible for this reason that iodine might be present in uranium minerals but it does not yet seem to have been detected. Iodine atoms would probably be built up by neutron capture on a slow time-scale, taking place in ' red giant ' stars.

Geochemistry

Iodine is among the rare elements in the Earth's crust, being estimated at 0·3 g/ton, and it is essentially a dispersed element.

The iodine content of igneous rocks is generally low and has not been extensively determined. The amounts recorded in rocks and in some of the minerals such as sodalite and the feldspars are never as high as 1 p.p.m. and usually much less. It may be possible that in spite of its large size (radius 2·2 Å) the iodine ion can replace Cl^- and OH^- in minerals such as the above, but of course only to a very small extent.

It appears that iodine is present in volcanic emanations as indicated by amounts exceeding 1 p.p.m. in fumarolic incrustations and in lava in the Naples–Vesuvius region.

Apart from the evaporites, the only independent iodine minerals are silver compounds, *viz.*, iodoargyrite AgI, iodoembolite Ag(Cl,Br,I), and miersite (Ag,Cu)I.

Iodine in Solution

Mineral springs, in some cases probably connected with magmatic activity, sometimes contain iodine, generally as iodide, but the free element has been claimed. Such springs and wells are quite numerous and widespread, being met with for example in Switzerland, Saratoga, U.S.A., California, U.S.A., Indonesia, Japan, Italy, and the U.S.S.R. The iodine contents range from 30–150 p.p.m. and up to 100 p.p.m. is also found in some oil-well brines.

The iodine content of sea water is 0·05 p.p.m. and this is the greatest potential source of the element, but it is not at present practicable to extract it directly. The average iodine content of river water is about 0·0018 p.p.m. so that the concentration factor for sea water is over 27—a much lower value than that for bromine (*q.v.*). It is interesting that the iodine content of the Dead Sea water is extremely low.

Iodine in the Atmosphere

This has been the subject of much study on account of the significance of iodine for human health. It is clear that a high proportion of this iodine is derived from sea spray. The content is 0·1–1·4 mg/m^3. This iodine finds its way into rainwater and figures of 0·001–0·003 p.p.m. are quoted for Switzerland. For the Eastern U.S.A. it has been shown that the iodine content of rainwater decreases to about one-thirtieth of the value on the Atlantic coast on proceeding inland to the region of the Great Lakes.

There is a certain amount of evidence, not entirely conclusive, that free iodine vapour is present in the atmosphere. It may be derived from marine iodine and also to a minor extent from industrial processes and the burning of kelp. It may be observed that the means exist in the atmosphere, particularly in its upper regions, for the oxidation of iodides by means of nitrogen oxides, by ozone, or possibly by ordinary oxygen in the presence of sunlight.

Iodine in Soils

This is related closely to airborne iodine from the ocean and a remarkable degree of concentration seems to be attained. The average value for soils of about 2 p.p.m. represents a seven-fold concentration as compared with ordinary rocks (both igneous and sedimentary). In far inland regions the content is low, in oceanic islands very high, in some cases as much as 70 p.p.m. Apart from geography, the iodine content of soils depends on the time factor so that very ancient soils (> 20000 years old) are noticeably rich in the element and post-glacial soils (< 20000 years old) are perceptibly deficient. This effect has been related to the incidence of iodine-deficiency diseases (mainly goitre) in certain countries.

Iodine in Evaporites

In general the well-known oceanic salt deposits do not contain noteworthy amounts of iodine and in view of the high solubility of iodides, which would be the last to separate and the first to be leached out, this is not surprising. An exception is, however, provided by the famous caliche deposits of the Chilean desert region already alluded to (Chapter 16). In these deposits iodide is present together with considerable amounts of iodate. An example quoted is 330 p.p.m. $NaIO_3$ and 470 p.p.m. NaI, but in other cases nearly all the iodine is present in the form of iodate. The iodate content may attain 2% of the nitrate present in the caliche. The origin of this large concentration of iodine is obscure.

Goldschmidt suggested four possible modes of origin:

(*a*) Accumulation of airborne oceanic iodine in connexion with alkaline soils.
(*b*) High oxidative weathering of marine sediments rich in iodine.
(*c*) Fractional leaching by ground water of salts in desert soils.
(*d*) Concentration of iodine from volcanic exhalations.

As previously mentioned (Chapters 16 and 19) the explanation of the very high oxidation state of these deposits presents considerable difficulty. Independent iodine minerals found in or near these deposits include lautarite $Ca(IO_3)_2$, dietzeïte $Ca_2(IO_3)_2(CrO_4)$, bellingerite $2Cu(IO_3)_2.2H_2O$, and swartzenbergite $Pb_5(IO_3)Cl_3O_3$.

Biogeochemistry

Iodine provides one of the classical examples of the concentration of an element by biological processes. Land and freshwater plants only contain 0·01–1·0 p.p.m. iodine but marine plants have long been noteworthy for their iodine content. It is the brown algae, particularly species of *Laminaria*, *Fucus*, and *Sargassum*, which are enriched in iodine to the greatest extent and the amount varies from 1000 to as much as 17000 p.p.m. on the dry weight. According to Vinogradov marine diatoms (Caspian Sea) may contain up to 620 p.p.m. In the case of animals some sponges actually contain 2% of iodine on dry weight.

In terrestrial animals iodine is concentrated in particular in the thyroid gland which contains iodine compounds of which the most important are 3 : 5-di-iodotyrosine and the 3 : 5-di-iodo-4-phenoxy derivative of this substance which is called thyroxine. These substances are elaborated from the iodide in the blood and represent a most remarkable degree of concentration of the element.

Sources

The world production of iodine is of the order of 2000 tons per annum of which at least 1200 tons are obtained as a by-product from the Chilean nitrate industry. This output could undoubtedly be increased ten-fold if necessary. Other minor sources are from iodiferous spring and well waters and oil-well brines. There are localities in the U.S.A., mainly in California, in the U.S.S.R. near the Caspian Sea, in Indonesia, Japan, and Italy. The production of iodine from ‘ kelp ’—the ash of seaweeds—in Western France, Scotland, Ireland, Norway, Japan, etc., is obsolete or obsolescent.

CHAPTER 21

The Metals of Group VII

We have here to consider three elements, the first of which, manganese, is both abundant and widespread and of considerable geochemical interest, the second, technetium, which does not occur on the Earth, and the third, rhenium, which is a very rare and highly dispersed element.

Manganese

Only the single stable nuclide ^{55}Mn exists in Nature. It is to be supposed that manganese, as one of the metals of the 'iron peak' of abundance, would be synthesized by the nuclear processes occurring at very high temperatures after the α-process is ended and when the evolution of a star is very rapid. At this stage there is for a time a statistical equilibrium between nuclei, protons, and neutrons. On this basis Hoyle *et al.* have shown that the relative abundance observed is in reasonable agreement with that predicted for conditions when $T = 3{\cdot}78 \times 10^9\,°K$ and the density is of the order of $10^5\,g/cm^3$.

There are certain stars known as 'metallic line stars' in which manganese appears to be superabundant. The reason for this is not yet known.

Geochemistry

In the Earth's crust manganese comes eighth in order of abundance as far as metals are concerned and twelfth for the elements as a whole. The estimated abundance is 1000 g/ton and this is not very different from the value which has been quoted for the Earth as a whole.

Analyses of meteorites show that the division of manganese between the metallic, sulphide, and silicate phases is in the ratio 1 : 10 : 26. Manganese could therefore be expected to be a lithophile element with some thiophile tendencies. This is largely borne out in practice.

As a transitional element manganese is well known for its multiplicity of oxidation states. As far as natural chemistry is concerned these are probably confined to 2, 3, and 4. The element is essentially cationic and the highly oxidized anions MnO_4^{2-} and MnO_4^- are unknown in Nature. In general it may be said that Mn^{2+} ions are characteristic of igneous rocks, particularly the deep-seated types, whilst the sediments generally show a higher oxidation state, Mn^{4+} being characteristic. The Mn^{3+} ion seems to occur in only a limited number of minerals.

The ionic radius of manganese as Mn^{2+} is 0·80 Å which is only slightly larger than Fe^{2+}, which is 0·74 Å. Considerable interchange between the two ions in crystal lattices is therefore possible. The ions of calcium, magnesium, and several other metals also come fairly near to manganese in size and the element is therefore widely present in minerals, particularly silicates, as a trace or minor element. The ability of manganese to enter into silicate minerals causes it to be found in amounts approaching the average crustal abundance (1000 g/ton) in all types of igneous rocks. On the whole manganese does not become enriched at any particular stage of magmatic crystallization and there are no manganese ore bodies of magmatic type.

Independent manganese minerals are not usually found in igneous rocks; they are more characteristic of hydrothermal and metasomatic deposits.

Manganese can also form a few sulphide minerals. The monosulphide MnS is analogous to the sulphides of zinc and cadmium. The mineral MnS_2 does not seem to be fully analogous to pyrite FeS_2—instead of being semi-metallic it may be fully ionic in structure.

Manganese-bearing minerals of primary origin predominantly contain the Mn^{2+} ion and, in the process of weathering, this passes into solution as $Mn(HCO_3)_2$. A few manganous compounds are found in hydrothermal veins. The stability of such compounds is probably related to their solubility (thus the tungstate minerals hübnerite and wolfram are very stable) or more generally to the hydrogen ion concentration of their environment which has an important bearing on the oxidation potential required to convert Mn^{2+} to Mn^{3+} or Mn^{4+}. It is at this stage that there is a decided parting of the geochemical ways of manganese and iron. In the case of the Mn^{2+} ion the oxidation potential in acid solution is quite high at + 1·51 V at pH 4 and below and 0 V decreasing to less than − 0·4 V at pH values above 8. In the case of iron the potential required at pH 2·5 for Fe^{2+} to Fe^{3+} is only 0·77 V, and at pH 6 has fallen to only − 0·1 V. It is evident from this that ferrous ions are more readily oxidized than manganous ions and that the latter will tend to be stable over a wider range of pH, hence the Mn^{2+} ions from weathering processes remain mobile for greater distances than Fe^{2+}

ions. The higher manganese hydroxides which are produced by oxidation are not stable but break down into the hydrated oxide $MnO_2.nH_2O$ and this substance separates as a negative colloid. Thus the ' oxidate ' sediments of manganese are mainly formed in the slightly alkaline conditions of the sea and commonly contain adsorbed cations such as K^+ and Ba^{2+}. It is from such deposits that the minerals pyrolusite and psilomelane are largely derived.

These redox relationships are of considerable importance geochemically and would be relatively more so if it were not for the superabundance of iron which also undergoes analogous changes (*q.v.*). The varying colours of the different types of manganese minerals are of some interest. The pink of the Mn^{2+} ion is evident in rhodonite and rhodo-

MANGANESE MINERALS

Name	*Formula*	*Occurrence*
Alabandite	MnS	Hydrothermal veins
Hauerite	MnS_2	Rare mineral in gypsum and sulphur deposits
Manganosite	MnO	In H.T. veins, etc.
Pyrochroite	$Mn(OH)_2$	Low-temperature H.T. deposits
Hausmannite	Mn_3O_4	Bedded ore deposits
Psilomelane	$(Ba,H_2O)_2Mn_5O_{10}$	
Pyrolusite	MnO_2	
Manganite	$MnO.OH$	
Galaxite	$MnAl_2O_4$	H.T. deposits
Bixbyite	$(Mn,Fe)_2O_3$	Metamorphic ores
Scacchite	$MnCl_2$	Fumarolic deposits
Rhodonite	$MnSiO_3$	H.T. and metasomatic minera
Tephroite	Mn_2SiO_4	
Sussexite	$(Mn,Mg)_2B_2O_5.H_2O$	
Rhodochrosite	$MnCO_3$	
Braunite	$Mn^{2+}Mn_6^{3+}SiO_{12}$	Bedded ore deposits
Alleghanyite	$Mn_5Si_2O_8(OH)_2$	
Glaucochroite	$(Mn,Ca)_2SiO_4$	
Pyroxmangite	$(Mn,Fe)SiO_3$	Metamorphic deposits
Spessartine	$Mn_3Al_2Si_3O_{12}$	
Pennantite	$Mn_9Al_6Si_5O_{20}(OH)_{16}$	
Stewartite	$Mn_3(PO_4)_2.4H_2O$	Oxidized zones of H.T. veins
Natrophilite	$NaMnPO_4$	
Mallardite	$MnSO_4.7H_2O$	
Hübnerite	$MnWO_4$	H.T. veins

chrosite and may also be the cause of the colour of rose quartz. Another variety of quartz, amethyst, is generally supposed to owe its colour to manganese—the reddish-violet tint being perhaps due to Mn^{3+} ions.

Under special conditions of low pH and lack of oxygen in deep seas and lakes, manganous ions remain in solution—unlike iron which commonly precipitates as sulphide in the black bottom muds.

Biogeochemistry

The manganese content of soils is variable in the range 200–5000 p.p.m. and it is frequently concentrated in horizons rich in organic matter where it is presumably immobilized by complex formation.

In certain exceptional acid soils—as in the Everglades of Florida, U.S.A.—the manganese ions have been largely leached out. On the other hand in alkaline soils manganese may be held as insoluble oxides. Manganese is one of the essential minor elements of plants and there are recognized manganese deficiency diseases which may develop in soils of both of the types mentioned above, showing that it is evidently the Mn^{2+} ion which is absorbed by the plants. The function of manganese in plants is uncertain but it is known that an excess can have toxic effects.

Manganese appears to be necessary to animals also. It may have some analogous function to iron in connexion with haemoglobin. There are manganese-containing blood pigments in certain molluscs—*e.g.*, *Pinna squamosa*. In man it is suggested that manganese may be important in the functioning of intracellular respiratory enzymes.

Production

A number of manganese minerals are available as ores, including pyrolusite, psilomelane, hausmannite, and braunite. For chemical uses the ore (usually pyrolusite) should be of high MnO_2 content but for some metallurgical purposes, such as smelting to manganiferous iron, it can be as low as 5–10%.

Typical Deposits

Nikopol, U.S.S.R. Bedded ores about 6ft thick in sandy Oligocene clay, mainly pyrolusite with earthy manganese (wad), manganite, and iron oxides. The deposit is believed to have resulted from microbiological activity, precipitating the manganese oxides in a littoral zone. The reserves are claimed to be some hundreds of millions of tons.

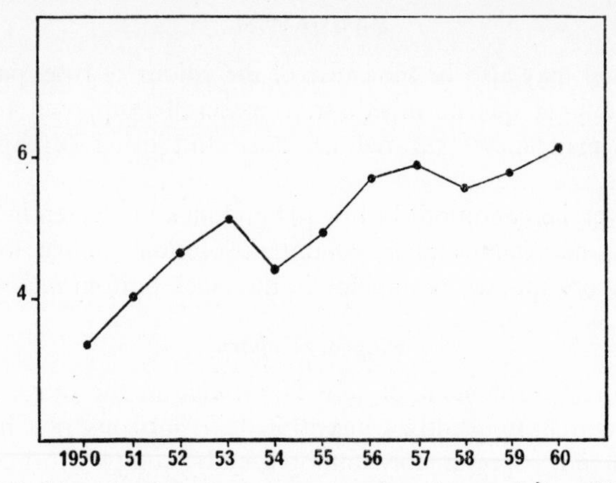

FIG. 21.1. World production of manganese ore (metal content), in millions of tons.

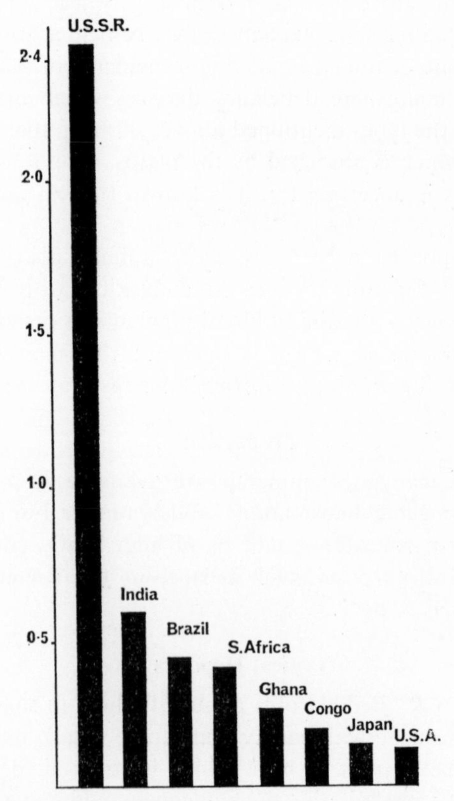

FIG. 21.2. Chief producers of manganese ore (metal content), 1959, in millions of tons.

Tchiaturi, Caucasus, U.S.S.R. Bedded ores, mainly pyrolusite and psilomelane in horizontal beds of marly sand of Eocene age. The beds are 6–10ft thick and extend 19 miles by 5–6 miles. Reserves of over 150 million tons are claimed.

Central Provinces, India. The deposits, which are extensive, are believed to have resulted from oxidative weathering of pre-Cambrian schists which contained spessartine and rhodonite. The ore minerals are mainly pyrolusite and psilomelane.

Ghana. Massive and concretionary bodies of ore in ancient metamorphic rocks. The reserves are said to be at least 10 million tons of ore.

Great Britain. Traces of manganese minerals are widespread in metalliferous areas but workable ore has only been raised in the past in a very few places. There was a small output from North Wales and for a short period from the iron ore mines of West Cumberland where some hausmannite was extracted.

Ireland. Small manganese deposits have been reported, as at Glandore, West Cork, but they are of little economic significance.

Technetium

The claims that element number 43 had been isolated from terrestrial sources have not been substantiated. There are no stable nuclides of the element. The isotopes ^{97}Tc and ^{99}Tc have the longest half-lives.

In 1952 Merrill established the presence of the lines of ^{99}Tc in the spectra of S-type stars. This is the longest-lived isotope which can be formed by neutron capture (half-life $2{\cdot}1 \times 10^5$ years). Its abundance is comparable with that of neighbouring elements. Since the S-stars are about 1% as abundant as the M-stars it follows that if all the red giants pass through the S-stage and continue as M-giants for 10 million years, then the duration of the S-stage is about 100000 years. The various interesting problems posed by this discovery are not fully solved but it must be said that it provides a very valuable piece of evidence in favour of the theories of element building in stars.

The mechanism proposed for the technetium synthesis is

$$^{98}Mo + n = {}^{99}Mo$$
$$^{99}Mo = {}^{99}Tc + \beta + \nu$$

Cameron, however, has proposed a mechanism involving photo-disintegration. It is now evident that the name of this element, proposed by Paneth when it was believed to be a purely artificial product, is really a misnomer. It remains to be seen whether some more appropriate designation will be introduced.

Rhenium

There are two natural isotopes of this element:

$$^{185}\text{Re} = 37{\cdot}07\,\%$$
$$^{187}\text{Re} = 62{\cdot}93\,\%$$

It is claimed that ^{187}Re is a long-lived β-emitter with a half-life of 4×10^{12} years, decaying to ^{187}Os. It would appear to be an element which would be synthesized by neutron capture on a slow time-scale.

Geochemistry

Rhenium was discovered in 1925 and as might be supposed it is a rare element. The crustal abundance of rhenium is estimated at 0·001 g/ton, which is comparable with that of the platinum metals. Its cosmic abundance, however, places it near to metals like silver and antimony.

In general it is dispersed but some concentration of the element occurs, particularly in the mineral molybdenite. It should be noted that ReS_2 has the same layered lattice structure as MoS_2. Amounts of the order of 20 p.p.m. are reported for some samples of molybdenite and there is one exceptional occurrence in Northern Sweden of a hydrothermal vein with molybdenite which carries 2500 p.p.m. of Re. Osmiridium associated with olivine rocks has been reported to contain rhenium up to 1 p.p.m.

The behaviour of rhenium in secondary processes is not yet known with any certainty. Goldschmidt suggested the possibility of the formation of the ReO_4^- ion.

CHAPTER 22

The Iron Group

The three metals, iron, cobalt, and nickel, which constitute this portion of Group VIII are conveniently treated together. In their natural chemistry they exhibit some interesting resemblances and also some marked differences. Cosmically, iron stands on a peak of the abundance curve of the elements and the values fall away rather steeply on each side. In consequence, iron is considerably more abundant than the elements which precede and succeed it in order of atomic number. Geochemically, iron, cobalt, and nickel are markedly siderophile so that the values for crustal abundances are unlikely to be a true measure of their actual terrestrial values.

Iron

There are four stable isotopes of the element in Nature:

$$^{54}Fe = 5{\cdot}84\% \qquad ^{57}Fe = 2{\cdot}17\%$$
$$^{56}Fe = 91{\cdot}68\% \qquad ^{58}Fe = 0{\cdot}31\%$$

There are no certain indications of any perceptible variation in isotopic composition of terrestrial and meteoric iron samples.

It is considered that the atoms of iron would be synthesized in the so-called ' equilibrium process ' occurring at the phase of stellar evolution when the temperature is in excess of 3×10^9 °K and the density perhaps as high as 10^5 g/cm³. This phase is supposed to precede a supernova explosion when many of the products of nuclear reactions are blown off into space, eventually perhaps to aggregate as some of the material of another star. Iron stands at a peak of nuclear stability and this is consistent with the high abundance of the element cosmically which is about thirty times greater than that of manganese which precedes it and about sixty times greater than that of cobalt which follows it.

Geochemistry

This extensive subject will be best dealt with under several subheadings.

(1) General geochemical character. Iron is of high abundance in the Earth. For the crust the value is 5 % and for the Earth as a whole, as deduced from physical properties, nature of meteorites, etc., it is quoted as 35 % by weight. It would appear that iron is in fact a typical siderophile element. But the affinity of iron for sulphur is considerable, large masses of iron sulphide minerals exist and the thiophile character of the element cannot be ignored. In the crust as a whole iron is in combination with oxygen and enters freely into silicate minerals; it is therefore also lithophile. Iron is an essential element in living organisms and must be reckoned as biophile also. It is thus a ubiquitous element. It does not, however, make any appreciable show in the atmosphere, except in compounds present in volcanic dust, and for various reasons its content in the hydrosphere is rather low.

(2) Forms in which iron exists. The element is sufficiently stable to exist in the free state and its compounds may be in either of two oxidation states, both of which can form readily under natural conditions. Hence the compounds of iron in Nature are abundant, widespread, and numerous.

(*a*) *Free metallic iron.* The occurrence of iron or of highly ferriferous alloys in meteorites is well known and has been referred to (Chapter 2). The siderites or iron meteorites consist essentially of a nickel–iron alloy and frequently contain certain highly reduced accessory minerals, particularly schreibersite $(Fe,Ni,Co)_3P$, troilite FeS, cohenite Fe_3C, and graphite. More rarely they contain such compounds as daubréelite $FeCr_2S_4$ and moissanite SiC.

The nickel–iron exists as two alloys. Kamacite has a body-centred α-iron structure and more than 5 % nickel with cobalt not usually over 6 %. Taenite may contain 13–48 % nickel and has the face-centred γ-iron structure. The total metal content of siderites is about 98 % and as previously noted (Chapter 2) they are thought to represent core material from a disintegrated member of the solar system. It is for this reason that they are generally held to present a sample of the sort of material of which the Earth's own core is composed.

It is estimated that the true metallic meteorites represent only 8 % of the total of meteorites which reach the Earth. It is to be noted that only the metallic types are likely to be recognizable as unusual substances whereas the silicate meteorites with their stony appearance require expert recognition. Most of the more famous meteorites are metallic. Some have had religious or mystical significance attached to them and their celestial origin early recognized. Probable examples are the ' Ka'bah ' at Mecca, venerated by Moslems, and ' the image which fell down from Jupiter ' (*Acts*, xix. 35).

The occurrence of native iron of terrestrial origin is rather local, which suggests that comparatively unusual conditions are required for its formation and exposure at the surface.

Some basaltic rocks contain minute grains of iron—*e.g.*, in Antrim, N. Ireland. The most interesting occurrence, however, is probably at Ovifak, Disco Island, Greenland, where masses of the metal many tons in weight have been weathered out of the basalt. The evidence seems to favour the view that the iron oxide (magnetite?) minerals of the original magma were reduced by carbonaceous material during ascent from lower levels of the Earth. It has on the other hand been claimed that some native iron has been erupted as such from great depths of the Earth.

Other localities for terrestrial iron are near Weimar, Thuringia, Germany, in Auvergne, France (in trachyte), at the Uil River, Urals, U.S.S.R., and Cameron, Missouri, U.S.A. (in coal measures).

Terrestrial irons are low in nickel and cobalt and are free from the typical meteoritic minerals like schreibersite. It is probable that the mode of origin of the iron is different in different localities.

(*b*) *Primary oxide and sulphide minerals.* As previously observed, at an early stage in the consolidation of an igneous magma there separate out certain refractory minerals of which magnetite Fe_3O_4 is the one carrying most iron.

According to Goldschmidt, changes corresponding to the following overall reaction occur by the action of water (or steam) on an ultra-basic (olivine) magma:

$$\begin{matrix} 3Fe_2SiO_4 \\ + \\ 3Mg_2SiO_4 \end{matrix} + 2H_2O = 2Fe_3O_4 + 6MgSiO_3 + 2H_2$$

This could evidently account for the separation of magnetite in the gabbroic stage rather than at the earlier peridotitic stage.

The resulting crystals may settle down in the magma and may be further concentrated by a sort of ' filter-pressing ' process. The exact mechanism by which such deposits have been emplaced in connexion with igneous rock masses is, however, open to several interpretations. The most famous examples of this kind are the magnetite ores of Northern Sweden, which are possibly the largest high-grade deposits of iron ore in the world.

There are also magmatic concentrations of titaniferous magnetite (containing ilmenite $FeTiO_3$). Not all magnetite deposits have originated in the above fashion, however. In some cases they seem to have resulted from metasomatic replacement in varied rocks, including schists, gneisses, and granites.

Sulphide ore bodies may consist to a large extent of pyrite FeS_2, but possibly carrying variable amounts of other metals (Cu, Au, etc.). They are not primarily sources of iron but may be valuable ores of sulphur and various non-ferrous metals. The famous Rio Tinto deposits in Spain have produced over 200 million tons of ore and have as much in reserve. They have been claimed as of magmatic origin but this is not certain.

Many deposits of primary iron minerals are undoubtedly of hydrothermal origin, that is they have been deposited from solutions.

Other sulphide bodies will be referred to below in connexion with cobalt and nickel. Iron compounds in hydrothermal deposits are only likely to be of economic interest if the deposit is very extensive. Very many valuable deposits are, as observed above, metasomatic replacements of some kind of rock and this may have involved the action of solutions, vapours, or both.

(*c*) *Primary silicate minerals*. These minerals are extremely numerous, particularly in igneous rocks. They are characteristic of the more basic types and the iron content generally falls with increasing silica content but the relationship between the two is not simple. It is notable that the ratio of ferrous to ferric iron tends to decrease as indicated in the following table:

	Dunites	*Gabbros*	*Diorites*	*Granites*
Silica (%)	40	48	57	70
Fe^{2+}	4·4	4·8	3·5	1·4
Fe^{3+}	2·0	2·2	2·2	1·1
Fe^{2+}/Fe^{3+}	2·2	2·2	1·6	1·3

In spite of the large number of iron silicate minerals there are in fact few of these which are simple silicates of iron alone.

The Fe^{2+} ion, radius 0·74Å, is somewhat larger than the Mg^{2+} ion (0·66Å), and although diadochy between the two is not uncommon, yet the complete replacement of magnesium by ferrous iron is rather rare, the most important example being the case of the orthosilicate series Mg_2SiO_4–Fe_2SiO_4. The two iron pyroxenes $FeSiO_3$ are not characteristic of magmatic rocks and seem to have been formed under low-temperature conditions. They are comparatively rare. The mineral hypersthene never contains more than 90% $FeSiO_3$, the remainder being $MgSiO_3$. It appears that $FeSiO_3$ is not stable at high temperatures and an artificial melt of this composition separates:

$$2FeSiO_3 = Fe_2SiO_4 + SiO_2$$

Other typical iron-bearing silicates of igneous rocks are quoted in the list below. The Fe^{2+} ion cannot replace Ca^{2+} in feldspars for thermochemical reasons but it is possible for Fe^{3+} to replace Al^{3+} in these minerals. Thus ferriferous orthoclase has been reported containing up to 2·88% Fe_2O_3. This is a rare occurrence, however, and it seems probable that the pink colour so often encountered in feldspars is due to exsolution of Fe^{3+} ions and decomposition to Fe_2O_3. The iron silicates of sedimentary rocks are mainly glauconite and chlorite. They characteristically impart a green colour to sediments which contain them. In both cases they contain both ferrous and ferric ions. Chloropal is a ferriferous clay mineral in which Fe^{3+} largely replaces Al^{3+}. Pure ferric silicates are of doubtful occurrence in Nature. In metamorphic rocks the minerals cordierite, anthophyllite, tremolite, biotite, and glaucophane are the chief silicates containing iron.

(*d*) *Secondary oxygen compounds of iron.* Under this heading we include oxides, hydroxides, and oxy-salts other than silicates. They are numerous and both ferrous and ferric salts are widely represented. This is to be expected in view of the relatively low redox potential involved in the action:

$$Fe(OH)_2 + OH^- = Fe(OH)_3 + e \quad (-0{\cdot}56V)$$

and also the oxidation of the ions in solution:

$$Fe^{2+} = Fe^{3+} + e \quad (+0{\cdot}77V)$$

Furthermore, the first can take place almost up to the neutral point and is well within the pH range of ordinary natural waters. On the other hand, reducing conditions at lower pH values generally favour the reverse process. In the weathering of rocks iron passes into solution mainly as $Fe(HCO_3)_2$ and perhaps to a minor extent as $FeSO_4$. As the solution becomes diluted and the pH rises to that of average river waters oxidation and hydrolysis soon lead to the deposition of ferric hydroxide, and various hydrated ferric oxide minerals are formed and they constitute an important group of sedimentary iron ores. Colloidal

$Fe(OH)_3$ tends to adsorb PO_4^{3-} ions so that many of these sedimentary ores are phosphatic.

As noted in the previous chapter, manganese is rather more stable in solution so that, in general, iron deposits at an earlier stage and even before the sea is actually reached. The solubility of ferric hydroxide in sea water is very low and the ocean contains very little iron in the ionic state. Under the reducing conditions which obtain in the black bottom muds of deeper waters, iron passes into the ferrous state and is precipitated as ferrous sulphide, which is the principal cause of the dark colour. It is from such accumulations that the pyrite content of many shales is derived. Sedimentary ores like the German *kupferscheifer* and the Mount Isa deposits in Australia contain considerable amounts of iron sulphide minerals.

Many iron salts are of low solubility and quite readily form crystalline minerals. Hydrothermal veins in their upper oxidized zones contain many iron compounds. The end-product is of course ferric oxide in anhydrous or hydrated forms. The rusty red appearance of exposures of rocks in mining areas is often cited as evidence of mineralized zones and the terms ‘ red gossan ’ and ‘ iron hat ’ are well known as names for the oxidized outcrop of mineral veins.

In sedimentary rocks generally, iron is found in the form of ferric oxide to a very large extent. The lower oxidation state is not stable on the Earth's surface and hence the upper layers of the crust are typically reddish or brownish in colour whereas the deeper-seated rocks are darker and tend to greenish and bluish-grey tints, due in the main to ferrous rather than ferric compounds. The significance of iron in respect of the inorganic cycle of oxygen has been referred to in Chapter 8.

(3) Biogeochemistry of iron. Iron is necessary and important to both plants and animals. The elaboration of chlorophyll requires the presence of iron although the prosthetic group of the substance contains only magnesium. In the case of higher animals the blood pigment haemoglobin contains a porphyrin complex in which iron supplies the central atom. Iron may also be present in some enzymes. It is present also in cell cytochrome which is involved in cell oxidation. The iron bacteria constitute a number of species which, when developing in a medium containing iron salts, accumulate a sheath of ferric hydroxide. This sheath does not form part of their cell structure. They belong mostly to the genera *Leptothrix*, *Gallionella*, and *Spirophyllum*. In the case of the last it has been shown that iron is essential as $Fe(HCO_3)_2$.

IRON MINERALS

Minerals of Mainly Magmatic Origin

Iron	Fe
Pyrrhotine	FeS (non-stoichiometric)
Pyrite Marcasite	FeS_2
Wüstite	FeO
Magnetite	Fe_3O_4
Fayalite	Fe_2SiO_4
Riebeckite	$Na_2Fe_3{}^{2+}Fe_2{}^{3+}Si_8O_{22}(OH)_2$
Aegirine	$NaFe^{3+}Si_2O_6$
Olivine	$(Mg,Fe)_2SiO_4$
Hypersthene	$(Mg,Fe)SiO_3$
Hedenbergite	$CaFeSi_2O_6$
Actinolite	$Ca_2(Mg,Fe)_5Si_8O_{22}(OH)_2$
Biotite	$K_2(Mg,Fe,Al,Fe^{3+})_{4-6}(Si,Al)_8O_{20}(OH)_4$
Augite	$(Ca,Mg,Fe^{2+},Fe^{3+},Al)(Si,Al)O_3$
Hornblende	$(Ca,Na,K)_{2-3}(Mg,Fe,Fe^{3+},Al)_5(SiAl)_8O_{22}(OH)_2$

Minerals of Hydrothermal Deposits

Achavalite	$FeSe$
Löllingite	$FeAs_2$
Arsenopyrite	$FeAsS$
Hercynite	$FeAl_2O_4$
Siderite	$FeCO_3$
Vivianite	$Fe_3(PO_4)_2.8H_2O$
Phosphosiderite	$FePO_4.2H_2O$
Scorodite	$FeAsO_4.2H_2O$
Melanterite	$FeSO_4.7H_2O$
Butlerite	$Fe^{3+}SO_4(OH).2H_2O$

Minerals of Sedimentary Rocks

Hematite	Fe_2O_3
Siderite	$FeCO_3$
Goethite Lepidocrocite	$FeO(OH)$
Pyrite	FeS_2

Minerals of Metamorphic Rocks

Cordierite	$(Mg,Fe)_2Al_4Si_5O_{18}$
Epidote	$Ca_2(Al,Fe)_3Si_3O_{12}(OH)$
Garnet	$Fe_3Al_2(SiO_4)_3$

Minerals of Evaporites and Fumarolic Deposits

Lawrencite	$FeCl_2$
Douglasite	$K_2FeCl_4.2H_2O$
Rinneite	$K_3NaFeCl_6$
Ludwigite	$(Mg,Fe)_2Fe^{3+}BO_5$

The carbon dioxide requirements of the organism are supplied from this source and it thrives at only 6 °C. Energy is obtained from the reaction

$$4FeCO_3 + O_2 + 6H_2O = 4Fe(OH)_3 + 4CO_2$$
$$(\Delta H = -81\,\text{kcal}, \qquad \Delta F = -40\,\text{kcal})$$

Assuming an efficiency of 5 % the synthesis of 1 g of organic carbon by the organism would require therefore 448 g of ferrous ions with the resulting deposition of about 900 g of ferric hydroxide. It is thus easy to see why these bacteria can be important agents in the deposition of hydrated ferric oxides of the bog iron ore type.

Production of Iron Ores

Iron is a very abundant metal of which the extraction presents certain technical difficulties. Consequently, iron minerals suitable as ores must be abundant, accessible, rich in the metal, and as free as possible of certain objectionable impurities such as sulphur, phosphorus, and titanium. Therefore most economically important deposits are oxides and, less commonly, carbonates. Magnetite Fe_3O_4 is the most desirable ore. Hematite Fe_2O_3 is perhaps the most widely worked but the hydrated oxides, loosely called limonite, are very important. Good ores are found in all parts of the world. A few typical deposits are described below.

Kiruna District, North Sweden. The deposit consists of a body of very high grade magnetite enclosed in pre-Cambrian rocks and associated with a syenite porphyry. There is secondary apatite and some pyroxene. It is believed that the ore, probably a magmatic differentiate, was injected in a molten condition. The total reserves of the area are many millions of tons.

Lake Superior Region, U.S.A. These are probably the largest and richest haematite deposits in the world. There are several ‘ iron ranges ’ and they all lie in pre-Cambrian rocks. They are considered to be weathered portions of former sedimentary iron ore beds from which silica, etc., has been leached out by circulating waters.

Lorraine, Luxembourg, Europe. These are very large sedimentary iron deposits interbedded in Jurassic rocks. The ore is soft earthy oolitic limonite and hematite with some siderite and iron silicates. The grade is low but the deposits are extensive and relatively easily worked, partly by open-cast methods. The reserves are about 5000 million tons.

Urals, U.S.S.R. The most important deposits may be at Mount

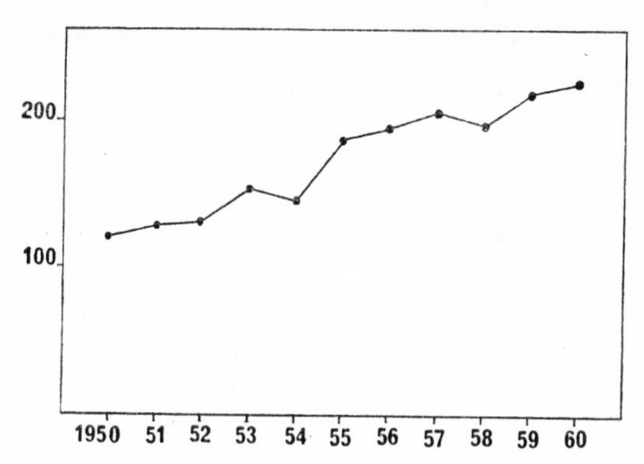

FIG. 22.1. World production of pig iron, in millions of tons.

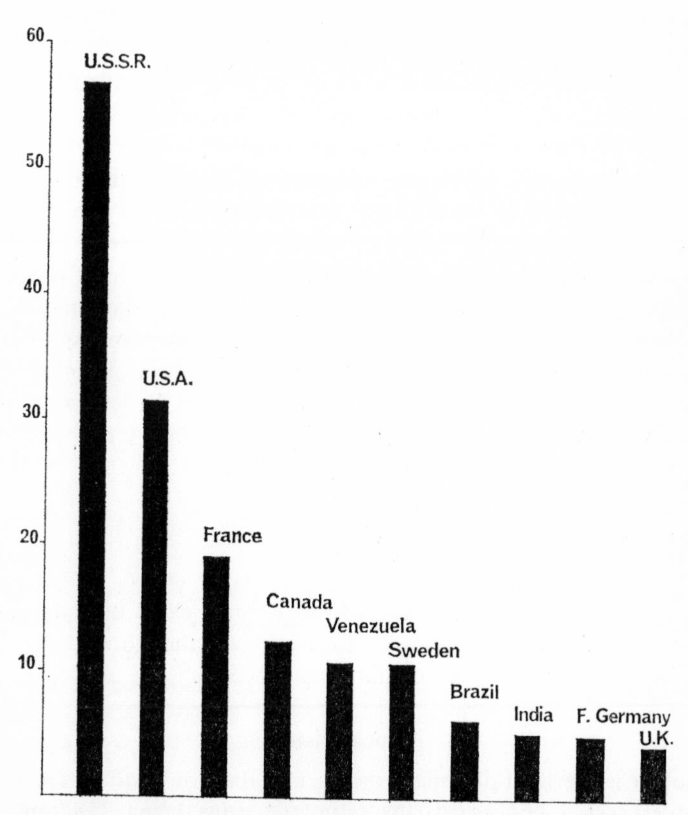

FIG. 22.2. Chief producers of iron ore (metal content), 1959, in millions of tons.

Magnitnaya where the ore is magnetite of contact-metasomatic origin. The reserves are said to be very large.

Great Britain. The most important production now is from low-grade Jurassic sedimentary ores of oolitic siderite type in several of the Midland counties—Lincoln, Rutland, Northampton, and Cleveland, Yorkshire. The ores can generally be worked open-cast and the reserves may be about 3000 million tons.

The high-grade hematite ores of West Cumberland and North West Lancashire are declining in output. They are mainly replacement deposits, probably of magmatic origin, in Carboniferous limestone. The ore is often crystalline or reniform and of high quality. Interesting accessory minerals including barytes, calcite, and fluorite have made the district famous for mineralogists.

Ultimate Forms

The total output of metallic iron and its alloys since 1900 may well be of the order of 6×10^9 tons. Much of this is probably no longer in the metallic form. Corrosion is comparatively rapid in the moist temperate countries which are the heaviest users of the metal and the end-product is in general a hydrated oxide, mainly lepidocrocite FeO(OH). Most of the iron compounds used will also tend to form a similar product. The oxide is not very mobile and there must be countless local accumulations all over the civilized world where engineering constructions have for various reasons been abandoned. As the rate of corrosion is dependent to some extent on surface area, it follows that many of the modern thin-metal constructions will be less permanent than the massive cast-iron products of the last century and the high-quality hand-forged objects of medieval times.

Cobalt

Only one stable nuclide, ^{59}Co, of this element exists in Nature. As in the case of iron, cobalt is one of the elements thought to have been synthesized mainly by the ‘ equilibrium process ’.

Geochemistry

Cobalt is the least abundant member of the iron triad as far as the Earth’s crust is concerned, the estimated value being 23g/ton. On account of the siderophile character of the element it should be enriched in the core of the Earth and for the planet as a whole the estimate is 0·2 % or 2000 g/ton.

The number of independent cobalt minerals is not very large. It has little tendency to enter silicate structures but on account of similarity of atomic and ionic radii to iron it is fairly commonly present as a 'concealed' element in sulphide minerals. Most of the workable deposits of cobalt are in fact subsidiary to copper and the metal is mainly obtained from the large copper-producing regions. Most of the natural compounds exhibit the lower oxidation state Co^{2+} but a few oxide minerals seem to contain the Co^{3+} ion. The much higher potential (1·84 V) needed to effect this oxidation than in the case of iron, explains the paucity of natural cobaltic compounds.

COBALT MINERALS

Name	*Formula*
Linnaeite	Co_3O_4
Cattierite	CoS_2
Smaltite	$CoAs_2$
Cobaltite	CoAsS
Heterogenite	CoO.OH
Sphaerocobaltite	$CoCO_3$
Lusakite	$(Fe,Co,Mg)_4Al_{18}Si_8O_{46}(OH)_2$
Erythrite	$Co_3(AsO_4)_2.8H_2O$
Roselite	$Ca_2(Co,Mg)(AsO_4)_2.2H_2O$
Bieberite	$CoSO_4.7H_2O$

Most of these minerals are found in hydrothermal deposits.

Biogeochemistry

Cobalt is mainly of interest as an essential trace element. It probably enters into the structure of certain enzymes. It is for instance known that glycylglycinase is activated by cobalt.

Vitamin B_{12} is a co-ordination complex containing cobalt with a principal valency of 3. As in other analogous compounds the cobalt is in the centre of a tetrapyrrole structure. Ruminants, for physiological reasons, require relatively more cobalt to elaborate the vitamin than do other kinds of animals and they are markedly sensitive to cobalt deficiency in soils over which they graze. Large areas in New Zealand and Western Australia are noted for natural cobalt deficiency. In such regions the soil may contain 1–4 p.p.m. whereas healthy soils normally carry 4–40 p.p.m. of cobalt. An organic compound of cobalt, $Na_2Co(CNS)_4.8H_2O$, is reported to occur in Katanga, Congo.

Production

Congo. Oxidized cobalt–copper ores are obtained in the Katanga mining field.

Northern Rhodesia. Concentrated ores contain Cu 3 %, Co 0·14 %,

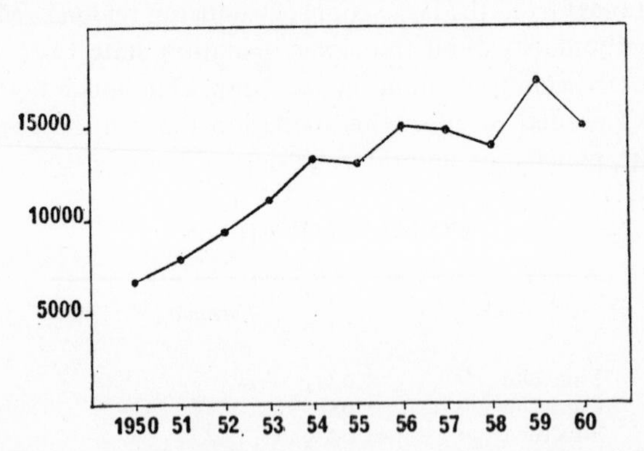

FIG. 22.3. World production of cobalt ore (metal content), in tons.

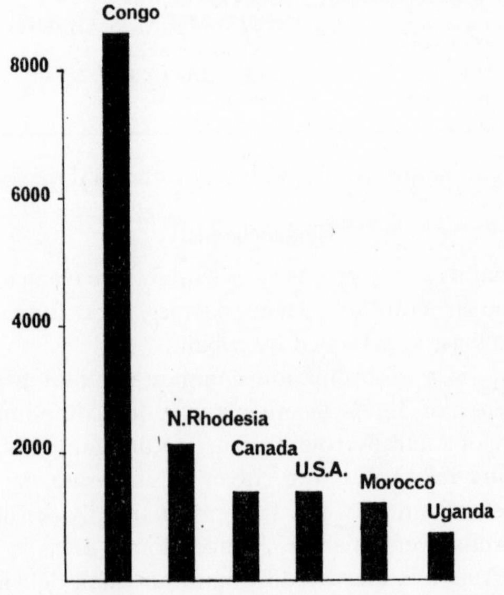

FIG. 22.4. Chief producers of cobalt ore (metal content), 1959, in tons.

Fe 2·5 % with sulphur, silica, etc. In the above cases the cobalt is largely obtained from the slags resulting from a suitable process of smelting the ore for copper.

Canada. Cobalt–silver arsenide ores are worked in North Eastern Ontario and there is also considerable by-product cobalt from the nickel–copper ores of Sudbury, Ontario.

Nickel

There are five stable isotopes of this element in Nature:

^{58}Ni = 67·76 % ^{62}Ni = 3·66 %
^{60}Ni = 26·16 % ^{64}Ni = 1·16 %
^{61}Ni = 1·25 %

There is no indication that the isotopic composition of meteoric nickel differs from the terrestrial.

As previously noted for iron and cobalt, nickel is one of the elements which should be formed by the 'equilibrium process' in stars. The predicted abundance is, however, considerably lower than that observed. It is probable that nickel isotopes are produced to a considerable extent also by neutron capture on a slow time-scale.

The occurrence of nickel in iron meteorites has already been noted in this chapter.

Geochemistry

With a crustal abundance of 80 g/ton, nickel is considerably more abundant than cobalt. It is a siderophile element and the estimated abundance for the Earth as a whole is 2·7 % (27000 g/ton).

Nickel occurs very markedly in sulphides and is associated with the earlier products of magmatic crystallization, *viz.*, the ultrabasic rocks. The nickel content of igneous rocks decreases progressively with increasing silica content, as shown below:

	Dunites	*Gabbros*	*Diorites*	*Granites*
Ni (p.p.m.)	1500	150	30	9
SiO_2 (%)	40	48	57	70

Probably most of the nickel is contained in sulphide accessory minerals such as pyrrhotine. Some nickel minerals are found in hydrothermal deposits and it shows marked resemblance to copper in this respect. Valuable concentrations of nickel are, however, very local.

Nickel minerals are fairly numerous and there are some hydrous silicate minerals analogous to serpentine, chlorite, and montmorillonite. In weathering processes the Ni^{2+} ion is stable and not prone to oxidation like Fe^{2+} and Co^{2+}. Nickel is largely retained in minerals of the above types and lateritic weathering of serpentine rocks may cause considerable enrichment in nickel.

It is probable that very little nickel reaches the ocean.

There is very little evidence that the element has any biological significance.

NICKEL MINERALS

Name	*Formula*
Millerite	NiS
Heazlewoodite	Ni_3S_2
Polydymite	Ni_3S_4
Vaesite	NiS_2
Niccolite	$NiAs$
Chloanthite	$NiAs_2$
Gersdorffite	$NiAsS$
Pentlandite	$(Fe,Ni)_9S_8$
Bravoite	$(Ni,Fe)S_2$
Bunsenite	NiO
Zaratite	$Ni_3CO_3(OH)_4.4H_2O$
Garnierite	$(Ni,Mg)_3Si_2O_5(OH)_4$
Annabergite	$Ni_3(AsO_4)_2.8H_2O$
Morenosite	$NiSO_4.7H_2O$

Production

The famous deposits of Sudbury, Ontario, Canada, produce more than half of the world's nickel and formerly produced as much as 90% of it. They are disseminated ore bodies in quartz diorite, massive sulphide bodies along faults or breccia zones, and sulphide stringer zones. The deposits are associated with a great elliptical-shaped basic intrusion of norite–micropegmatite which is zoned in a manner indicating differentiation. The main ore is a complex of sulphides containing particularly pyrrhotine, chalcopyrite, pentlandite, and minor contents of bismuth, tellurium, gold, silver, and the platinum metals. It is not now considered to be a magmatic segregation but rather a replacement deposit caused by hydrothermal solutions. The ore averages 1·5% nickel and the reserves are estimated at over 400 million tons of ore. In the Petsamo region, U.S.S.R., a nickel–copper deposit generally similar to the Sudbury type has a fairly large output and is said to have

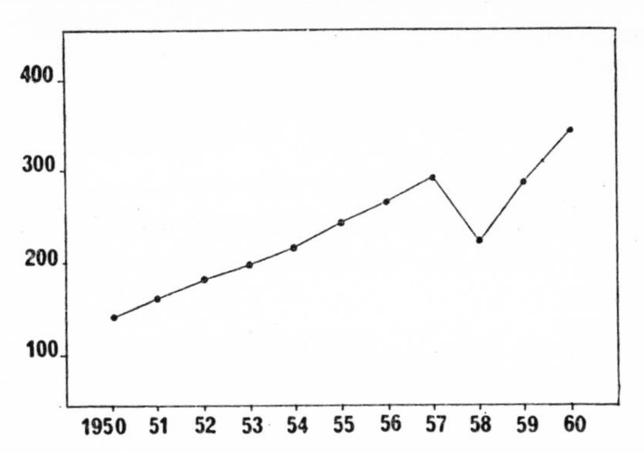

FIG. 22.5. World production of nickel ore (metal content), in thousands of tons.

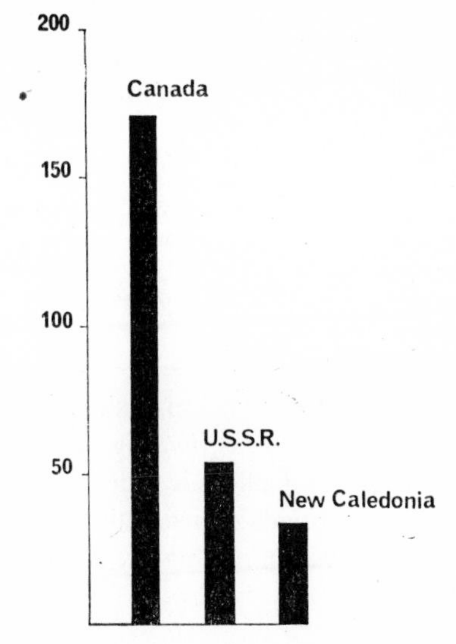

FIG. 22.6. Chief producers of nickel, 1959, in thousands of tons.

reserves of 4 million tons of ore carrying 1·6% nickel. In New Caledonia there is a growing production from deposits which consist of residual enrichments of nickel silicates.

Great Britain. There have been very small outputs of nickel ores from time to time in the past, mainly from localities in the South of Scotland. They were never of any great significance.

CHAPTER 23

The Platinum Metals

General

These six metals of Group VIII belong to two different periods but they are conveniently treated together as their natural chemistry is similar and not very extensive. The number of nuclides is quite large and only rhodium is without isotopes. They are tabulated below.

Ruthenium	^{96}Ru = 5·68%, ^{98}Ru = 2·22%, ^{99}Ru = 12·81%, ^{100}Ru = 12·70%, ^{101}Ru = 16·98%, ^{102}Ru = 31·34%, ^{104}Ru = 18·27%
Rhodium	^{103}Rh = 100%
Palladium	^{102}Pd = 0·8%, ^{104}Pd = 9·3%, ^{105}Pd = 22·6%, ^{106}Pd = 27·2%, ^{108}Pd = 26·8%, ^{110}Pd = 13·5%
Osmium	^{184}Os = 0·018%, ^{186}Os = 1·59%, ^{187}Os = 1·64%, ^{188}Os = 13·3%, ^{189}Os = 16·1%, ^{190}Os = 26·4%, ^{192}Os = 41·0%
Iridium	^{191}Ir = 38·5%, ^{193}Ir = 61·5%
Platinum	^{190}Pt = 0·012%, ^{192}Pt = 0·78%, ^{194}Pt = 32·8%, ^{195}Pt = 33·7%, ^{196}Pt = 25·4%, ^{198}Pt = 7·23%

It will be observed that as many as four pairs of isobars exist among these elements. In spite of numerous investigations there is no reliable evidence that any of the above nuclides is radioactive. They are all rare elements but their cosmic abundances greatly exceed their terrestrial crustal abundances by factors approximately as indicated below:

	Ru	Rh	Pd	Os	Ir	Pt
Cosmic/Terrestrial	690	150	16	1500	850	160

It seems probable that the synthesis of the various atoms of these elements could take place both by the s-process and the r-process of neutron capture. The notable discrepancies between terrestrial and cosmic abundances are no doubt due to the peculiarities of the mode of occurrence in the Earth. It is unlikely that the values quoted have a very high degree of accuracy.

General Geochemical Character

The geochemistry of the platinum metals is easily summed up. They are both relatively and absolutely rare in the accessible parts of the

Earth. Their distribution in meteorites indicates that they have a very marked siderophile character and the compounds they can form in Nature are very few—being restricted, apart from alloys, to a few binary compounds with sulphur and the metalloids. They do not form simple cations and the conditions under which they can exist in complex ions scarcely exist on the Earth. They are associated with deep-seated magmatic (peridotitic) rocks and their weathered counterparts (serpentines).

They do not enter into silicate minerals and probably occur only in the elemental state or as the simple compounds mentioned above.

MINERALS OF THE PLATINUM METALS

Name	*Formula*
Potarite	PdHg
Stannopaladinite	Pd_3Sn_2
Iridosmine, Osmiridium	Ir,Os
Niggliite	Pt,Sn
Platinum	Pt
Norilskite	(Pt,Fe,Ni)
Platiniridium	(Ir,Pt)
Laurite	RuS_2
Arsenopalladinite	Pd_3As
Stibiopalladinite	Pd_3Sb
Cooperite	PtS
Sperrylite	$PtAs_2$
Braggite	(Pt,Pd,Ni)S

As a result of the lanthanide contraction the atomic radii of these metals lie between 1·32 Å and 1·38 Å, and they are so closely similar in their chemical character that in practically all of the above minerals more than one and frequently the whole of the six metals are actually present. The chemical resistance, high melting points, and high densities are properties which largely explain why the metals of this family are mainly associated with the early magmatic rocks and why in the weathering process they form resistates well adapted for accumulation in placer deposits. Although scarce, the actual localities for platinum metals are rather widespread.

The slight thiophile character, particularly of ruthenium, palladium, and platinum, leads to their occurrence in sulphide ore bodies in certain

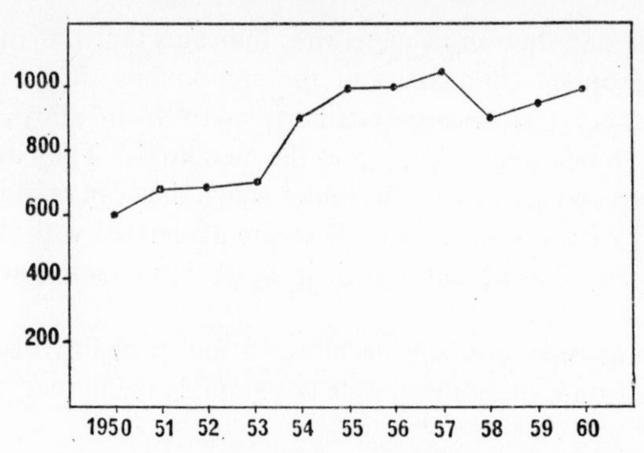

FIG. 23.1. World production of platinum metals, in thousands of troy ounces. (10^6 troy oz = 31 tons.)

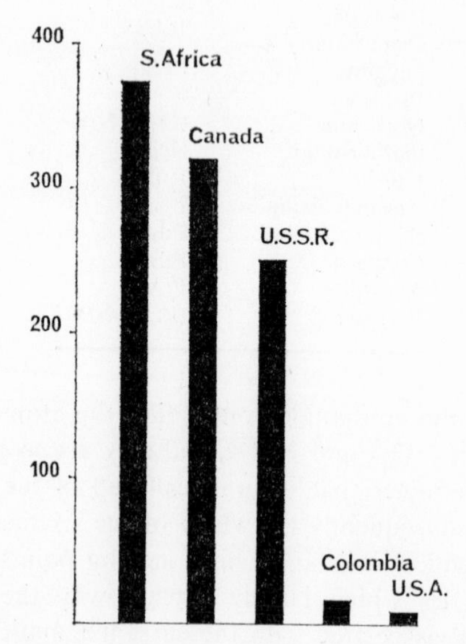

FIG. 23.2. Chief producers of platinum metals, 1959, in thousands of troy ounces.

cases such as in the Sudbury complex (Chapter 22). The mineral molybdenite MoS_2 can also carry amounts up to 1g/ton of platinum metals.

There is no present indication that any of these elements has any biological significance.

Production of the Platinum Metals

South Africa. Associated with the Bushveld igneous complex (see Chapter 19) are two main groups of platiniferous concentrations:

(*a*) Remarkable ultrabasic ‘ pipes ’ of dunite or hornolite rock up to 60ft diameter and 1000ft deep. The central part of the pipes is richest and averages 3–60g/ton.
(*b*) Chromitite segregations in the lower part of the differentiated ‘ Merensky Reef ’. The platinum band is about 12in. thick and averages 15–18g/ton. These are the only large-scale ‘ straight ’ platinum ores worked at present.

Canada. Platinum is mainly obtained from the Frood mine, Sudbury, Ontario (see Chapter 22) where in the massive copper–nickel sulphide body of the lower levels a platinum content of about 1·5g/ton is obtained together with about 1g/ton of the other platinum metals.

U.S.S.R. The metal is won from placers on both sides of the Northerly Ural Mountains where small masses of dunite contain magmatic concentrations of platinum metals. The weathering of these bodies has produced the placers in both old and new river beds. The chief areas are the basin of the Iss and the Tura rivers and the Nizhni Tagil district.

South America. There is a waning production from placer deposits derived from weathering of ultrabasic material in Colombia.

CHAPTER 24

The Rare Earth Elements Scandium, Yttrium, and the Lanthanides

The fifteen elements to be considered here belong to three different periods, but all to the same group of the Periodic Table. They are so closely related in their geochemistry, however, that they may be advantageously considered together. The essential nuclear data for these elements follow.

Name	*Symbol*	*Mass Numbers of Isotopes*
Scandium	Sc	45 = 100%
Yttrium	Y	50 = 100%
Lanthanum	La	138 = 0·089%, 139 = 99·911%
Cerium	Ce	136 = 0·193%, 138 = 0·25%, 140 = 88·48%, 142 = 11·07%
Praseodynium	Pr	141 = 100%
Neodymium	Nd	142 = 27·13%, 143 = 12·2%, 144 = 23·87%, 145 = 8·3%, 146 = 17·18%, 148 = 5·72%, 150 = 5·60%
Prometheum	Pm	145 ?
Samarium	Sm	144 = 3·16%, 147 = 15·07%, 148 = 11·27%, 149 = 13·84%, 150 = 7·47%, 152 = 26·63%, 154 = 22·53%
Europium	Eu	151 = 47·77%, 153 = 52·23%
Gadolinium	Gd	152 = 0·20%, 154 = 2·15%, 155 = 14·73%, 156 = 20·47%, 157 = 15·68%, 158 = 24·87%, 160 = 21·9%
Terbium	Tb	159 = 100%
Dysprosium	Dy	156 = 0·0524%, 158 = 0·0092%, 160 = 2·294%, 161 = 18·88%, 162 = 25·53%, 163 = 24·97%, 164 = 28·18%
Holmium	Ho	165 = 100%
Erbium	Er	162 = 0·136%, 164 = 1·56%, 166 = 33·41%, 167 = 22·94%, 168 = 27·07%, 170 = 14·88%
Thulium	Tm	169 = 100%
Ytterbium	Yb	168 = 0·14%, 170 = 3·03%, 171 = 14·31%, 172 = 21·82%, 173 = 16·13%, 174 = 31·84%, 176 = 12·73%
Lutecium	Lu	175 = 97·4%, 176 = 2·6%

One of the most striking characters of the lanthanides proper is in their relative abundances and Fig. 24.1 illustrates this. Two points will be at once apparent:

(1) There is a close parallelism between the cosmic and terrestrial abundance values, showing that little or no differentiation of these elements takes place in geochemical processes.
(2) The elements of even atomic number are always relatively more abundant.

There also follows another point and that is that the elements of even

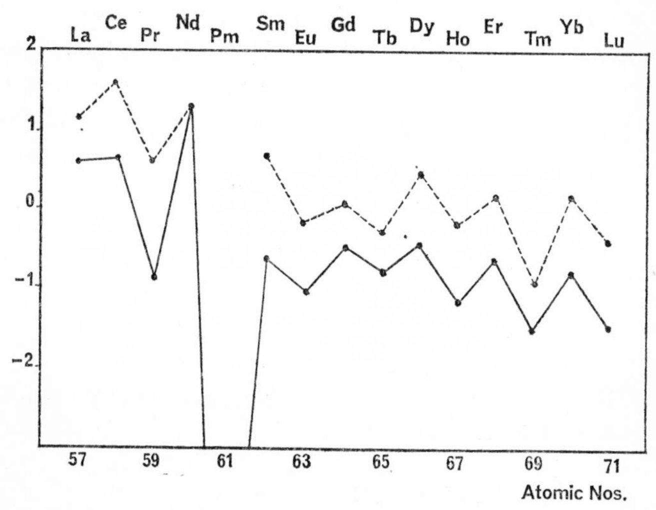

FIG. 24.1. Relative abundance of the lanthanides (logarithmic values). Dotted line, cosmic. Continuous line, terrestrial.

atomic number are also those with the largest numbers of isotopes. Thus:

Element	La	Ce	Pr	Nd	Pm	Sm	Eu	Gd	Tb	Dy	Ho	Er	Tm	Yb	Lu
Z	57	58	59	60	61	62	63	64	65	66	67	68	69	70	71
No. of isotopes	2	4	1	7	?	7	2	7	1	7	1	6	1	7	2

It is evident that the even-proton nuclides have a greater range of stable isotopes and can be synthesized by a greater variety of nuclear processes.

The large number of nuclides of this group (fifty-five) is reflected in the number of processes by which they can be built up. It appears that these will be both the s-process and the r-process of neutron capture. Some of them are also probably fission products of heavy nuclei.

Samarium and gadolinium are notable for high-capture cross-sections for slow neutrons.

Radioactivity among the Lanthanides

^{138}La is feebly radioactive with a branched decay to:

(*a*) ^{138}Ba by *K*-electron capture (half-life 7×10^{10} years).
(*b*) ^{138}Ce by β-emission (half-life $1{\cdot}2 \times 10^{12}$ years).

^{140}La has been detected in snow following atomic test explosions—it is one of the fission products.

If any promethium isotopes exist in Nature their content must be very small and probably beyond the limits of detection. ^{145}Pm as obtained artificially has the longest half-life, 30 years, and decays to ^{145}Nd by orbital electron capture.

^{147}Sm is naturally radioactive. It is an α-emitting nuclide which decays with a half-life of $6{\cdot}7 \times 10^{11}$ years, the decay product being ^{143}Nd. Samarium appears to be the only element having an α-emitting isotope and also a number of stable isotopes, and it is the lightest element that decays by α-particle emission. The rate at which radiogenic heat is produced in the Earth's crust by the radioactivity of samarium is relatively small and of little significance.

Lutecium also has a radioactive isotope, ^{176}Lu, which has branched decay, 33% by β-emission to ^{176}Hf (half-life $2{\cdot}4 \times 10^{10}$ years) and 67% by *K*-electron capture to ^{176}Yb.

Geochemistry of the Rare Earth Elements

It will now be convenient to include scandium and yttrium in our studies of the group and employ the somewhat wider terminology ‘ rare earths ’, rather than ‘ lanthanides ’. The crustal abundance of these elements is estimated as follows:

Element	Sc	Y	Ce	Pr	Nd	Sm	Eu	Gd	Tb	Dy	Ho	Er	Tm	Yb	Lu
(g/ton)	5	40	46	6	24	7	1	6	0·9	5	1	3	0·2	3	0·8

The whole family is characteristically lithophile and they are probably mainly concentrated in granitic magmas and their related products. It is here that the interesting effect of the decreasing size of the ions with increasing atomic number is encountered.

Geochemically, the significance of the lanthanide contraction, which is a consequence of the special electronic configurations of these elements, is mainly that the ionic sizes (for the common trivalent cations) average 0·96 Å and this also includes yttrium and scandium. The largest of the ions, La^{3+}, is about 18 % larger than the average and this is within the tolerance for diadochic substitution under high-temperature (magmatic) conditions. It therefore follows that the rare earths show only slight differences in their geochemistry and in most of the minerals at least small amounts of every one of the elements is likely to be present.

There is, however, some tendency for the yttrium–erbium group to predominate in certain minerals such as xenotime YPO_4 and for the cerium–lanthanum group to predominate in others such as monazite $CePO_4$. Analyses of these two minerals as quoted by Dana are given below.

	Xenotime	*Monazite*
Locality	N. Carolina, U.S.A.	Bihar, India
$(Y,Er)_2O_3$	56·81%	1·15%
$(La,Nd)_2O_3$	0·93%	32·72%
Ce_2O_3	—	22·0%

The trivalent lanthanide ions are of such size that they are unacceptable into crystal lattices in place of Al^{3+} (radius 0·51 Å), nor are they likely to deputize for Mg^{2+} (0·66 Å) or ferrous iron (0·74 Å) to any appreciable extent. Hence the lanthanides do not separate to any extent in the earlier stages of magmatic crystallization. On the whole they tend to form independent minerals in the later stages—*e.g.*, as minor accessory minerals in granitic rocks and more particularly in the pegmatites of granites and syenites.

There is also some degree of concealment of lanthanide ions in alkaline earth minerals and in some lead minerals. It is possible that samarium, europium, and ytterbium may be present in such minerals as their divalent ions. An interesting case is provided by strontianite from Strontian, Argyllshire, Scotland (the original locality) which is bluish-green in colour, and contains europium and samarium. The lead mineral pyromorphite has also been recorded not infrequently with considerable rare earth content (*circa* 1000 p.p.m.). It appears that europium also deputizes for calcium in sphene $Ca(TiO)SiO_4$. The silicate allanite which is also an accessory rock mineral may contain trivalent rare earths replacing calcium.

The possibility that cerium might exist in the tetravalent state in Nature has been considered and there appears to be some evidence for this. Goldschmidt thought it might occur in this form in minerals of the thorite group ($ThSiO_4$) along perhaps with uranium.

On account of the general tendency for the rare earths to separate in the later stages of crystallization, particularly in pegmatites, it is not surprising that they are quite often associated with other rarer elements like thorium, uranium, niobium, and tantalum.

The number of rare earth minerals is quite large and includes complex oxides, fluorides, carbonates, silicates, tantalates, phosphates, and arsenates. There are no sulphides or sulphates, however. They are in general minerals of pegmatites or minor accessory minerals in granitic rocks and occasionally detrital minerals of sands.

RARE EARTH MINERALS

Name	*Formula*
Yttrocrasite	$Yt_2Ti_4O_{11}$?
Knopite	$(Ca,Ti,Ce)_2O_3$
Davidite	$(Fe,U,Ce)(TiFe)_3O_7$
Tysonite	$(Ce,La, etc.)F_3$
Cerfluorite	$(Ca,Ce)F_{2-3}$
Yttrofluorite	$(Ca,Yt)F_{2-3}$
Lanthanite	$(La,Ce)_2(CO_3)_3.9H_2O$
Tengerite	$BeYtCO_3(OH)_3.H_2O$
Ancylite	$Sr_3Ce_4(CO_3)_7(OH)_4.3H_2O$
Bastnaesite	$(La,Ce)FCO_3$
Parisite	$Ca(Ce,La)_2(CO_3)_3F_2$
Röntgenite	$Ca_2(Ce,La)_3(CO_3)_5F_3$
Törnebohmite	$Ce_3Si_2O_8OH$
Thortveitite	$(Sc,Yt)_2Si_2O_7$
Thalenite	$Yt_2Si_2O_7$
Cerite	$Ce_4Si_3O_{12}.3H_2O$?
Gadolinite	$Be_2FeYt_2Si_2O_{10}$
Kainosite	$Ca_2(Ce,Yt)_2Si_4O_{12}CO_3.H_2O$
Fergusonite	$(Yt,Er)(Nb,Ta)O_4$
Pyrochlor	$(Ca,Na,Ce)(Nb,Ti,Ta)_2(O,OH,F)_7$
Euxenite	$(Yt,Er,Ce,La,U)(Nb,Ti,Ta)_2(O,OH)_6$
Polycrase	$(Yt,Er,Ce,La,U)(Ti,Nb,Ta)_2(O,OH)_6$
Samarskite	$(Yt,U,Fe,Th)(Nb,Ta)_2O_6$
Xenotime	$YtPO_4$
Monazite	$(La,Ce)PO_4$
Retzian	$Ca_5CeMn_6(AsO_4)_3(OH)_{16}$

Secondary Processes

In the processes of weathering and sedimentation, the slight differentiation which occurs in the primary crystallization is to a large extent reversed and all of the lanthanides with scandium and yttrium tend to come together again. This is indicated by Fig. 24.2.

Small amounts, a few p.p.m. in most cases, have been found in

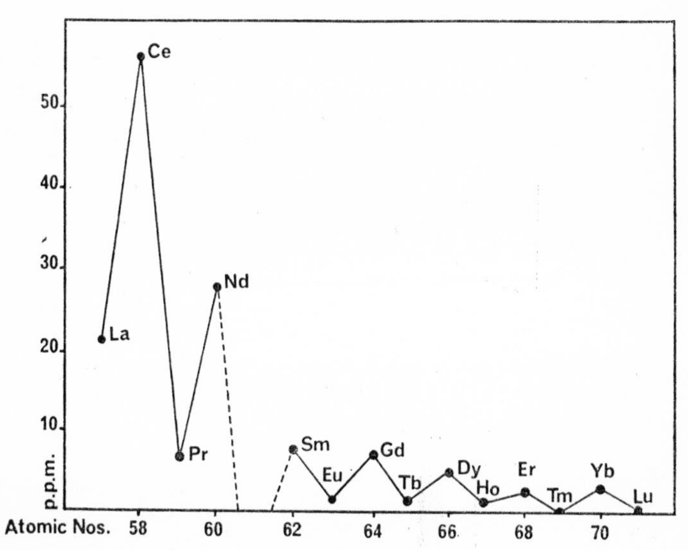

FIG. 24.2. Average content of the lanthanide oxides M_2O_3 in clay sediments (after Goldschmidt).

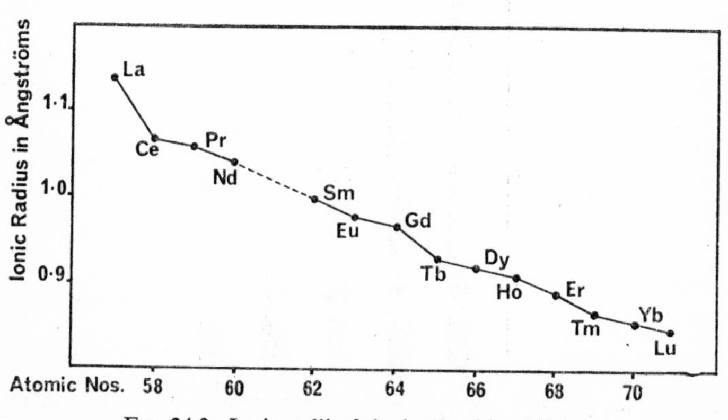

FIG. 24.3. Ionic radii of the lanthanides, M^{3+}.

various kinds of biological material but there is no evidence that any of the rare earth elements has biological significance.

Production of the Rare Earths

The various rare earth minerals are, as noted above, either accessory rock minerals or else major constituents of pegmatites.

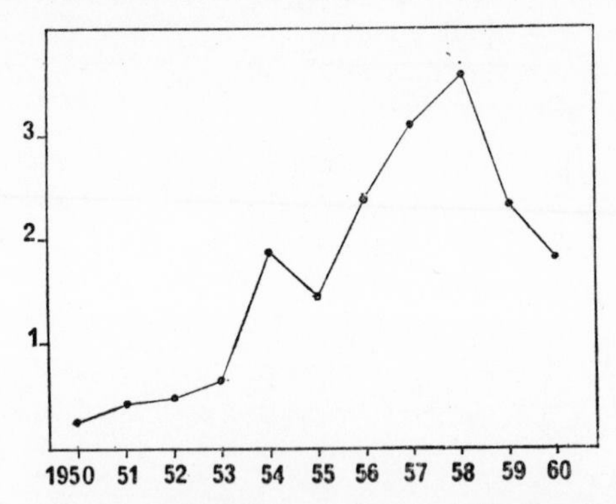

FIG. 24.4. World production of monazite and rare earth minerals, in thousands of tons.

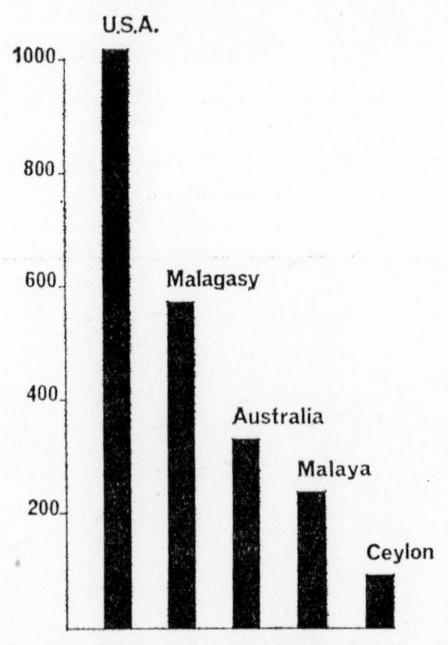

FIG. 24.5. Chief producers of thorium and rare earth minerals in 1959, in tons (information incomplete).

Some of the most famous localities, including the one at Ytterby, are in Southern Scandinavia, generally pegmatites associated with granites or syenites. Considerable production now takes place from residual concentrations mainly of monazite in beach sands in Ceylon and India.

The mineral bastnaesite occurs in quantity at Nipton, San Bernadino County, California, U.S.A. in a large vein where it is associated with calcite, barytes, and other minerals. The ore can be concentrated up to 72 % rare earth oxides and the reserves are claimed to be 25 million tons of ore. Many of the deposits and others are also of interest as sources of thorium and will be referred to in Chapter 25.

CHAPTER 25

The Actinides

There are no stable nuclides among these elements but several are sufficiently long-lived to occur in perceptible quantities in Nature. Several of the short-lived decay products are found in Nature only in association with their parent elements. For this reason, radium, which does not belong here, but in Group II, will also be considered.

HALF-LIVES OF THE MORE STABLE ACTINIDES AND RELATED ELEMENTS

Name	*Nuclide*	*Half-life*
Polonium	^{210}Po	138 days
Radon	^{222}Rn	3·8 days
Radium	^{226}Ra	1 622 years
Actinium	^{227}Ac	27·7 years
Thorium	^{229}Th	7 340 years
	^{230}Th	8×10^4 years
	^{232}Th	$13{\cdot}89 \times 10^9$ years
Protoactinium	^{231}Pa	34 300 years
Uranium	^{234}U	$2{\cdot}52 \times 10^5$ years
	^{235}U	$7{\cdot}07 \times 10^8$ years
	^{238}U	$4{\cdot}5 \times 10^8$ years

Of these nuclides only ^{232}Th, ^{235}U, and ^{238}U can be expected in primitive terrestrial material. The others are found to be present merely because they are being continuously formed in the radioactive decay processes of their parent elements.

Uranium and thorium are elements which by reason of their instability cannot, it is thought, have been formed by neutron capture on a slow time-scale, but only by the r-process which involves neutron capture at such a great rate that a nucleus can capture another neutron before it undergoes decay.

Neutron-rich nuclides are thus built up which decay by α-emission. If the neutron flux is of sufficient density elements up to ^{254}Cf or even higher will be built up. These will decay by spontaneous fission with a half-life of 55 days. The exponential decline of radiation from a supernova in about this period is cited by Hoyle *et al.* as evidence that in a supernova the neutron-flux is such that the region of lead and bismuth is by-passed and addition of neutrons occurs up to a mass number of about 260 when the nuclides become unstable against fission. Some of these nuclides will decay to thorium and uranium. The r-process is more favourable to the formation of ^{235}U than to ^{238}U by a factor of about 1·6 because of the greater number of progenitors of ^{235}U. The half-lives of the two isotopes being known, it is possible to calculate the time required at these rates of decay to change the relative abundance to the terrestrial value of 1 : 140. This time, according to Hoyle, is $6{\cdot}6 \times 10^9$ years—which is the ' age ' of terrestrial uranium, and presumably that of the supernova event that produced it. If allowance is made for the possible contribution by the many supernovae distributed uniformly in time, then the age is about 10^{10} years. Further consideration on these lines of the three long-lived radioactive elements, ^{235}U, ^{238}U, and ^{232}Th, has led Fowler and Hoyle to conclude that the age of our galaxy is between 1·2 and $2{\cdot}0 \times 10^{10}$ years.

Thorium

Thorium is over three times as abundant in the Earth's crust as the so-called ' common ' element tin, the crustal abundance being estimated at about 10g/ton. Thorium generally exists in Nature as its tetravalent ion and it does not form many independent minerals. The radius of the Th^{4+} ion is 1·02Å so that it is not far in size from U^{4+} (= 0·97Å), from many of the lanthanides and also from Ca^{2+}. The separation of thorium takes place in the later stages of magmatic crystallization and it behaves in fact much like the rare earths and uranium, with which elements it is frequently associated.

THORIUM MINERALS

Name	*Formula*
Thorianite	ThO_2
Thorite	$ThSiO_4$
Blomstrandine	$(Yt,Th)(Nb,Ti)_2(O,OH)_6$
Samarskite	$(Yt,U,Fe,Th)(Nb,Ta)_2O_6$
Cheralite	$(Ca,Th,Ce_2)(PO_4)_2$

The independent thorium minerals are characteristically present in pegmatites. Much thorium in the Earth's crust must, however, be carried in such important accessory rock minerals as zircon, monazite, allanite, apatite, and sphene. In the latter minerals the tetravalent thorium is substituting for divalent calcium.

The thorium content of igneous rocks ranges from as much as 25 p.p.m. in granitic rocks down to 5 p.p.m. or less in gabbros and dunites.

Uranium

The crustal abundance of uranium, estimated at 2 g/ton, is higher than that of several of the 'rarer' elements like tungsten and molybdenum.

Uranium differs from thorium in its tendency to exist in the upper layers of the Earth in the hexavalent state, particularly as the very stable UO_2^{2+} ion. In the minerals from deep-seated sources where the oxidation potential is lower it tends to be mainly as the U^{4+} ion. The uranyl compounds in general have a higher solubility than the tetravalent series and are in consequence more mobile in Nature. There are apparently many more minerals of uranium than of thorium, but some consideration should be given to the assiduity of prospectors for the element, for the demand for thorium has never been as high as that for uranium.

The number of minerals which contain at least 1 % of uranium is about 100 (see the list below).

The redox potential in respect of $U^{4+} \rightarrow U^{6+}$ is of the same order as that for the $Fe^{2+} \rightarrow Fe^{3+}$ conversion and therefore oxidation states of natural uranium compounds will vary in somewhat the same way as those of iron.

General Geochemistry of Thorium and Uranium

In the discussion which follows it will be convenient to treat of both of the elements together. The analogy between lanthanides and actinides in Nature is apparent by the fractionation of the elements which occurs to some extent due to varying oxidation states. Thus in the lanthanides cerium may form the Ce^{4+} ion and in this state be differentiated to some extent from most of the other elements of the group. Similarly uranium tends to differentiate from the other actinides when in its hexavalent state (both as U^{6+} and UO_2^{2+}).

URANIUM MINERALS

Name	*Formula*
Uraninite	UO_2
Becquerelite	$7UO_3.11H_2O$
Billietite	$BaU_6O_{13}.11H_2O$
Davidite	$(Fe,U,Ce)(Ti,Fe)_3O_7$
Rutherfordine	UO_2CO_3
Sharpite	$(UO_2)_6(CO_3)_5(OH)_2.7H_2O$
Bayleyite	$Mg_2UO_2(CO_3)_3.18H_2O$
Liebigite	$Ca_2U(CO_3)_4.10H_2O$
Andersonite	$Na_2CaUO_2(CO_3)_3.6H_2O$
Rabbittite	$Ca_3Mg_3(UO_2)_2(CO_3)_6(OH)_4.18H_2O$
Soddyite	$(UO_2)_5Si_2O_9.6H_2O$
Coffinite	$USiO_4$?
Sklodowskite	$Mg(UO_2)_2Si_2O_7.6H_2O$
Uranotile	$Ca(UO_2)_2Si_2O_7.6H_2O$
Euxenite	$(Yt,Er,Ce,La,U)(Nb,Ti,Ta)_2(O,OH)_6$
Samarskite	$(Yt,U,Fe,Th)(Nb,Ta)_2O_6$
Tobernite	$Cu(UO_2)_2(PO_4)_2.12H_2O$
Salléeite	$Mg(UO_2)_2(PO_4)_2.12H_2O$
Autunite	$Ca(UO_2)_2(PO_4)_2.12H_2O$
Parsonite	$Pb_2UO_2(PO_4)_2.2H_2O$
Renardite	$Pb_2(UO_2)_4(PO_4)_2(OH)_4.7H_2O$?
Dumontite	$Pb_2(UO_2)_3(PO_4)_2(OH)_4.3H_2O$
Nováčekite	$Mg(UO_2)_2(AsO_4)_2.12H_2O$
Uranospinite	$Cu(UO_2)_2(AsO_4)_2.12H_2O$
Zeunerite	$Cu(UO_2)_2(AsO_4)_2.16H_2O$
Zippeïte	$(UO_2)_3(SO_4)_2(OH)_2.8H_2O$
Carnotite	$K_2(UO_2)_2(VO_4)_2.3H_2O$

Some conclusions have been drawn from the facts of distribution of thorium and uranium in igneous rocks and minerals.

The average Th/U ratio is approximately 3·5:1. In respect of individual minerals it seems, however, that the earlier-formed species like apatite and zircon may have a ratio of about unity whereas in the later-formed minerals like monazite the ratio may be 100:1 yet in xenotime it may be only 0·01:1. It seems, on the whole, that the minerals of the later stages (mainly pegmatitic) have been subjected to a greater variety of conditions of deposition, leading to marked fractionation of the two actinides. In the rather rare cases of hydrothermal deposition, uranium predominates or is present exclusively. It is probable that uranium can be transported as a simple or complex uranyl ion and subsequently reduced to the U^{4+} state.

Thorium and Uranium in Sedimentary Rocks

On account of the ease and delicacy of radiation methods of determination, there is now a considerable body of information about the

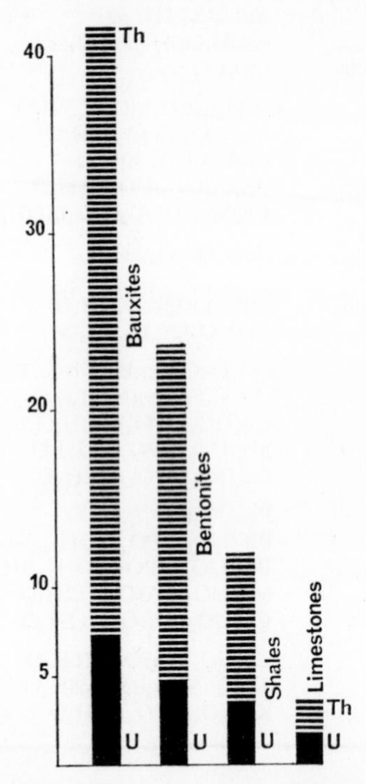

FIG. 25.1. Thorium and uranium content of sedimentary rocks in p.p.m.

occurrence of radioactivity in sedimentary rocks, particularly in connexion with petroleum prospecting. The highest values are found in residual rocks like bauxite and bentonite but there is marked differentiation in favour of uranium in chemically precipitated rocks like limestones which seems to indicate the greater mobility of uranium ions in sedimentary processes. The relative values are indicated in Fig. 25.1.

Natural Waters

There is a very low thorium–uranium content in most natural waters but the values for uranium alone are known with reasonable accuracy.

Very rarely do concentrations exceed 0·001 p.p.m. in fresh waters and for ocean water the value is about 0·003 p.p.m. Thorium contents are in general vanishingly small.

Biological Aspects

It appears that certain plants and animals contain thorium and uranium in amounts above the average for sea water and normal sediments. It is, however, unlikely that these elements are biologically significant. Where there is an appreciable content in carbonaceous material, such as uranium in certain asphalts, it is probably due to secondary processes.

Production of Thorium

The most important sources of the element are closely related to those of the rare earth elements previously discussed. Minerals which are, or have been, exploited include:

Monazite—Nigeria, Ceylon, Congo, Malaya, Egypt, Madagascar, Korea, and U.S.A.

Thorite—Nigeria.

Bastnaesite—U.S.A.

The ThO_2 content of these minerals varies from about 10 % (bastnaesite) to 72 % (thorite). For strategic reasons several countries now withhold output figures and the information is incomplete (see Figs. 24.4 and 24.5).

Production of Uranium

The main ores used as sources of radium became important for the parent element from 1939 onward. Production figures are not complete but several of the major producers again issue them. There have been marked fluctuations in the relative importance of the different producers. The more important are summarized below.

Congo. Deposits at Shinkolobwe (Kasolo) are the most important. High-grade pitchblende (amorphous uraninite) with torbernite and kasolite in veins, stockworks, and disseminations in dolomitic limestones are mainly worked by open cuts. The ores are probably connected with the same magmatic source as the associated copper minerals.

Canada, Great Bear Lake. The deposits are said to be replacement lodes and stockworks in pre-Cambrian sedimentary and volcanic rocks. The complex of minerals includes pitchblende, native silver, pyrite, and chalcopyrite. The pitchblende is thought to be a colloidal deposit in cavities connected with the hydrothermal stage of a magmatic episode associated with the related granites.

Czechoslovakia, Joachimstahl. The classical source used by Madame Curie. It consists of fissure veins with silver, copper, lead, zinc, bismuth, etc., ores. They are of hydrothermal origin.

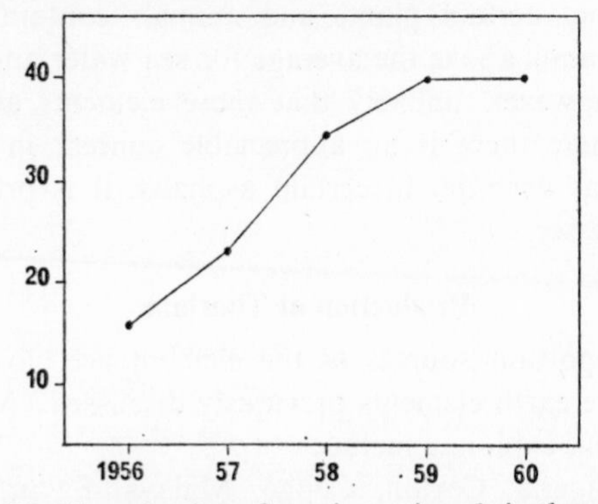

FIG. 25.2. World production of uranium minerals in thousands of tons.

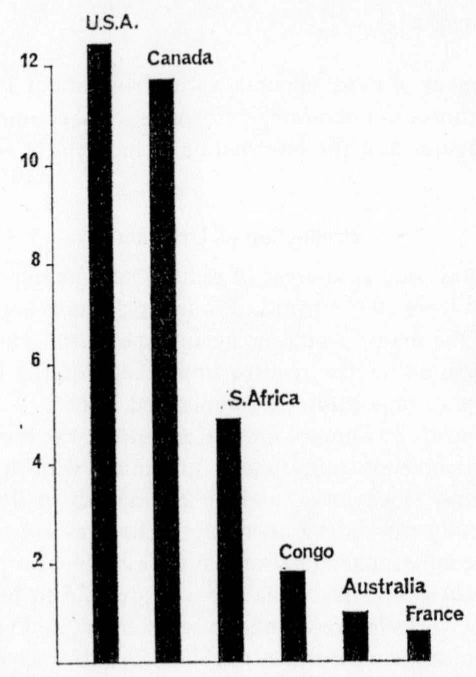

FIG. 25.3. Chief producers of uranium minerals, 1959, in thousands of tons.

U.S.A., Colorado Plateau. The uranium ore is carnotite with associated vanadium minerals. It occurs disseminated in sandstones. Extraordinary concentrations of ore were found in petrified tree logs but the biological origin of the deposits is nevertheless uncertain.

Great Britain. Small concentrations of uranium ores have been located in recent years in many places but they are of doubtful economic interest. Small amounts of pitchblende were obtained at St. Ives in Cornwall in connexion with the tin veins of the area. Torbernite and autunite are also found at Gunnislake in the same county.

Radium

Prior to 1939 radium was the principal element of interest in the ores of uranium, and ores were preferred of low thorium content. In good-quality pitchblendes the radium content is about 320 mg/ton of uranium metal. However, in ordinary geochemical processes radium, an alkaline earth metal, becomes separated to some extent from its parent element. Thus certain geologically young minerals may contain radium—*e.g.*, barytes and pyromorphite. The ionic radius of radium (1·52 Å) is such as to allow it to replace barium or lead with comparative ease. In sedimentary processes the intermediate element ionium or ^{230}Th is of significance. With its considerable half-life of 80000 years, it can accumulate in deep-sea sediments. The age of these sediments has been determined in suitable cases by means of their ionium content.

In respect of radium it has been found that its accumulation in the abyssal red clay reaches a maximum concentration at a depth in the clay of about 10 cm and thereafter declines. Studies of this kind have been used to investigate rates of deposition and other oceanographic problems.

Selected References

G. R. Burbidge, E. M. Burbidge, W. A. Fowler, and F. Hoyle, *Revs. Mod. Phys.* **28, 29** (1956–1957).

R. A. Lyttleton, *The Modern Universe*. Hodder and Stoughton (1956). *The Universe. Scientific American*, Bell (1958).

G. P. Kuiper, *The Earth as a Planet*. University of Chicago Press (1958).

H. C. Urey, *The Planets, Their Origin and Development*. Yale University Press (1932).

H. C. Aller, *The Abundance of the Elements*. Interscience Publishers (1961).

V. V. Cherdyntsev, *Abundance of Chemical Elements* (Trans. W. Nichiporuk). University of Chicago Press (1961).

K. Rankama, *Isotope Geology*. Pergamon Press (1956).

Physics and Chemistry of the Earth, Vols. I–IV. Pergamon Press (1956–1961).

B. Mason, *Principles of Geochemistry* (2nd edn). John Wiley (1958).

V. M. Goldschmidt, *Geochemistry* (Ed. A. Muir). Oxford University Press (1958).

K. Rankama and Th. G. Sahama, *Geochemistry* (2nd edn). University of Chicago Press (1950).

A. A. Smales and L. R. Wager (Ed.), *Methods in Geochemistry*. Interscience Publishers (1960).

I. I. Ginzburg, *Principles of Geochemical Prospecting* (Trans. V. P. Sokoloff). Pergamon Press (1960).

E. O'F. Walsh, *Introduction to Biochemistry*. English Universities Press (1961).

M. Stephenson, *Bacterial Metabolism*. Longmans (1952).

H. Lees, *Biochemistry of Autotrophic Bacteria*. Butterworth (1955).

J. D. Bernal, *The Physical Basis of Life*. Routledge (1951).

A. I. Oparin, *Life, its Nature, Origin and Development* (Trans. A. Synge). Oliver & Boyd (1961).

A. I. Oparin, *Origin of Life on the Earth* (Trans. A. Synge). Oliver & Boyd (1960).

J. F. Kirkaldy, *General Principles of Geology*. Hutchinson (1958).

E. B. Branson, W. A. Tarr, and W. D. Keller, *Introduction to Geology*. McGraw-Hill (1952).

A. Harker, *Petrology for Students* (8th edn). Cambridge University Press (1955).

L. G. Berry and B. Mason, *Mineralogy*. W. H. Freeman & Co. (1959).

E. S. Dana, *The System of Mineralogy* (7th edn), Vols. I and II. John Wiley (1955).

M. H. Hey, *Chemical Index of Minerals*. British Museum (1955).

J. W. Mellor, *Comprehensive Treatise on Inorganic Chemistry*. Vols. I–XVI and Supplements. Longmans (1927–1962).

A. F. Wells, *Structural Inorganic Chemistry*. Oxford University Press (1962).
W. Francis, *Coal, its Formation and Composition*. Edward Arnold (1961).
A. M. Bateman, *Economic Mineral Deposits*. John Wiley (1958).
S. J. and M. G. Johnstone, *Minerals for the Chemical and Allied Industries*. Chapman & Hall (1961).
Statistical Summary of the Mineral Industry. Overseas Geological Surveys. H.M. Stationery Office (1954–1959).
Future of Non-Ferrous Mining in Great Britain and Ireland. Institution of Mining and Metallurgy (1959).

Index